B-Complex Vitamins - Sources, Intakes and Novel Applications

Edited by Jean Guy LeBlanc

Published in London, United Kingdom

IntechOpen

Supporting open minds since 2005

B-Complex Vitamins - Sources, Intakes and Novel Applications
http://dx.doi.org/10.5772/intechopen.95718
Edited by Jean Guy LeBlanc

Contributors
Muhammad Sarwar, Muhammad Haroon Sarwar, Muhammad Farhan Sarwar, Lisa J. Hill, Adrian C. Williams, Bastian Lubis, Putri Amelia, Muhammad Akil, Vincent Viandy, Aznan Lelo, Abbas Shams, Marcela Bermudez Echeverry, Silvia Honda Takada, Bruna Petrucelli Arruda, Debora Sterzeck Cardoso, Pamela Pinheiro Martins, Juliane Midori Ikebara, Aline V. Sousa-Santos, Victor R.C. Torres da Silva, Özlem Çoban, Hilal H. Al-Shekaili, Clara van Karnebeek, Blair R. Leavitt, Fatma Nur Parın, Jagdish Singh, Quratulain Babar, Ayesha Saeed, Muhammad Farrukh Tahir, Anwar Ali, Shridhar C. Ghagane, Akshata Sangoli, Rajendra B. Nerli

First published in London, United Kingdom, 2022 by IntechOpen
IntechOpen is the global imprint of INTECHOPEN LIMITED, registered in England and Wales, registration number: 11086078, 5 Princes Gate Court, London, SW7 2QJ, United Kingdom
Printed in Croatia

British Library Cataloguing-in-Publication Data
A catalogue record for this book is available from the British Library

Additional hard and PDF copies can be obtained from orders@intechopen.com

B-Complex Vitamins - Sources, Intakes and Novel Applications
Edited by Jean Guy LeBlanc
p. cm.
Print ISBN 978-1-83969-797-5
Online ISBN 978-1-83969-798-2
eBook (PDF) ISBN 978-1-83969-799-9

For EU product safety concerns:
IN TECH d.o.o., Prolaz Marije Krucifikse Kozulić 3, 51000 Rijeka, Croatia, info@intechopen.com or visit our website at intechopen.com.

Meet the editor

Dr. Jean Guy LeBlanc has an MSc and BSc in Biochemistry from the Université de Moncton, Canada, and a Ph.D. in Biochemistry from the National University of Tucuman, Argentina, where he is also a post-graduate teacher in the Faculty of Biochemistry, Chemistry and Pharmacy. He is also a Principal Researcher for CERELA-CONICET, Argentina. His main areas of study include the use of lactic acid bacteria to increase bioactive compounds (vitamins, digestive enzymes, antioxidants, etc.) for the fermentation of foods or as biopharmaceuticals to treat and prevent vitamin deficiencies, inflammatory diseases, and some types of cancer. Dr. LeBlanc has published 110 peer-reviewed articles and 33 book chapters and edited 8 books. He has also participated in 143 works in scientific meetings.

Contents

Preface

Although B-complex vitamins have been extensively studied, novel sources, updated intake values and chemical compositions, and their role in numerous diseases are still being discovered. The objective of this book is to provide the most up-to-date information on the effects of vitamin B deficiency from retrospective, perspective and prospective points of view. Specific chapters have been written to describe as well as the roles of niacin (vitamin B3), pyridoxine (vitamin B6), folate (vitamin B9), and vitamin B12 in numerous disorders. The book also presents novel applications of B-complex vitamins, such as thiamin in patients with critical conditions, dietary supplements in the prevention of renal stones, and treatment of COVID-19.

This book project contains twelve chapters dealing with different aspects of B-complex vitamins including their sources, intakes and novel applications.

This multi-topic book is divided into two sections. The first section comprises nine different chapters discussing sources, intakes and deficiencies of B-complex vitamins while the second section consists of three chapters discussing the novel applications of B-complex vitamins.

Jean Guy LeBlanc
CERELA-CONICET,
San Miguel de Tucuman,
Argentina

Section 1

B-Complex Vitamins - Sources, Intakes and Deficiencies

Chapter 1

Poverty and Pellagra's Penumbras

Adrian C. Williams and Lisa J. Hill

Abstract

Pellagra has largely been forgotten. This is unfortunate as important lessons are to be learnt about the diseases and social and economic consequences of poverty – and for the root cause of poverty (and of affluence) – that involve dietary nicotinamide and nicotinamide adenine dinucleotide (NAD) homeostasis. NAD disruption can occur not only from poor diet but from increased consumption from genotoxic, infectious and metabolic stresses. NAD deficiency is closely linked to poor physical and intellectual development, premature ageing and diseases of ageing. Acute infections, many with NAD-consuming toxins, that may differentially affect the NAD-depleted, now include COVID-19. Some Covid manifestations, such as myoclonic encephalopathy and "Long Covid," resemble pellagra clinically and biochemically as both have disturbed nicotinic and tryptophan metabolism. Symbionts that supply nicotinic acid, such as TB and some gut micro-organisms, can become dysbiotic if the diet is very deficient in milk and meat, as it is for 1–2 billion or more. High doses of nicotinamide lead to inhibition of NAD-consuming enzymes and excessive induction of nicotinamide-n-methyl transferase (NNMT) with consequent effects on the methylome: this gives a mechanism for an unrecognised hypervitaminosis-B3 with adverse effects of nicotinamide overload for consumers on a high meat diet with "fortified" foods and "high energy" drinks. Methods of measuring NAD metabolism routinely for screening the populations at risk of deficiency and in metabolically ill or infectious disease patients should be developed urgently. Successful intervention should improve human capital and prevent many aspects of poverty, reduce discrimination and even the drive to emigrate.

Keywords: nicotinamide, tuberculosis, meat transitions, ageing, neurodegeneration, dementia, obesity, cancer, ACE2, Covid-19, NAD, Long Covid

1. Introduction

The 4 "D's" of Dementia, Dermatitis, Diarrhoea and Death are taught to medical students but the interesting history of pellagra and its wider phenotype is largely forgotten [1–4]. A characteristic blank facies and a festinating gait, fasciculation of the tongue or myoclonic encephalopathy are classic features of well-known neurological diseases that were actually first described in the pellagra epidemics. Many other close mimics of neurodegenerative and neuropsychiatric disease, including frank psychoses were seen. Pellagrins harboured dysbiotic infections and succumbed to acute infections explaining the high mortality, the gut manifestations and the high incidence of tuberculosis (TB). Pellagra was widely believed to be hereditary and certainly ran in families. Transgenerational effects created vicious cycles of ill health and poor brain development and further poverty in a man-made economic and market failure that caused a nutritional and metabolic trap.

Sufferers were exposed to considerable discrimination and "Othering", whether as poor whites or blacks, that attracted the attention of eugenicists and sterilisation programmes as their fertility was high (when short of outright starvation), yet this situation is cured by a simple dietary intervention.

2. History and background

The 18thC European epidemics, as described by Gaspar Casal (1735), affected poor Mediterranean peasants on monophagic maize based polenta diets and little meat. The peasants themselves were all too aware of the condition and the relationship to lack of animal products, such as milk, meat and butter. Earlier cases must have existed, perhaps called leprosy in biblical and in earlier times (true leprosy "disappears", like TB, on a high meat diet or nicotinamide administered as an antibiotic). Central American peasants in the New World largely avoided pellagra by a cultural evolutionary approach that involved eating and growing maize with beans and cooking with alkali releasing nicotinamide – but these cultural adaptations were not transported with the plant (that compared with other cereals is low in both nicotinamide and tryptophan) in the Columbian exchange.

Casal agreed about the important role for diet and only later were genetic or infectious, from rotten maize, aetiologies favoured. The early 20th C American epidemic, that killed hundreds of thousands, predominantly affected poor blacks (and whites) working as semi-slave sharecroppers thrown in to poverty with the collapse of the cotton market and the loss of their own farms and hunting rights to plantations, eating maize, molasses (rum) and small quantities of low-quality pork. Joseph Goldberger working in the 1920's after ground-breaking epidemiological and experimental work showed once again that diet was the crucial factor [5–7]. The discovery of nicotinic acid by Elvehjem and a role for tryptophan led to the cure of patients by Spies and the prevention of others through supplementation programmes in milled bread in the 1930s-40's.

Pellagra was long believed to be degenerative in the dehumanising sense of the term. Recent research is clear that Homo sapiens evolved on a high meat diet: pellagra can therefore be seen as an atavistic example of human evolution in reverse gear. This does not downplay the role of dietary balance with plant foods and their contribution to our cooking and (agri-)culture and consciousness given their psychoactive, poisonous and medicinal properties [8–11]. Still, typical early modern and modern societies, unlike hunter-gatherers that share meat, have ruling "meat elites" that will go to almost any lengths to obtain it from wars or if necessary (in the past), human sacrifice creating their own dietary habitat and stratified classes of cognitive, creative and social capital in fragile social contracts [12–14].

Pellagrins, living in a very low meat habitat, were seen a different race having a different physiognomy that crossed colour lines even though it was an archetypal disease of poverty, rather like TB with which it is associated. Inferior cognition, antisocial and addictive behaviours led to discrimination as the "Butterfly caste." This iconic example of cultural and retaliatory "honour" wars may be a denominator common to other disadvantaged groups and their identity politics (or the drive to migrate), that can distract from the underlying economic and dietary issues that need to be faced – and resolved by (meat) redistribution [15–17]. Pellagra casts a long shadow. Some historians date many of the tensions, racial discrimination and stereotyping with insulting epithets between poor whites and blacks from these times as in "The Mind of the South" with distant echoes in segregation, incarceration and apartheid around the world, such as in South Africa (another pellagra zone), to this day [15, 18].

3. Pellagra summary

Pellagra was a systems failure causing premature ageing and widespread neurodegeneration or dysfunction with evidence of mitochondrial failure, oxidative stress and proteinopathy. Poor intellectual development and dementia were key features as were many neuropsychiatric effects and poor social behaviour. Severe forms resembled Jacob-Creutzfeld disease (still misdiagnosed at times), with myoclonic encephalopathy and common biochemical features with NAD disturbances [19]. Gut infections and a high incidence of TB were major features. There were many undiagnosed and untreated cases as the exaggerated sunburn rash (Casal's necklace) was often not present or as noticeable ("pellagra sine pellagra") particularly in those with pigmented skin.

4. Modern copy-cats

Pellagra demonstrates that a disorder that mimicked many neurodegenerative conditions, as now classified, can have a single and simple dietary cause even when there is evidence for (epi-) genetic involvement, dysbiotic microbiomes, mitochondrial and oxidative stress, and proteinopathy (**Figure 1**). This is not surprising as NAD is so central to metabolism in a "NAD" world (**Figure 2**) [20–23]. Dietary nicotinamide backed-up by the degradation of tryptophan on the kynurenine and "immune tolerance" pathway are the precursors to NAD. NAD(H) is critical to mitochondrial energetics as NADH, other dehydrogenase reactions, anabolism (as NADP), and NAD consumer pathways [24]. Stress from chemical or microbial toxins, requiring DNA or tissue repair by poly ADP ribose polymerases (PARPs) and Sirtuins could have the same pathological result by consuming NAD.

Some cancer and age-related antagonistic pleiotropy-type or somatic mutations clearly interact with NAD metabolism and may respond to nicotinamide supplementation or restriction later in life [25–27]. Others such as high energy neurones in the frontal cortex or in dopaminergic neurones may suffer if there

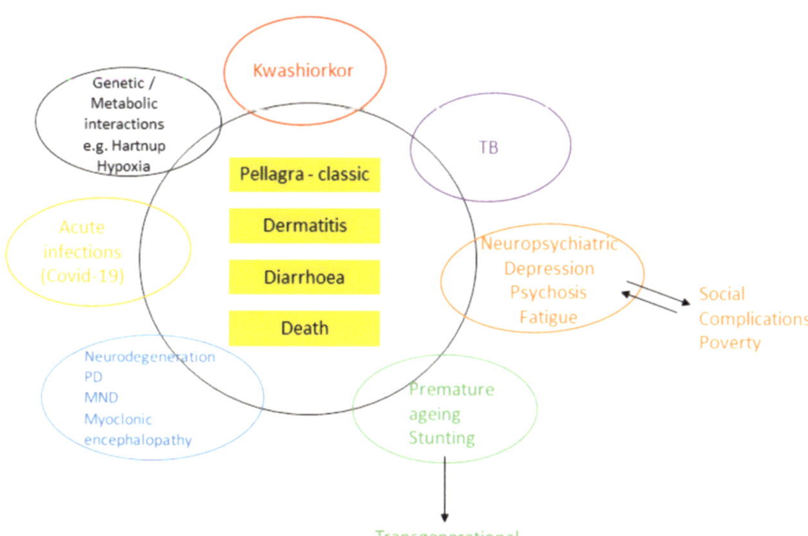

Figure 1.
Pellagra's penumbras. Classical pellagra encompassed multi-organ involvement and disturbed symbiotic and social relationships in the context of poverty. NAD deficiency may involve an even wider multifactorial phenotype and involve excess consumption and nicotinamide overload.

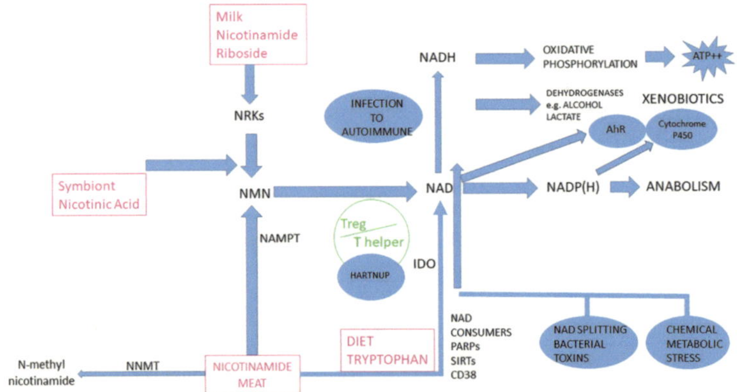

Abbreviations NMN=Nicotinamide mononucleotide; NAMPT= Nicotinamide phosphoribosyl-transferase; IDO= Indoleamine 2,3-dioxygenase; NNMT= Nicotinamide N-methyl-transferase; NRK= Nicotinamide riboside-kinase; PARP=Poly ADP-ribose polymerases; SIRTs=Sirtuins; CD38= Cyclic ADP ribose-hydrolase; AhR= Aryl hydrocarbon receptor.

Figure 2.
NAD is so central to our metabolism and our relationships with the outside milieu that it is appropriate to call this perspective an "NAD world" with many opportunities for lost homeostasis that could lead to the prevention of at the least many diseases of poverty.

are "too many mouths to feed" and need supplements [28–30]. Epigenetic developmental origins of health and disease (DOHaD) or somatic mutations may also be helped by a steady satisfactory dose of nicotinamide throughout lives and across generations avoiding various trade-offs such as poor repair or "disposable soma's" to allow high fertility and unite downstream mechanisms: such as reactive oxygen species, mitochondrial failure, DNA methylation and protein misfolding or physiological and immune collapse with the beneficial effects of calorie restriction, ketogenic diets and exercise [31–37]. Pellagra probably used all these mechanisms although this is best documented for mitochondrial, oxidative stress, and amyloidosis.

5. Test and trace

Pellagra may be being missed even with classical presentations let alone "pellagra sine pellagra" and when hidden in the pellagra penumbra where NAD deficiency may exacerbate other conditions. Alcoholism is a known risk factor and some cases treated rightly for thiamine deficiency with multivitamins may be obscuring cases with a pellagrous element contributing to lack of awareness of the condition. Pellagra may be endemic in the millions in poverty who are meat and milk deprived masquerading as Kwashiorkor ("juvenile pellagra") or "environmental enteropathy" or as poor cognition or general ill-health and susceptibility to adverse effects of infection or trauma. A community screening test that would not be difficult to develop should be a priority and where found family and other contacts should be traced as they will be at risk [38–40].

6. TB known cons but some surprising pros

Tuberculosis, common diarrhoeal illnesses and very high death rates from acute infections, such as smallpox and measles, decrease markedly as societies modernise and increase their meat and nicotinamide intake. Nicotinamide and its analogues, such as Isoniazid, are TB antibiotics and many bacterial toxins (including TB's), interact with NAD-consumer pathways so being NAD replete would improve host resistance making this less of a mystery [41–47]. Intriguingly this is more complex as TB excretes nicotinic acid (used as a test for pathogenic forms for many years), suggesting that when diet is poor, but not too poor, a low population of TB can act as a helpful symbiont, as may some gut organisms in stark contrast to acute infections [48–53]. High dietary dosage, as much as improved hygiene and less cross-infection, will lead to an "absence of TB", and other "Old Friends". TB and even BCG vaccination have important roles in educating the immune system particularly the T cell population and reducing the over-reaction to otherwise harmless antigens characteristic of auto-immune and allergic disease [54].

7. Covid-19 discovers our Achilles heel

Covid-19 probably interferes with the tryptophan uptake pathway and therefore T cell and Interferon responses via a chaperone mechanism with the Angiotensin-converting enzyme (ACE2) receptor through which it enters cells. This amino-acid uptake mechanism also malfunctions with mutations that lead to Hartnup disease that includes a pellagra-like syndrome [55]. Some clinical manifestations of Covid, such as on gut, skin and cognition, or "long Covid" (whose multi-organ symptoms as happened to pellagrins were often doubted), could be "formes fruste" or new versions of pellagra as is supported by documented abnormal tryptophan and nicotinamide metabolism [56–59]. At risk groups such as the elderly or those in poor countries on poor diets may be at risk as they start off from an NAD depleted state. It remains to be seen if supplementation would help before, during or after this or other acute infections.

8. Acute infections and switch to auto-immunity

"Meat transitions" appear to lead to a reduction in many infections and almost simultaneously (as in the late 19th C UK), trigger a demographic and epidemiological switch toward infertility and auto-immune, allergic and other diseases of modernity. Immune intolerance with changes in T cell subset ratios result from inhibition of Indoleamine 2,3-dioxygenase (IDO) and the tryptophan to kynurenine "immune tolerance" pathway as no longer necessary to supply nicotinamide [60, 61]. This immune tolerance extends to the foetus, where it was originally discovered, so may be partly responsible for declines (and occasional reversals), in fertility with modernity – higher doses also affecting cognition and educational levels contributing to a non-coercive form of population control even if not always welcome [25, 62–67]. Dietary modification of nicotinamide or tryptophan in diet, perhaps in concert with other vitamins such as Vitamin D, could affect the incidence of auto-immune conditions such as Multiple Sclerosis [68].

9. Ageing gracefully

Meat transitions and "modernity" have, as if by magic, reduced the incidence of premature ageing, dementia, and death [69–73]. The extraordinary and fast increases in longevity and the fall of age adjusted incidence of dementia have no convincing explanation, and cannot be genetic or related to modern medicine even if antibiotics and vaccinations are part of the answer. In all species there are well described links between NAD metabolism and ageing, alongside resistance to infection, and premature death as was the rule with pellagra – so no magic is required.

NAD levels fall with age and with many diseases of ageing so could respond to nicotinamide supplementation [74–80]. Much has been written about paleo-diets as if major adaptations have not occurred since such as the co-evolved and convergent evolution of genetic (lactase persistence), and cultural adaptations (fermented milk as yoghurt and cheeses and cereals or beer (supplying potent nicotinamide-riboside [81]), and extra amylases for starches as well as careful cooking of maize with alkali as "nixtamal" that may justify the term "Paleofantasy". Yet, there may be something in the elderly being more reliant on the ancestral partly abandoned high meat diet as these and other genetic adaptations may be attenuated once past the reproductive peak [82].

10. Poverty, inequality and discrimination

Early pellagra-ologists, such as Lombroso, may have been right in sensing that pellagrins were atavistic examples of degeneration in the 19th C sense of the term given that increasing meat intake was an important step in our evolution from more herbivorous primates [83]. However this increase was tempered by a move down the food chain in the Mesolithic with more plant based foods, perhaps to increase our fertility, and this move accelerated with the Neolithic agricultural revolution [84, 85]. This was the start of the mixed blessings of "Cereal-ization" and "Calorie-ization" that continues to this day – except for those getting richer where "Meatification" is the rule as observed by Engel and his law [86]. Furthermore for 95% of our evolution as hunter-gatherers we shared meat or individuals were shunned without mercy (even though successful hunters may have used meat to obtain extra mates), so the non-egalitarian meat variances that developed recently are surprising and extreme with 100 fold variances across the globe between rich and poor [64, 87–90]. The poor particularly in poor countries as a consequence face an adverse metabolic and transgenerational NAD headwind that may come to define poverty [91]. Countries that do well ("rosbifs"), by contrast have a healthier anabolic NAD metabolism whereas collapses of empires have been linked to poor diet and uncontrolled "catabolism" unless a meat "safety net" is built [92–95].

Inequalities of meat intake between classes and countries may need to be fairer for everyone's safety. These extremes are traceable back to 17th C common pastureland "enclosure" movements and 19th C colonialism with the creation of the "third world." All these and other mechanisms channel meat to the wealthy. The New World originally had few natural animal domesticates but this was corrected by the Columbian exchange enabling the rise of the West [96]. The global South was also unlucky in its meat supply particularly in Africa where a lack of animal domesticates and an abundance of human and veterinary infections in the tsetse fly belt such as trypanosomiasis and rinderpest and were prone to pellagra. Darker skin colour reduces the diagnostic help from the sunburn of pellagra but is a mixed blessing if it is acting as a warning (including allowing self-treatment), of the more serious cognitive effects of nicotinamide deficiency [38].

Much discrimination may be against groups previously or currently at risk from nicotinamide deficiency and conversely many who claim supremacy or that they are part of a meritocracy may have always had a better diet with more meat and nicotinamide. Engel first pointed out that all groups will increase meat intake once they can afford it and should be seen as an essential need for personal and national prosperity [97] as we elaborate in our companion chapter. Dominance in primates has been linked to high serotonin levels and may be the basis of "Biopower and Biopolitics" of "Superior" humans on the better diet [98, 99].

11. Meat markets

The meat market and food supply chain has become very unequal and may be driving inequality affecting disease and demographic transitions and all working through differential doses of micronutrients relative to caloric intake (**Figure 3**) [100–104]. This dysfunctional market may be having profound effects on planetary and human health including those related to the commercial determinants of disease and the double burden of both wasting and obesity with epigenetic effects playing themselves out over individual's lifetimes and across generations [105–108]. There is a case for de-commodifying meat (and fruit and vegetables), as was true in our ancestral state, enabling healthy living for all whatever their income, gender, or ethnic status [109–111].

12. Too much of a good thing?

A state of hypervitaminosis-B3 is also possible. Many conditions common with affluence and greed with a high meat intake (let alone nicotinamide supplementation), are linked to induction of the enzyme NNMT that includes many cancers, Parkinson's disease and obesity and other aspects of the metabolic syndrome,

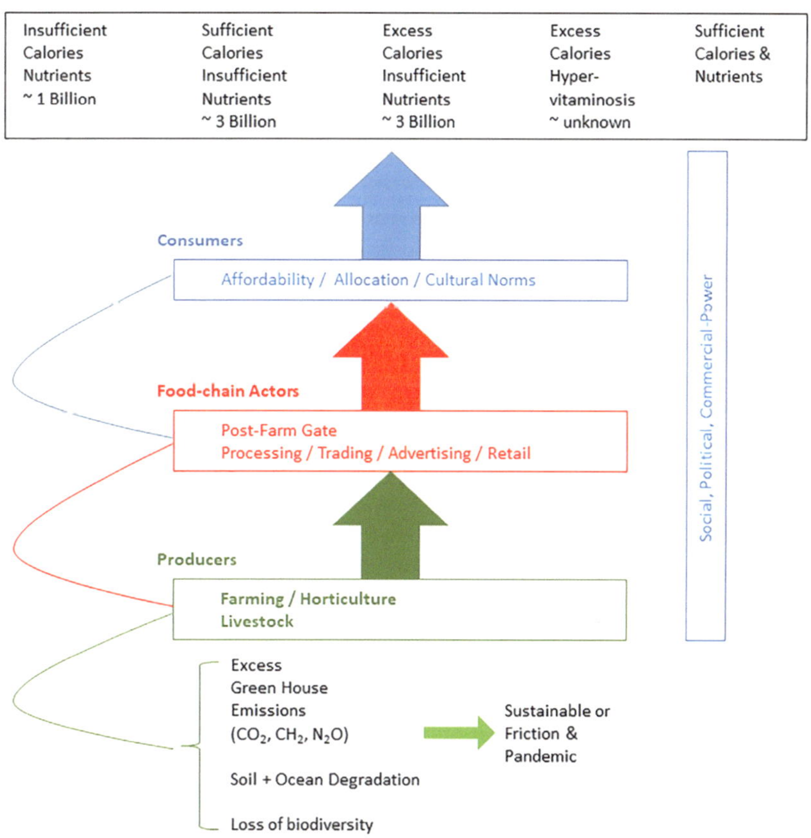

Figure 3.
A solution needs to be found between a more even supply of animal products to the rich and the NAD deficient poor with better farming techniques and supply chains with less waste. Rather than exacerbate climate change this could help through increased human and social capital and reduced risk of zoonoses and political friction.

including diabetes [112–117]. NNMT detoxifies nicotinamide and is induced by high doses (being absent in herbivores), but consumes valuable methyl groups and nicotinamide overload might over-inhibit NAD-consumer enzymes that are metabolic master molecules (14–16) [118]. Nicotinamide's methylated derivative resembles the dopaminergic neurotoxin MPTP and may, like nicotinamide, be a "double-edged sword". When nicotinamide fortification was introduced in the 1940's (in processed breads), this was never universal and never monitored or aimed at eliminating even classical pellagra worldwide. Furthermore any adverse effects in countries where meat and milk were in ample supply was never investigated even as manufacturers of foods and "high energy" drinks (such as Red Bull) added far more than was necessary for supplementation. Short term there are few signs of toxicity but any adverse effects of nicotinamide overload may be long-term side-effects and harder to spot.

13. Poverty prevention: looking upstream

Pellagra's history is well worth remembering given that nobody systematically makes sure that it or "pellagra sine pellagra" was eliminated. In addition several acquired infectious or metabolic or multifactorial genetic diseases may be ameliorated by temporary or permanent adjustments in the dose. Many in poverty are at risk given that variances of meat intake are now extreme and their acute infections will cause further NAD depletion in a vicious cycle. Pellagra should have remained a public health concern, doubtless helped by supplementation, but not helped by never being a universal policy or, as a multi-organ disease not being owned by any single specialty. Nicotinamide's toxicity in high meat economies was never monitored over the long-term and is not short of potential mechanisms through affecting NAD-Consumer controlled metabolism directly or via the methylome.

Recent poverty literature rarely indexes pellagra even though it is a paradigm of how economics and both material absolute and more relational social and cultural needs overlap let alone giving a mechanism for the extraordinarily strong correlations between income and life expectancy across countries and, to some extent, relative measures of income inequality within countries [119, 120]. Pellagra is a proven pathology of poverty that could make disputed behavioural mechanisms, such as high-fat high-calorie "fast food" or addictions or poor (economic) habits, or lack of exercise or even stress and psychosocial processes and other "life style drifts" secondary phenomena. The North–South geographies of poverty and ill-health is both explained and our responsibility to repair it given as reparations for exacerbation in colonial exploitative times. The "Mediterranean paradox" when, despite considerable income inequality, there is less than expected health inequality is also more understandable as a healthier omnivorous diet is more affordable here than elsewhere – and not considered as much of a mark of status or "Bourdieuian" distinction [121, 122].

Pellagra links poverty and poor diet to poor human and social capabilities and, we propose, if this need for meat was corrected in a "Moral economy" – as opposed to the worst of neoliberalism and demonisation of the poor and "Precariats" born of austerity – would allow for more equality of opportunity and education across lives and across generations with less discrimination. It would also create a new dawn for researchers and policy makers [123–128].

Many were surprised at how previous well-meaning policy interventions in the developed and less developed world had failed to reduce health inequalities that included welfare systems, such as the free at the point of need NHS. Since the introduction of the NHS the UK has seen an increase in many measures of health inequality including longevity [129–131]. Low income increases exposure to toxins and pollutants and accidents that may be partly to blame (as may poor access compared with the middle classes) but diet that is often squeezed by costs of housing and other understandable social and less materials needs for entertainment and respect is contender. Diet is thought to be at least as important a risk factor for modern diseases as smoking and those that defined poverty as "a family is poor if it cannot afford to eat" may not have been so wide of the mark particularly if transgenerational teratogenic "cycles of disadvantage" are taken in to consideration as many reports on child development and maternal health even 50 years ago emphasised [132–137]. Sorting diet was largely beyond the reaches of the NHS as it is a preventive factor that starts in very early life whereas medical interventions, important as they are, happen late and that may explain their lack of impact on inequality.

Preston curves more optimistically suggest that modest increases in income improves health and happiness quickly with a low ceiling effect and diminishing returns and this is compatible with a climb up Engels curve rather than with improvements in hygiene, health, education or technology [138–144]. Such a materialist effect as the poor eating more meat could be part of a new win-win "Enlightenment" [145]. Dietary dosage or nicotinamide supplements may in addition need to be boosted when individuals have certain mutations or are under stress, whether genotoxic or anoxic/metabolic – or restrained if there really is a hypervitaminosis B3 contributing to diseases of affluence. High meat diets and being NAD-Replete may even help solve the dangers of antibiotic resistance and the emergence of superbugs [146]. The environmental cost of optimising meat intake would be mitigated by affluent countries eating and wasting less but sharing more. The meat supply needs to be safe with the poor not having to rely on "bush meat" or risk food poisoning or old and new zoonoses, such as COVID-19, that are a danger to all [88, 147–149]. Supplementation of nicotinamide alone may however not be enough as animal products contain other helpful micronutrients such as iron and sources of methyl-groups such as choline and vitamin B12.

14. Conclusion

We should imagine along with John Lennon (1971) "No need for greed or hunger – A brotherhood of Man" and allow a return to our meat and micronutrient sharing roots. The immediate need is for further study using real world data by measuring Nicotinamide/NAD/NNMT and tryptophan metabolism. Monitoring should happen widely in populations at risk of both deficiency and excess from cradle to grave and in those with a variety of established diseases or trauma or asymptomatic mutations. Interventional studies should provide firm evidence from better nutritional research that does not assume that meat in moderation is toxic to planetary or human health [150].

Acknowledgements

We would like to thank Queen Elizabeth Hospital Birmingham Charity for funding this study.

Conflict of interest

The author(s) declared no potential conflicts of interest with respect to the research, authorship, and/or publication of this article.

Author details

Adrian C. Williams[1]* and Lisa J. Hill[2]

1 Department of Neurology, University Hospitals Birmingham NHS Foundation Trust, Birmingham, United Kingdom

2 School of Biomedical Sciences, Institute of Clinical Sciences, University of Birmingham, United Kingdom

*Address all correspondence to: adrian.williams1949@gmail.com

References

[1] Rajakumar K. Pellagra in the United States: A historical perspective. South Med J. 2000;93(3):272-277.

[2] Williams AC, Ramsden DB. Pellagra: A clue as to why energy failure causes diseases? Med Hypotheses. 2007;69(3): 618-628.

[3] Harris S. Clinical Pellagra: C. V. Mosby Company; 1941.

[4] Roe DA. A Plague of Corn: The Social History of Pellagra: Cornell University Press; 1973.

[5] Carrasco D. City of Sacrifice: The Aztec Empire and the Role of Violence in Civilization: Beacon Press; 2000.

[6] Cole J. Assessing the calorific significance of episodes of human cannibalism in the Palaeolithic. Sci Rep. 2017;7:44707.

[7] Kraut AM. Goldberger's War: The Life and Work of a Public Health Crusader: Farrar, Straus and Giroux; 2021.

[8] Anderson E. Plants, Man and Life: University of California Press; 1969.

[9] Pendell D, Snyder G. Pharmako Poeia: Plant Powers, Poisons, and Herbcraft: North Atlantic Books; 2010.

[10] Pollan M. This Is your Mind on Plants: Penguin Publishing Group; 2021.

[11] Standage T. A History of the World in Six Glasses: Doubleday Canada; 2010.

[12] Etheridge EW. The Butterfly Caste: A Social History of Pellagra in the South: Greenwood Publishing Company; 1972.

[13] Sunderland D. Social Capital, Trust and the Industrial Revolution: 1780-1880: Taylor & Francis; 2007.

[14] Florida R. The Rise of the Creative Class--Revisited: Revised and Expanded: Basic Books; 2014.

[15] Hartman A. A War for the Soul of America, Second Edition: A History of the Culture Wars: University of Chicago Press; 2019.

[16] Isenberg N. White Trash: The 400-Year Untold History of Class in America: Penguin Publishing Group; 2016.

[17] Kuppens T, Easterbrook MJ, Spears R, Manstead AS. Life at both ends of the ladder: Education-based identification and its association with well-being and social attitudes. Personality and Social Psychology Bulletin. 2015;41(9):1260-1275.

[18] Hochschild AR. Strangers in their Own Land: Anger and Mourning on the American Right: New Press; 2018.

[19] Zhou M, Ottenberg G, Sferrazza GF, Hubbs C, Fallahi M, Rumbaugh G, Brantley AF, Lasmezas CI. Neuronal death induced by misfolded prion protein is due to NAD+ depletion and can be relieved in vitro and in vivo by NAD+ replenishment. Brain. 2015;138(Pt 4):992-1008.

[20] Canto C, Menzies KJ, Auwerx J. NAD(+) metabolism and the control of energy homeostasis: A balancing act between mitochondria and the nucleus. Cell Metab. 2015;22(1):31-53.

[21] Imai S. "Clocks" in the NAD world: NAD as a metabolic oscillator for the regulation of metabolism and aging. Biochim Biophys Acta. 2010;1804(8): 1584-1590.

[22] Stromland O, Niere M, Nikiforov AA, VanLinden MR, Heiland I, Ziegler M. Keeping the balance in NAD metabolism. Biochem Soc Trans. 2019;47(1):119-130.

[23] Yaku K, Okabe K, Nakagawa T. NAD metabolism: Implications in aging and longevity. Ageing Res Rev. 2018;47:1-17.

[24] Imai S. The NAD World: A new systemic regulatory network for metabolism and aging--Sirt1, systemic NAD biosynthesis, and their importance. Cell Biochem Biophys. 2009;53(2):65-74.

[25] Austad SN, Hoffman JM. Is antagonistic pleiotropy ubiquitous in aging biology? Evol Med Public Health. 2018;2018(1):287-294.

[26] Kanakkanthara A, Kurmi K, Ekstrom TL, Hou X, Purfeerst ER, Heinzen EP, Correia C, Huntoon CJ, O'Brien D, Wahner Hendrickson AE, et al. BRCA1 deficiency Upregulates NNMT, which reprograms metabolism and sensitizes ovarian cancer cells to mitochondrial metabolic targeting agents. Cancer Res. 2019;79(23):5920-5929.

[27] Rodriguez JA, Marigorta UM, Hughes DA, Spataro N, Bosch E, Navarro A. Antagonistic pleiotropy and mutation accumulation influence human senescence and disease. Nat Ecol Evol. 2017;1(3):55.

[28] Bolam JP, Pissadaki EK. Living on the edge with too many mouths to feed: Why dopamine neurons die. Mov Disord. 2012;27(12):1478-1483.

[29] Nitrini R. Selective vulnerability of von Economo neurons in frontotemporal dementia. Dement Neuropsychol. 2008;2(3):164.

[30] Talhelm T, English AS. Historically rice-farming societies have tighter social norms in China and worldwide. Proceedings of the National Academy of Sciences. 2020;117(33):19816-19824.

[31] Fontana L, Partridge L. Promoting health and longevity through diet: From model organisms to humans. Cell. 2015;161(1):106-118.

[32] Fu H, Hardy J, Duff KE. Selective vulnerability in neurodegenerative diseases. Nature neuroscience. 2018;21(10):1350-1358.

[33] Gluckman P, Hanson MB, Hanson M, Gluckman PP. The Fetal Matrix: Evolution, Development and Disease: Cambridge University Press; 2005.

[34] Jasienska G, Bribiescas RG, Furberg AS, Helle S, Nunez-de la Mora a. human reproduction and health: An evolutionary perspective. Lancet. 2017;390(10093):510-520.

[35] Kaptijn R, Thomese F, Liefbroer AC, Van Poppel F, Van Bodegom D, Westendorp RGJ. The trade-off between female fertility and longevity during the epidemiological transition in the Netherlands. PLoS One. 2015;10(12): e0144353.

[36] Meij J, Van Bodegom D, Ziem J, Amankwa J, Polderman A, Kirkwood T, De Craen A, Zwaan B, Westendorp R. Quality–quantity trade-off of human offspring under adverse environmental conditions. Journal of evolutionary biology. 2009;22(5):1014-1023.

[37] Barker DJ. Fetal and infant origins of adult disease. Monatsschrift Kinderheilkunde. 2001;149(1):S2-S6.

[38] Williams AC, Hill LJ. The 4 D's of pellagra and Progress. Int J Tryptophan Res. 2020;13:1178646920910159.

[39] Liu Y, Clement J, Grant R, Sachdev P, Braidy N. Quantitation of NAD+: Why do we need to measure it? Biochim Biophys Acta Gen Subj. 2018;1862(12):2527-2532.

[40] Schultz MB, Lu Y, Braidy N, Sinclair DA. Assays for NAD(+)-dependent reactions and NAD(+)

metabolites. Methods Mol Biol. 2018;1813:77-90.

[41] Deng Q, Barbieri JT. Molecular mechanisms of the cytotoxicity of ADP-ribosylating toxins. Annu Rev Microbiol. 2008;62:271-288.

[42] Pajuelo D, Gonzalez-Juarbe N, Tak U, Sun J, Orihuela CJ, Niederweis M. NAD(+) depletion triggers macrophage Necroptosis, a cell death pathway exploited by mycobacterium tuberculosis. Cell Rep. 2018;24(2):429-440.

[43] Kwon WY, Suh GJ, Kim KS, Kwak YH. Niacin attenuates lung inflammation and improves survival during sepsis by downregulating the nuclear factor-kappaB pathway. Crit Care Med. 2011;39(2):328-334.

[44] Li WY, Ren JH, Tao NN, Ran LK, Chen X, Zhou HZ, Liu B, Li XS, Huang AL, Chen J. The SIRT1 inhibitor, nicotinamide, inhibits hepatitis B virus replication in vitro and in vivo. Arch Virol. 2016;161(3):621-630.

[45] Mahon RN, Hafner R. Immune Cell Regulatory Pathways Unexplored as Host-Directed Therapeutic Targets for Mycobacterium tuberculosis: An Opportunity to Apply Precision Medicine Innovations to Infectious Diseases. Clin Infect Dis. 2015;61Suppl 3:S200-216.

[46] Maslowski KM. Metabolism at the Centre of the host-microbe relationship. Clin Exp Immunol. 2019;197(2):193-204.

[47] Tak U, Vlach J, Garza-Garcia A, William D, Danilchanka O, de Carvalho LPS, Saad JS, Niederweis M. The tuberculosis necrotizing toxin is an NAD+ and NADP+ glycohydrolase with distinct enzymatic properties. Journal of Biological Chemistry. 2019;294(9): 3024-3036.

[48] Boshoff HI, Xu X, Tahlan K, Dowd CS, Pethe K, Camacho LR, Park TH, Yun CS, Schnappinger D, Ehrt S, et al. Biosynthesis and recycling of nicotinamide cofactors in mycobacterium tuberculosis. An essential role for NAD in nonreplicating bacilli. J Biol Chem. 2008;283(28): 19329-19341.

[49] Konno K, Kurzmann R, Bird KT, Sbarra A. Differentiation of human tubercle bacilli from atypical acid-fast bacilli, II. Clinical application. American Review of Tuberculosis and Pulmonary Diseases. 1958;77(4): 675-680.

[50] Leistikow RL, Morton RA, Bartek IL, Frimpong I, Wagner K, Voskuil MI. The mycobacterium tuberculosis DosR regulon assists in metabolic homeostasis and enables rapid recovery from nonrespiring dormancy. J Bacteriol. 2010;192(6): 1662-1670.

[51] Manca C, Koo MS, Peixoto B, Fallows D, Kaplan G, Subbian S. Host targeted activity of pyrazinamide in mycobacterium tuberculosis infection. PLoS One. 2013;8(8):e74082.

[52] Mc KD, Malone L, et al. The effect of nicotinic acid amide on experimental tuberculosis of white mice. J Lab Clin Med. 1948;33(10):1249-1253.

[53] Simmons JD, Peterson GJ, Campo M, Lohmiller J, Skerrett SJ, Tunaru S, Offermanns S, Sherman DR, Hawn TR. Nicotinamide limits replication of Mycobacterium tuberculosis and BCG within macrophages. J Infect Dis. 2019.

[54] Rook G. The Hygiene Hypothesis and Darwinian Medicine: Birkhäuser Basel; 2009.

[55] Ge XY, Li JL, Yang XL, Chmura AA, Zhu G, Epstein JH, Mazet JK, Hu B, Zhang W, Peng C, et al. Isolation and characterization of a bat SARS-like

coronavirus that uses the ACE2 receptor. Nature. 2013;503(7477):535-538.

[56] Baig AM, Khaleeq A, Ali U, Syeda H. Evidence of the COVID-19 virus targeting the CNS: Tissue distribution, host-virus interaction, and proposed neurotropic mechanisms. ACS Chem Neurosci. 2020;11(7):995-998.

[57] Blasco H, Bessy C, Plantier L, Lefevre A, Piver E, Bernard L, Marlet J, Stefic K, Benz-de Bretagne I, Cannet P, et al. The specific metabolome profiling of patients infected by SARS-COV-2 supports the key role of tryptophan-nicotinamide pathway and cytosine metabolism. Sci Rep. 2020;10(1):16824.

[58] Robertson J, Gostner JM, Nilsson S, Andersson L-M, Fuchs D, Gisslen M. Serum neopterin levels in relation to mild and severe COVID-19. BMC infectious diseases. 2020;20(1):942.

[59] Thomas T, Stefanoni D, Reisz JA, Nemkov T, Bertolone L, Francis RO, Hudson KE, Zimring JC, Hansen KC, Hod EA, et al. COVID-19 infection alters kynurenine and fatty acid metabolism, correlating with IL-6 levels and renal status. JCI insight. 2020;5(14).

[60] Williams AC, Hill LJ. Nicotinamide and demographic and disease transitions: Moderation is best. Int J Tryptophan Res. 2019;12:117864691 9855940.

[61] Kepert I, Fonseca J, Muller C, Milger K, Hochwind K, Kostric M, Fedoseeva M, Ohnmacht C, Dehmel S, Nathan P, et al. D-tryptophan from probiotic bacteria influences the gut microbiome and allergic airway disease. J Allergy Clin Immunol. 2017;139(5): 1525-1535.

[62] Bauer TM, Jiga LP, Chuang JJ, Randazzo M, Opelz G, Terness P. Studying the immunosuppressive role of indoleamine 2,3-dioxygenase: Tryptophan metabolites suppress rat allogeneic T-cell responses in vitro and in vivo. Transpl Int. 2005;18(1):95-100.

[63] Bongaarts J, Casterline J. Fertility transition: Is sub-Saharan Africa different? Population and Development Review. 2013;38:153-168.

[64] Chang RQ, Li DJ, Li MQ. The role of indoleamine-2, 3-dioxygenase in normal and pathological pregnancies. American Journal of Reproductive Immunology. 2017.

[65] Clark CM, Lawlor-Savage L, Goghari VM. The Flynn effect: A quantitative commentary on modernity and human intelligence. Measurement: Interdisciplinary Research and Perspectives. 2016;14(2):39-53.

[66] Harper S. Demography: A Very Short Introduction: Oxford University Press; 2018.

[67] Pearson CS. On the Cusp: From Population Boom to Bust: Oxford University Press; 2015.

[68] Sonner JK, Keil M, Falk-Paulsen M, Mishra N, Rehman A, Kramer M, Deumelandt K, Röwe J, Sanghvi K, Wolf L, et al. Dietary tryptophan links encephalogenicity of autoreactive T cells with gut microbial ecology. Nat Commun. 2019;10(1):4877.

[69] Caballero B. The Nutrition Transition: Diet and Disease in the Developing World: Elsevier Science; 2002.

[70] Caparros M. Hunger: The Oldest Problem: Melville House; 2020.

[71] Caulfield LE, de Onis M, Blossner M, Black RE. Undernutrition as an underlying cause of child deaths associated with diarrhea, pneumonia, malaria, and measles. Am J Clin Nutr. 2004;80(1):193-198.

[72] Clement J, Wong M, Poljak A, Sachdev P, Braidy N. The plasma

NAD(+) metabolome is Dysregulated in "Normal" aging. Rejuvenation Res. 2018.

[73] Deaton A. The Great Escape: Health, Wealth, and the Origins of Inequality: Princeton University Press; 2013.

[74] McReynolds MR, Chellappa K, Baur JA. Age-related NAD(+) decline. Exp Gerontol. 2020;134:110888.

[75] Rajman L, Chwalek K, Sinclair DA. Therapeutic potential of NAD-boosting molecules: The In vivo evidence. Cell Metab. 2018;27(3):529-547.

[76] Zhang M, Ying W. NAD(+) deficiency is a common central pathological factor of a number of diseases and aging: Mechanisms and therapeutic implications. Antioxid Redox Signal. 2019;30(6):890-905.

[77] Bonkowski MS, Sinclair DA. Slowing ageing by design: The rise of NAD(+) and sirtuin-activating compounds. Nat Rev Mol Cell Biol. 2016;17(11):679-690.

[78] Covarrubias AJ, Perrone R, Grozio A, Verdin E. NAD(+) metabolism and its roles in cellular processes during ageing. Nat Rev Mol Cell Biol. 2021;22(2):119-141.

[79] Hou Y, Lautrup S, Cordonnier S, Wang Y, Croteau DL, Zavala E, Zhang Y, Moritoh K, O'Connell JF, Baptiste BA. NAD+ supplementation normalizes key Alzheimer's features and DNA damage responses in a new AD mouse model with introduced DNA repair deficiency. Proceedings of the National Academy of Sciences. 2018:201718819.

[80] Ryu D, Zhang H, Ropelle ER, Sorrentino V, Mázala DA, Mouchiroud L, Marshall PL, Campbell MD, Ali AS, Knowels GM, et al. NAD+ repletion improves muscle function in muscular dystrophy and counters global PARylation. Sci Transl Med. 2016;8(361):361ra139.

[81] Mehmel M, Jovanović N, Spitz U. Nicotinamide Riboside-The Current State of Research and Therapeutic Uses. Nutrients. 2020;12(6).

[82] Rutledge GA, Cabral LG, Kuey BJ, Lee JD, Mueller LD, Rose MR. Hamiltonian patterns of age-dependent adaptation to novel environments. PLoS One. 2020;15(10):e0240132.

[83] Lee RB, DeVore I. Man the Hunter: Taylor & Francis; 2017.

[84] Henry AG, Brooks AS, Piperno DR. Plant foods and the dietary ecology of Neanderthals and early modern humans. J Hum Evol. 2014;69:44-54.

[85] Mellars P, French JC. Tenfold population increase in western europe at the neandertal-to-modern human transition. Science. 2011;333(6042): 623-627.

[86] Steckel RH, Rose JC, Foundation NS, Ohio State University. Columbus O. The Backbone of History: Health and Nutrition in the Western Hemisphere: Cambridge University Press; 2002.

[87] Henrich J. The Weirdest People in the World: How the West Became Psychologically Peculiar and Particularly Prosperous: Penguin Books Limited; 2020.

[88] Hickel J. The Divide: A Brief Guide to Global Inequality and its Solutions: William Heinemann; 2017.

[89] Milanovic B. Global Inequality: A New Approach for the Age of Globalization: Harvard University Press; 2016.

[90] Pilcher JM. Food in World History: Taylor & Francis; 2017.

[91] Williams AC, Hill LJ. Nicotinamide as independent variable for intelligence, fertility, and health: Origin of human

creative explosions? Int J Tryptophan Res. 2019;12:1178646919855944.

[92] Greer JM. The Long Descent: A User's Guide to the End of the Industrial Age: FOUNDERS HOUSE PUB LLC; 2019.

[93] Huntington E. Mainsprings of Civilization. 1945.

[94] Lowrey A. Give People Money: How a Universal Basic Income Would End Poverty, Revolutionize Work, and Remake the World: Crown; 2018.

[95] Wylie D. Starving on a Full Stomach: Hunger and the Triumph of Cultural Racism in Modern South Africa: University Press of Virginia; 2001.

[96] Crosby AW. The Columbian Exchange: Biological and Cultural Consequences of 1492: Greenwood Publishing Group; 2003.

[97] Prais SJ, Houthakker HS. The Analysis of Family Budgets: Cambridge University Press; 1971.

[98] Raleigh MJ, McGuire MT, Brammer GL, Yuwiler A. Social and environmental influences on blood serotonin concentrations in monkeys. Archives of general psychiatry. 1984;41(4):405-410.

[99] Cisney VW, Morar N. Biopower: Foucault and beyond: University of Chicago Press; 2015.

[100] Andiman WA. Animals Viruses and Humans, a Narrow Divide: How Lethal Zoonotic Viruses Spill Over and Threaten us: Paul Dry Books; 2018.

[101] Conway G. The Doubly Green Revolution: Food for all in the Twenty-First Century: Comstock pub. Associates; 1998.

[102] Denison RF. Darwinian Agriculture: How Understanding Evolution Can Improve Agriculture: Princeton University Press; 2016.

[103] Grace D, Lindahl J, Wanyoike F, Bett B, Randolph T, Rich KM. Poor livestock keepers: ecosystem-poverty-health interactions. Philos Trans R Soc Lond B Biol Sci. 2017;372(1725).

[104] Steel C. Hungry City: How Food Shapes our Lives: Vintage Books; 2013.

[105] Easterly W, Easterly WR. The White Man's Burden: Why the West's Efforts to Aid the Rest Have Done So Much Ill and So Little Good: Penguin Press; 2006.

[106] Farmer P, Sen A, Sen M. Pathologies of Power: Health, Human Rights, and the New War on the Poor: University of California Press; 2005.

[107] Lang T. Feeding Britain: Our Food Problems and How to Fix Them: Penguin Books Limited; 2020.

[108] Zotor FB, Ellahi B, Amuna P. Applying the food multimix concept for sustainable and nutritious diets. Proc Nutr Soc. 2015;74(4):505-516.

[109] Esping-Andersen G. The Three Worlds of Welfare Capitalism: Wiley; 2013.

[110] Evans G, Kantrowitz E. Socioeconomic status and health: The potential role of environmental risk exposure Annu rev public health 23: 303-331. Find this article online. 2002.

[111] Morris JN, Deeming C, Wilkinson P, Dangour AD. Action towards healthy living--for all. Int J Epidemiol. 2010;39(1):266-273.

[112] Gao Y, Martin NI, van Haren MJ. Nicotinamide N-methyl transferase (NNMT): An emerging therapeutic target. Drug Discov Today. 2021.

[113] Kraus D, Yang Q, Kong D, Banks AS, Zhang L, Rodgers JT,

Pirinen E, Pulinilkunnil TC, Gong F, Wang YC, et al. Nicotinamide N-methyltransferase knockdown protects against diet-induced obesity. Nature. 2014;508(7495):258-262.

[114] Neelakantan H, Brightwell CR, Graber TG, Maroto R, Wang HL, McHardy SF, Papaconstantinou J, Fry CS, Watowich SJ. Small molecule nicotinamide N-methyltransferase inhibitor activates senescent muscle stem cells and improves regenerative capacity of aged skeletal muscle. Biochem Pharmacol. 2019;163:481-492.

[115] Pissios P. Nicotinamide N-Methyltransferase: More than a vitamin B3 clearance enzyme. Trends Endocrinol Metab. 2017;28(5):340-353.

[116] Roberti A, Fernández AF, Fraga MF. Nicotinamide N-methyltransferase: At the crossroads between cellular metabolism and epigenetic regulation. Mol Metab. 2021;45:101165.

[117] Sperber H, Mathieu J, Wang Y, Ferreccio A, Hesson J, Xu Z, Fischer KA, Devi A, Detraux D, Gu H, et al. The metabolome regulates the epigenetic landscape during naive-to-primed human embryonic stem cell transition. Nat Cell Biol. 2015;17(12):1523-1535.

[118] Hwang ES, Song SB. Possible Adverse Effects of High-Dose Nicotinamide: Mechanisms and Safety Assessment. Biomolecules. 2020;10(5).

[119] Wilkinson R, Pickett K. The Spirit Level: Why Equality Is Better for Everyone: Penguin Books Limited; 2010.

[120] Greve B. Routledge International Handbook of Poverty: Taylor & Francis; 2019.

[121] Mackenbach JP. Health Inequalities: Persistence and Change in Modern Welfare States: Oxford University Press; 2019.

[122] Kulhánová I, Bacigalupe A, Eikemo TA, Borrell C, Regidor E, Esnaola S, Mackenbach JP, Consortium E. why does Spain have smaller inequalities in mortality? An exploration of potential explanations. The European Journal of Public Health. 2014;24(3):370-377.

[123] Stevenson J. Social Conditions in Britain between the Wars: Penguin; 1977.

[124] Sen A, Sen PEPAK. Commodities and Capabilities: North-Holland; 1985.

[125] Doyal L, Gough I. A Theory of Human Need: Palgrave Macmillan UK; 1991.

[126] Townsend P. International Analysis Poverty: Taylor & Francis; 2014.

[127] Shetty P, Emerton V. Nutrition through the Life Cycle: Royal Society of Chemistry; 2007.

[128] Klug F. A magna carta for all humanity: Homing in on human rights. Soundings. 2015;60(60):130-144.

[129] Black D, Townsend P, Davidson N. Inequalities in Health: The Black Report: Penguin; 1992.

[130] Acheson D, Office GBHMsS. Independent inquiry into inequalities in health: Report: Stationery Office; 1998.

[131] Bartley M. Health Inequality: An Introduction to Concepts, Theories and Methods: Wiley; 2016.

[132] Baird D. Epidemiology of congenital malformations of the central nervous system in (a) Aberdeen and (b) Scotland. Journal of Biosocial Science. 1974;6(2):113-137.

[133] Birch HG. Disadvantaged Children: Health, Nutrition and School Failure: Harcourt Brace Jovanovich; 1972.

[134] DHSS P. Health: Eating for Health. HMSO; 1978.

[135] Marshall WA. Human growth and its disorders: Academic Press Inc. (London) Ltd.; 1977.

[136] Rutter M, Madge N. Cycles of Disadvantage: A Review of Research: Ashgate Publishing Company; 1976.

[137] Swinburn BA, Kraak VI, Allender S, Atkins VJ, Baker PI, Bogard JR, Brinsden H, Calvillo A, De Schutter O, Devarajan R, et al. The global Syndemic of obesity, Undernutrition, and climate change: The lancet commission report. Lancet. 2019;393(10173):791-846.

[138] Caldwell JC. Mortality in relation to economic development. Bulletin of the World Health Organization. 2003;81: 831-832.

[139] Cassedy JH. The modern rise of population. JAMA. 1977;238(1):66-66.

[140] Fogel RW. The Escape from Hunger and Premature Death, 1700-2100: Europe, America, and the Third World: Cambridge University Press; 2004.

[141] Imamura F, Micha R, Khatibzadeh S, Fahimi S, Shi P, Powles J, Mozaffarian D, Global burden of diseases N, chronic diseases expert G. dietary quality among men and women in 187 countries in 1990 and 2010: A systematic assessment. The Lancet Global health. 2015;3(3):e132-e142.

[142] Lim SS, Vos T, Flaxman AD, Danaei G, Shibuya K, Adair-Rohani H, Amann M, Anderson HR, Andrews KG, Aryee M, et al. A comparative risk assessment of burden of disease and injury attributable to 67 risk factors and risk factor clusters in 21 regions, 1990-2010: a systematic analysis for the Global Burden of Disease Study 2010. Lancet (London, England). 2012;380 (9859):2224-2260.

[143] Masters WA, Rosenblum NZ, Alemu RG. Agricultural transformation, nutrition transition and food policy in Africa: Preston curves reveal new stylised facts. The Journal of Development Studies. 2018;54(5): 788-802.

[144] Vaupel JW, Villavicencio F, Bergeron-Boucher M-P. Demographic perspectives on the rise of longevity. Proc Natl Acad Sci U S A. 2021;118(9): e2019536118.

[145] Therborn G. Cities of Power: The Urban, the National, the Popular, the Global: Verso Books; 2021.

[146] Izcue A, Powrie F. Immunology: Malnutrition promotes rogue bacteria. Nature. 2012;487(7408):437-439.

[147] Cleaveland S, Laurenson MK, Taylor LH. Diseases of humans and their domestic mammals: Pathogen characteristics, host range and the risk of emergence. Philos Trans R Soc Lond B Biol Sci. 2001;356(1411):991-999.

[148] Friant S, Ayambem WA, Alobi AO, Ifebueme NM, Otukpa OM, Ogar DA, Alawa CBI, Goldberg TL, Jacka JK, Rothman JM. Eating Bushmeat improves food security in a biodiversity and infectious disease "hotspot". Ecohealth. 2020;17(1):125-138.

[149] Mason P, Lang T. Sustainable Diets: How Ecological Nutrition Can Transform Consumption and the Food System: Taylor & Francis; 2017.

[150] Spector T. Spoon-Fed: Why Almost Everything we've Been Told about Food Is Wrong: Random House; 2020.

Chapter 2

Folates: An Introduction

Abbas Shams

Abstract

Folate is a naturally occurring essential nutrient which is vital for DNA replication and a necessary substrate in various enzymatic reactions which are involved in synthesis of amino acids and vitamin metabolism. The synthetic and oxidized form of folate is folic acid, it is mainly incorporated into fortified foods and dietary supplements for preventive measures against folate deficiency. Folate deficiency has been linked with several abnormalities in both mother (anemia, peripheral neuropathy) and fetus (congenital abnormalities). Folic acid supplementation taken around the time of conception has been known to alleviate the risk of neural tube defects in the off springs. Optimal intake and absorption of folates is required for the maintenance of the human body's normal functioning and keeping the genomic integrity intact.

Keywords: folate, folic acid, neural tube defects, folate deficiency, congenital abnormalities

1. Introduction

Around 90 years ago anemic pregnant women in India were treated with a yeast extract used by a physician named Lucy Wills [1]. Later, the active compound in the yeast was identified as folate and at that time it was named as anti-anemia factor. The term folates refer to the group of various forms of water-soluble vitamin B9 namely folic acid, dihydrofolate (DHF), tetrahydrofolate (THF), 5,10-methyl tetrahydrofolate and 5-methyl tetrahydrofolate (5-MTHF) [2]. Folate cannot be synthesized by human body due to which we have to obtain it from exogenous diet sources such as beans, citrus fruits, leafy green vegetables, brewer's yeast, and cow's liver. It is available as folic acid in fortified foods, supplements, and multivitamins. It is available as folic acid in its synthetic oxidized form having only one glutamate residue in fortified in fortified foods, supplements, and multivitamins. Folate is essential for the normal functionality of human body as it provides one-carbon groups for biosynthesis of nucleotides, amino acids metabolism, and methylation of DNA [2]. A number of scientific studies have showed that folic acid is effective in the prevention of neural tube defects (NTD's) [3, 4]. NTD's are congenital malformations of structures of central nervous system which occur due to the failure of neural tube closure after conception between 21 and 28 days [5]. Hence optimal intake of folate during early pregnancy is necessary for the developing fetus; there is a higher risk of NTD'S occurrence in offspring during early pregnancy due to folate deficiency. Recently research have shown that normal folate metabolism reduces the concentration of blood homocysteine which in high concentrations may increase the risk of stroke and coronary heart diseases [6]. Studies suggests that during early pregnancy low maternal blood folate might be related to behavioral disorders in childhood and folate may also be a factor in cognitive functions such as

Alzheimer's disease [7, 8]. In order to prevent these disorders and complications it is recommended to keep optimal blood folate levels and homocysteine levels below the accepted cutoff values, the recommended daily allowance of folate is 300 μg/day while there is a recommendation of 400 μg/day for women of reproductive age for prevention of the risk of NTD'S occurrence [9].

1.1 Structure of folate

The term folate represents the group of B vitamins (Vitamin B9) which have similar biological activity to folic acid. There are three major components in the parent structure of folic acid: a pteridine ring that can be either oxidized or reduced, coupled with para-aminobenzoic acid (PABA) through a methylene bridge, which is bound to glutamic acid or polyglutamate by a γ-peptide link [10]. Structure of an oxidized form of folate is illustrated in **Figure 1**, which can be converted to DHF after being reduced at the double bond present at the N-8 position. Further reduction at N-5 double bond leads to the formation of THF and in this state the N-5 of the pteridine moiety and N-10 of the PABA group may act as acceptors of single carbon units [11].

1.2 Folate absorption and one-carbon metabolism

The difference between synthetic folic acid and naturally occurring is that the former is the oxidized monoglutamate form while the latter being the reduced polyglutamate form. Dietary forms of folates are predominantly reduced polyglutamates in order get transported across the intestinal wall these polyglutamates need to be hydrolyzed via glutamate carboxypeptidase II (GCPII) inside the gut into monoglutamate forms which can cross the cell membranes [12]. There are three different types of proteins that helps in transport of monoglutamate folate form across cell membrane, such as proton-coupled folate transporter (PCFT), reduced folate carrier (RFC) and folate receptor proteins (FRα & FRβ). Intestinal uptake of folate appears to be PCFT dependent which is a pH dependent transporter because even if RFC is expressed in the intestine, mutations in the genes encoding for the RFC are not found to be linked with deficiency in folate absorption from intestine; the optimal pH required for folate transport is pH 5.5 [13]. RFC employs reduced folate carrier (RFC) as anion exchanger, cellular uptake of folate is dependent on RFC (supports low- affinity high-capacity uptake system) and folate receptor proteins (supports high- affinity low-capacity uptake system) while folate transport across blood brain barrier appears to require both PCFT and FRα [14]. Both the PCFT and RFC are

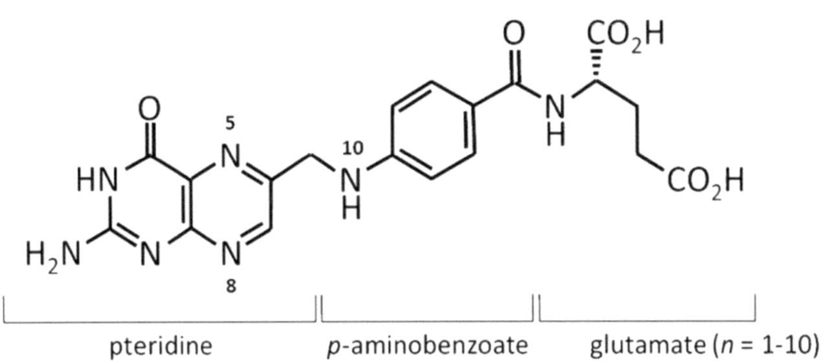

Figure 1.
The structure of folic acid, N-5, N-8, and N-10 are one carbon unit and/or hydrogen acceptors, glutamate may be n = 1 (monoglutamate form) or n = 2–10 (polyglutamate form [11].

different in their specificities towards reduced and oxidized folate species and their expression inside gut, mutations in the genes encoding transporter proteins, leads to complications in the dietary folate uptake from intestine such mutations at the gene ALC46A1 (encodes PCFT) on chromosome 17q11.2, its treatment involves parenteral folate administration as orally administration of folate has been successful in some cases and it is really important to maintain the CSF (cerebrospinal fluid) folate above the levels associated with deficiency which are 15 ng/mL [15, 16].

The set of biochemical reactions mediated by folate cofactors is known as one carbon metabolism because there is transfer of one carbon units inside the enterocytes, firstly conversion from monoglutamate folic acid to dihydrofolate (DHF) occurs and then DHF is converted into tetrahydrofolate (THF) via dihydrofolate reductase (DHFR) which is NADPH dependent enzyme. Once reduction into THF occurs, the N-5 and/or N-10 serve as acceptors for one carbon units and these carbons are transferred at varying oxidation states which depends on their sources and the enzymes which catalyzes the reactions. In the next step THF is converted into 5,10-methylene tetrahydrofolate, where one carbon unit is accepted by THF from serine hydroxymethyl transferase (SHMT), vitamin B6 has a role of cofactor for SHMT, and glycine is released as the final product. Part of 5,10-methylene tetrahydrofolate can be consumed for the synthesis of thymidine by donating its methylene unit or can be used in the de novo purine synthesis where it undergoes oxidation to 10-formyl-THF. Another part of 5,10-methylene tetrahydrofolate leads to the production of 5-methyl-THF in a reduction catalyzed by methylenetetrahydrofolate reductase (MTHFR).

5-MTHF serves as a substrate for the formation of methionine by donating its methyl group to homocysteine (Hcy) and converting it to methionine in a reaction catalyzed by methionine synthase which is a B-12 dependent enzyme. In mammalian cells methionine can be regenerated from homocysteine in a manner which is independent of folate and instead betaine (product of degradation of choline) is used via an enzyme called betaine homocysteine methyltransferase (BHMT), the expression of BHMT is limited only to liver and kidney [17]. Methionine has a vital role in protein synthesis as it can be converted to S-adenosylmethionine (SAM) which is a universal methyl donor and methylates major biomolecules such as adrenaline, carnitine or phosphatidylcholine, SAM is converted into S-adenosylhomocysteine (SAH) which undergoes hydrolysis to produce homocysteine for beginning a new cycle as shown in **Figure 2**. Along with methionine synthase pathway, homocysteine also goes into the transsulfration pathway where in a cystathionine-β-synthase (CBS) catalyzed reaction, it is converted into cystathionine (Cys), cystathionine can be hydrolyzed into cysteine by cystathionine-γ-lyase enzyme as shown in **Figure 2**.

One-carbon metabolism is compartmentalized, and it takes place in compartments inside cell, such as cytosol and mitochondria with nucleus having very low levels of folates [19]. These two subcellular compartments (cytosolic and mitochondrial) have different redox states due to which inside the mitochondria generation of formate from serine and glycine is favored while in cytoplasmic folate metabolism there is incorporation of one-carbon units derived from mitochondrial produced formate into 10-formyl-THF and then to 5,10-methylene-THF, and 5-MTHF as well as synthesis of purines, thymidine along with homocysteine remethylation to form methionine [14]. Some of the metabolic steps such as the interconversion of serine and glycine occurs in both cytosolic and mitochondrial compartments [20]. In nucleus dUMP (deoxyuridine monophosphate) is converted to deoxythymidine monophosphate (dTMP) by thymidylate synthase (TYMS).

Folate one-carbon metabolism depends majorly on factors such as dietary folate intake and genetic polymorphism in the associated genes due to which any

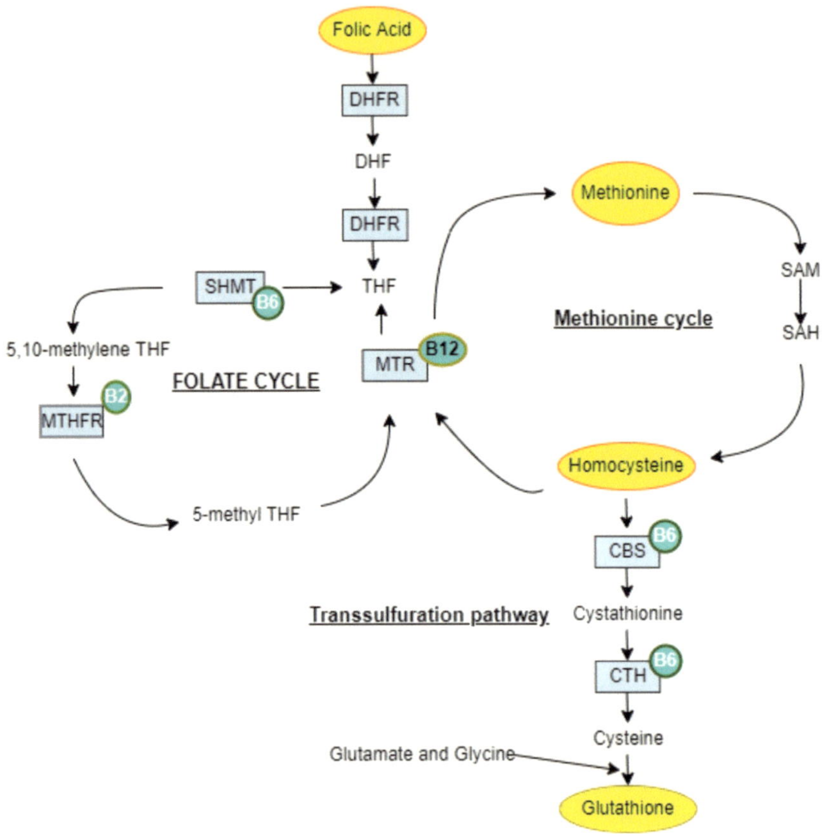

Figure 2.
Folate one-carbon metabolism [18].

disturbances in folate metabolism can lead to disruptions in regulation of DNA synthesis, abnormal homocysteine levels followed by subsequent pathological consequences. Deleterious mutations in associated genes such as a 19 bp deletion in gene encoding DHFR and single base substitutions in gene encoding MTHFR such as C677T and G1958A results in the formation of a thermolabile MTHFR enzyme and ultimately increasing levels of unmetabolized folic acid (UMFA) in blood, hyperhomocysteinemia and underlying health conditions such as cardiac complications and CNS malformations [21, 22].

2. Different species of folate

As we discussed in the introduction that folate is an umbrella term used for a diverse forms of water-soluble Vitamin B9, we will focus on two major folate species the first being the synthetic form called folic acid and the second one called 5-methyl tetrahydrofolate (5-MTHF) which is the active form.

2.1 Folic acid

Folic acid is synthetic form and oxidized state of folate, it is mainly incorporated in fortified foods and dietary supplements for preventive measures against folate deficiency especially during pregnancy. Folic acid is different from naturally occurring folate as shown in **Figure 3** as it has a monoglutamate residue unlike the natural

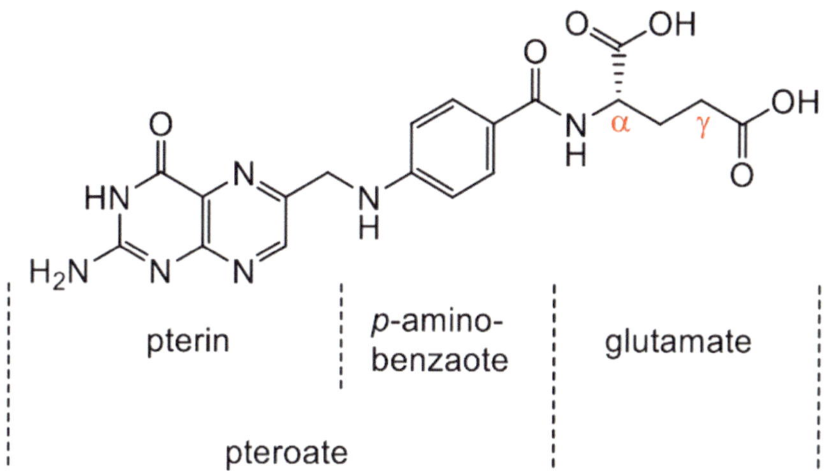

Figure 3.
Structure of folic acid [23].

folates which have polyglutamate residues. In order to be active metabolically folic acid needs to be converted into its reduced form THF, the oxidized form of folate has the advantage of being more heat stable as compared to the natural food form which is more heat labile and light sensitive [24].

Moreover, folic acid is 70% more bioavailable as compared with the other polyglutamates which needs to be hydrolyzed to monoglutamate form in order to get absorbed, it is absorbed more readily because of its monoglutamate nature, there is very minute probability of any adverse effects or hypersensitivity thus making it the best choice for food fortification. Folic acid is vital for necessary metabolic functions and its optimal intake is recommended to maintain normal folate status. In Europe, the recommended daily allowance (RDA) for folic acid is 170–300 μg/day for women and for men it is 200–300 μg/day [25]. Folic acid is the major component in prenatal nutrition as folic acid requirement is increased during pregnancy and lactation in order to prevent NTD's and other congenital complications.

2.2 5-MTHF: the active form of folic acid

5-Methyltetrahydrofolate (5-MTHF), also known as 'Levomefolic acid' accounts for 98% of the entire circulating folate, catalyzed by Methylene tetrahydrofolate reductase (MTHFR) and has several important roles which includes methylation,

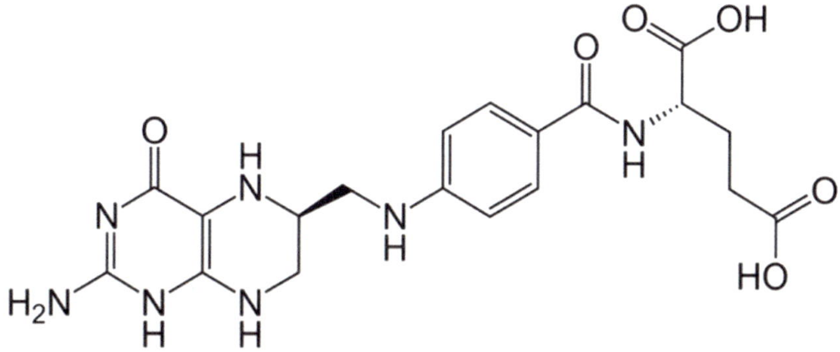

Figure 4.
Structure of 5-MTHF [26].

conversion of homocysteine to methionine, production of serotonin and melatonin. It is also responsible for the synthesis of DNA, the chemical structure of 5-MTHF is illustrated in **Figure 4**.

The L-5-methyl-tetrahydrofolate (L-5-MTHF) is present as the primary form of dietary folate and is also the only form (under the folate category) which is found normally in the systemic circulation. Thus, 5-MTHF is transported into peripheral tissues and is used for the maintenance of cellular metabolism. The crystalline form of L-5-MTHF is available as calcium salt (Metafolin (R)) and is popularly used in the commercial market as a supplement [27]. The data generated by Verhaar and colleagues, [28] indicated that administration of 5-MTHF can restore the endothelial function in hypercholesterolemia patients and this could probably because 5-MTHF has the potency to alter the cellular oxidative metabolism. These findings also tell us about the beneficial exploitation of oral folic acid therapy, which can be helpful in the reduction of cardiovascular disease risk.

It has been proven that 5-MTHF which occurs naturally is more advantageous to the body when compared to synthetic form of folic acid. This is due to the higher absorption rate of the former even when there is a change in the pH of the gastrointestinal tract. Another reason for this property of 5-MTHF is that its overall bioavailability is unaltered in the presence of metabolic defects. The usage of 5-MTHF contrary to folic acid, potentially decreases the masking of certain hematological symptoms of vitamin B12 deficiency. The 5-MTHF is also capable of minimizing the interaction with drugs that cease the action of dihydrofolate reductase and thereby is successful in overcoming metabolic defects caused by methylene tetrahydrofolate reductase polymorphism [29].

In the human body the folate molecule acts as a one-carbon unit carrier, thus it serves as an essential component for the biosynthesis of nucleotides and also in the process of DNA replication. 5-MTHF has a vital role in methionine cycle as it donates a –CH3 group for homocysteine methylation to produce methionine and any imbalance in this process will act as a contributing factor to hyperhomocysteinemia [30].

2.3 Folate deficiency

Folate is an essential nutrient vital for DNA replication and it is a necessary substrate in various enzymatic reactions which are involved in synthesis of amino and vitamin metabolism. The demand for folate increases during pregnancy because it is required for nourishment of the fetus. Folate deficiency has been linked with several abnormalities in both mother (anemia, peripheral neuropathy) and fetus (congenital abnormalities). Folic acid supplementation taken around the time of conception has been known to alleviate the risk of NTD's in the offspring [2].

The term 'folate deficiency anemia' is given to the medical condition in which there is a decrease in the number of red blood cells (anemia) in the blood. It is characterized by the presence of large-sized, abnormal RBC's (megaloblasts), which are formed due to disrupted DNA synthesis. Folate deficiency occurs when the body's demand for folate is not fulfilled, or when there is insufficient dietary intake or inadequate absorption of folate, and when the body loses more folate than usual.

Folate deficiency is also known to have many adversities, such as 'megaloblastic anemia', resulting from deranged cell maturation during erythropoiesis. NTDs, which is caused by failure of neural tube fusion during the early month of pregnancy. Another complication due to folate deficiency is hyperhomocysteinemia, which is a risk factor for cardiovascular and metabolic diseases especially in patients with end stage renal diseases. Folate deficiency usually occurs due to insufficient dietary intake, but other possible reasons are intestinal malabsorption, deranged folate metabolism and elevated demand of folate during pregnancy, or even due to chronic alcoholism. Dietary folate intake can be assessed by the plasma folate concentration, and it is the most widely used method. For long-term status and tissue folate stores, the erythrocyte folate proved to be the best indicator [31].

Both in animals and humans, the methylation status of certain genes at birth can be modified due to folate deficiency, with probable pathogenic and tumorigenic effects in the offspring. The metabolic network of folates can be modified when there is a presence of pre-existing genetic polymorphisms and will also increase the risk of cancer, which include childhood leukemias. The protective effects of folic acid might be dosage dependent, as excessive (hyper) folic acid may have the adverse effect of nourishing certain types of tumors. Thus, it was concluded that the right amount of folate was required for the maintaining normal functioning of the human body and keeping the genomic integrity intact [32].

Abbreviations

DHF	dihydrofolate
THF	tetrahydrofolate
5-MTHF	5-methyl tetrahydrofolate
5,10-MTHF	5,10-methyl tetrahydrofolate
MTHFR	methylene tetrahydrofolate
NTD'S	neural tube defects
RDA	recommended daily allowance
UMFA	unmetabolized folic acid
PABA	para amino benzoic acid
GCPII	glutamate carboxypeptidase II
PCFT	proton-coupled folate transporter
RFC	reduced folate carrier
CSF	cerebrospinal fluid
DHFR	dihydrofolate reductase
NADPH	reduced nicotinamide dinucleotide phosphate
SHMT	serine hydroxymethyl transferase
BHMT	betaine homocysteine methyl transferase
SAM	S-adenosyl methionine
SAH	S-adenosyl homocysteine
CBS	cystathionine-β-synthase
CYS	cystathionine
TYMS	thymidylate synthase
DTMP	deoxythymidine monophosphate
DUMP	deoxyuridine monophosphate

Author details

Abbas Shams
University of Campania Luigi Vanvitelli, Naples, Italy

*Address all correspondence to: abbas.shams@biogem.it

References

[1] Wills L, Clutterbuck PW, Evans BDF. A new factor in the production and cure of macrocytic anaemias and its relation to other haemopoietic principles curative in pernicious anaemia. Biochemical Journal. 1937;**31**(11): 2136-2147. DOI: 10.1042/bj0312136

[2] Crider KS, Bailey LB, Berry RJ. Folic acid food fortification—Its history, effect, concerns, and future directions. Nutrients. 2011;**3**(3):370-384. DOI: 10.3390/nu3030370

[3] De Wals P, Tairou F, Van Allen MI, Lowry RB, Evans JA, Van den Hof MC, et al. Spina bifida before and after folic acid fortification in Canada. Birth Defects Research Part A: Clinical and Molecular Teratology. 2008;**82**(9): 622-626. DOI: 10.1002/bdra.20485

[4] Berry RJ, Li Z, Erickson JD, Li S, Moore CA, Wang H, et al. Prevention of neural-tube defects with folic acid in China. China-U.S. Collaborative Project for Neural Tube Defect Prevention. New England Journal of Medicine. 1999;**341**(20):1485-1490. DOI: 10.1056/ NEJM199911113412001

[5] Blencowe H, Cousens S, Modell B, Lawn J. Folic acid to reduce neonatal mortality from neural tube disorders. International Journal of Epidemiology. 2010;**39**(suppl_1):i110-i121. DOI: 10.1093/ije/dyq028

[6] Cui R, Iso H, Date C, Kikuchi S, Tamakoshi A. Dietary folate and vitamin b6 and B12 intake in relation to mortality from cardiovascular diseases: Japan collaborative cohort study. Stroke. 2010;**41**(6):1285-1289. DOI: 10.1161/ STROKEAHA.110.578906

[7] Schlotz W, Jones A, Phillips DI, Gale CR, Robinson SM, Godfrey KM. Lower maternal folate status in early pregnancy is associated with childhood hyperactivity and peer problems in offspring. Journal of Child Psychology and Psychiatry. 2010;**51**(5):594-602. DOI: 10.1111/j.1469-7610.2009.02182.x

[8] Smith AD. The worldwide challenge of the dementias: A role for B Vitamins and homocysteine? Food and Nutrition Bulletin. 2008;**29**(2_suppl 1):S143-S172. DOI: 10.1177/15648265080292S119

[9] De Benoist B. Conclusions of a WHO Technical Consultation on folate and vitamin B12 deficiencies. Food and Nutrition Bulletin. 2008;**29**(2_suppl1): S238-S244

[10] Shane B. Folate chemistry and metabolism. Clinical Research and Regulatory Affairs. 2001;**18**(3):137-159. DOI: 10.1081/CRP-100108170

[11] Froese DS, Fowler B, Baumgartner MR. Vitamin B12, folate, and the methionine remethylation cycle—Biochemistry, pathways, and regulation. Journal of Inherited Metabolic Disease. 2019;**42**(4):673-685. DOI: 10.1002/jimd.12009

[12] Zhao R, Matherly L, Goldman I. Membrane transporters and folate homeostasis: Intestinal absorption and transport into systemic compartments and tissues. Expert Reviews in Molecular Medicine. 2009;**11**:E4. DOI: 10.1017/S1462399409000969

[13] Zhao R, Diop-Bove N, Visentin M, Goldman ID. Mechanisms of membrane transport of folates into cells and across epithelia. Annual Review of Nutrition. 2011;**31**(1):177-201. DOI: 10.1146/ annurev-nutr-072610-145133

[14] Watkins D, Rosenblatt DS. Update, and new concepts in vitamin responsive disorders of folate transport and metabolism. Journal of Inherited Metabolic Disease. 2012;**35**:665-670. DOI: 10.1007/s10545-011-9418-1

[15] Qiu A, Jansen M, Sakaris A, Min SH, Chattopadhyay S, Tsai E, et al.

Identification of an intestinal folate transporter and the molecular basis for hereditary folate malabsorption. Cell. 2006;**127**(5):917-928. DOI: 10.1016/j.cell.2006.09.041

[16] Inoue K, Nakai Y, Ueda S, Kamigaso S, Ohta KY, Hatakeyama M, et al. Functional characterization of PCFT/HCP1 as the molecular entity of the carrier-mediated intestinal folate transport system in the rat model. American Journal of Physiology-Gastrointestinal and Liver Physiology. 2008;**294**(3):G660-G668. DOI: 10.1152/ajpgi.00309.2007

[17] Pajares MA, Pérez-Sala D. Betaine homocysteine S-methyltransferase: Just a regulator of homocysteine metabolism? Cellular and Molecular Life Sciences. 2006;**63**:2792-2803. DOI: 10.1007/s00018-006-6249-6

[18] Burgess K, Bennett C, Mosnier H, Kwatra N, Bethel F, Jadavji NM. The antioxidant role of one-carbon metabolism on stroke. Antioxidants. 2020;**9**(11):1141. DOI: 10.3390/antiox9111141

[19] Cook RJ, Blair JA. The distribution and chemical nature of radioactive folates in rat liver cells and rat liver mitochondria. Biochemical Journal. 1979;**178**(3):651-659. DOI: 10.1042/bj1780651

[20] Fox JT, Stover PJ. Folate-mediated one-carbon metabolism. Vitamins & Hormones. 2008;**79**:1-44. DOI: 10.1016/S0083-6729(08)00401-9

[21] Munshi A, Rebekah P, Tella S, Akka J. Variants of DHFR, MTHFD1, and MTHFR genes: Relation to CNS malformations. Journal of Obstetrics and Gynaecology of India. 2014; **64**(Suppl 1):24-25. DOI: 10.1007/s13224-013-0429-8

[22] Stanisławska-Sachadyn A, Brown KS, Mitchell LE, et al. An insertion/deletion polymorphism of the dihydrofolate reductase (DHFR) gene is associated with serum and red blood cell folate concentrations in women. Human Genetics. 2008;**123**:289-295. DOI: 10.1007/s00439-008-0475-y

[23] Graffner-Nordberg M, Kolmodin K, Åqvist J, Queener SF, Hallberg A. Journal of Medicinal Chemistry. 2001;**44**(15):2391-2402. DOI: 10.1021/jm010856u

[24] Seyoum E, Selhub J. Properties of food folates determined by stability and susceptibility to intestinal pteroyl polyglutamate hydrolase action. The Journal of Nutrition. 1998;**128**(11): 1956-1960. DOI: 10.1093/jn/128.11.1956

[25] de Bree A, van Dusseldorp M, Brouwer I, et al. Folate intake in Europe: Recommended, actual and desired intake. European Journal of Clinical Nutrition. 1997;**51**:643-660. DOI: 10.1038/sj.ejcn.1600467

[26] Kuznesof PM. 65th JECFA-Chemical and technical assessment (CTA) 2005

[27] Pietrzik K, Bailey L, Shane B. Folic acid and L-5-methyltetrahydrofolate. Clinical Pharmacokinetics. 2010;**49**: 535-548. DOI: 10.2165/11532990-000000000-00000

[28] Verhaar MC, Wever RM, Kastelein JJ, Van Dam T, Koomans HA, Rabelink TJ. 5-methyltetrahydrofolate, the active form of folic acid, restores endothelial function in familial hypercholesterolemia. Circulation, 1998;**97**(3);237-241

[29] Scaglione F, Panzavolta G. Folate, folic acid and 5-methyltetrahydrofolate are not the same thing. Xenobiotica. 2014;**44**(5):480-488. DOI: 10.3109/00498254.2013.845705

[30] Baierle M, Göethel G, Nascimento SN, Charão MF, Moro AM, Brucker N, et al. DNA damage in the

elderly is associated with 5-MTHF levels: A pro-oxidant activity. Toxicology Research. 2017;**6**(3):333-341. DOI: 10.1039/c6tx00247a

[31] Ndiaye NF, Idohou-Dossou N, Diouf A, Guiro AT, Wade S. Folate deficiency and anemia among women of reproductive age (15-49 years) in Senegal: Results of a national cross-sectional survey. Food and Nutrition Bulletin. 2018;**39**(1):65-74. DOI: 10.1177/0379572117739063

[32] Cantarella CD, Ragusa D, Giammanco M, et al. Folate deficiency as predisposing factor for childhood leukaemia: A review of the literature. Genes Nutr. 2017;**12**:14. DOI: 10.1186/s12263-017-0560-8

Chapter 3

Vitamin B9 in Dark Green Vegetables: Deficiency Disorders, Bio-Availability, and Fortification Issues

Jagdish Singh

Abstract

Folic acid is a B complex water-soluble vitamin that is essential to humans, and its deficiency can cause problems including neural tube defects as well as heart-related diseases. An important feature of such vitamins is that they are generally not synthesized by mammalian cells and therefore must be supplied in sufficient amounts in the diet. Folate is a generic term for compounds, possessing vitamin activity similar to that of pteroylglutamic acid, and is the form of the vitamin, which is naturally present in foods. The main dietary sources of folic acid are dark green and leafy vegetables such as spinach, asparagus, romaine lettuce, broccoli, bok choy, turnip green, beet, dried or fresh beans, and peas. The amount of folate that is absorbed and utilized physiologically varies among different food sources and different chemical forms of the vitamin. About 85% of folic acid is estimated to be bioavailable; however, the bioavailability of food folate is estimated at about 50% of folic acid. Several national health authorities have introduced mandatory food fortification with synthetic folic acid, which is considered a convenient fortificant, being cost efficient in production, more stable than natural food folate, and superior in terms of bioavailability and bio-efficacy. Presently, many countries affected by diseases associated with a lack of folic acid have made it mandatory to supplement foods with the vitamin. Considering the need, several analytical procedures were standardized to determine the presence of folic acid in different food matrices. The reported methods are simple, selective, robust, and reproducible and can be used in routine analyses.

Keywords: folic acid, vitamin B9, green leafy vegetables, bio-fortification, bio-availabilty

1. Introduction

Vitamin B9 also called folate or folic acid (FA) or pteroyl-L-glutamic acid is one of the eight water-soluble B vitamins. The chemical name of folic acid is *N*-[4-[[(2-amino-3,4-dihydro-4-oxo-6-pteridinyl)methyl]amino]benzoyl]-L-glutamic acid. The chemical formula of folic acid is $C_{19}H_{19}N_7O_6$ and its molecular weight is 441.4. The IUPAC name is (2*S*)-2-[[4-[(2-amino-4-oxo-3*H*-pteridin-6-yl) methyl amino] benzoyl] amino] pentanedioic acid. Folic acid appears as odorless orange-yellow needles. Folic acid is a synthetic form of folate, found in vitamin supplements and fortified foods. Its structure has been shown in **Figure 1**.

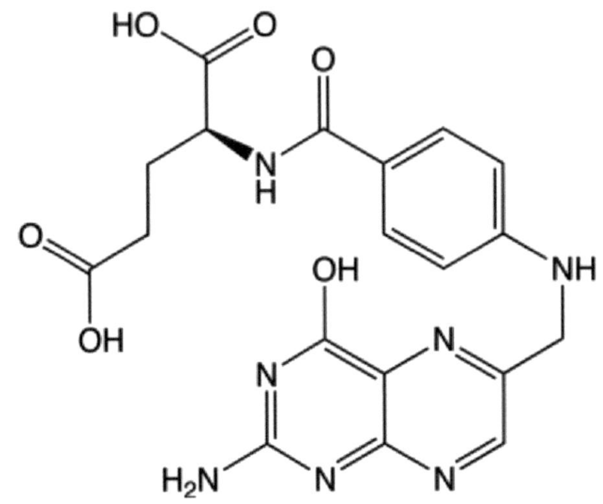

Figure 1.
Chemical structure of the folic acid molecule.

Folate, vitamin B12, and riboflavin have attracted scientific as well as health interest in recent years. Folate has a well-established role in preventing neural tube defects (NTDs); however, there are several other reports highlighting the potential role of folate and B-vitamins in protecting against several lifestyle diseases including heart-related cardiovascular diseases (especially strokes), certain types of cancers, cognitive impairment, and bone-related osteoporosis. Folic acid is involved in carbon transfer reactions of amino acid metabolism, in addition to purine and pyrimidine synthesis, and is essential for hematopoiesis and red blood cell production. It is found in many foods and particularly in leafy green vegetables that are essential for the critical biosynthetic pathways involving the transfer of methyl groups to organic compounds. Folate is important for a range of functions in the body. Folic acid (vitamin B9) works with vitamin B12 and vitamin C to help the body break down, use, and make new proteins. Folate is required in the synthesis of nucleic acids *viz.*, DNA and RNA and is also part of the protein metabolism. It helps in the degradation of homocysteine, which is a risk factor for heart disease. Folic acid is required for growth, reproduction (during gestation and lactation), and antibody formation. As a coenzyme, it is involved in glycine metabolism and is essential for the synthesis of purines, as well as pyrimidines. It plays a major role in cell division and protein synthesis. Its deficiency induces chromosomal abnormalities [1]. It is essential for the formation of RBCs and prevention of folate deficiency anemia [2]. Folic acid deficiency can lead to congenital malformations in the fetus (spina bifida, encephalocele, cleft palate, and hydrocephalus), as well as heart disease [3–5]. Deficiency of folate leads to megaloblastic anemia (a condition where there is a reduction in RBCs, and the red blood cells are larger in size than normal). Other symptoms include weakness, fatigue; irregular heartbeat, shortness of breath, hair loss, pale skin, mouth sores, etc. The nutrient is very important during early pregnancy to reduce the risk of birth defects of the brain and spine [1]. An important feature of this vitamin is that they are generally not synthesized by mammalian cells and therefore must be supplied in sufficient amounts in the diet [6, 7].

2. Recommended dietary allowance (RDA)

The RDA for males and females aged 15 years and older is 400 μg DFE day^{-1} (**Table 1**), whereas the RDA ranges from 65 to 300 μg DFE day^{-1} for ages between

Category	Age	RDA (μg day^{-1})
Infants	0–6 months	65
	6–12 months	80
	1–3 years	150
Children	4–6 years	200
	7–14 years	300
Adults	15+ years	400
Pregnancy		600
Lactation		500

Source: Miller [8].

Table 1.
Recommended dietary allowances (RDA) for folate.

0 and 14 years. The requirement of folate for pregnant and lactating women is comparatively higher, respectively, 600 and 500 μg DFE day^{-1}. The upper tolerable limit (UL) for folic acid has been reported as 1000 μg day^{-1}. Men and women of age 19 years and older should aim for 400 μg DFE. Those who are breastfeeding should aim to take around 500 μg per day. People who regularly drink alcohol should aim for at least 600 mcg DFE of folate daily since alcohol can impair its absorption. Higher daily doses (up to 4 mg) are recommended for women who have had a baby with a neural tube defect. Folate is essentially nontoxic, although there is some concern that high doses may mask pernicious anemia. The body absorbs folic acid from supplements and fortified foods better than the folate from naturally occurring foods.

3. Common symptoms of folate deficiency

A diet lacking folate or folic acid can lead to a folate deficiency. Inadequate levels of folate (vitamin B9) and vitamin B12 during pregnancy have been reported to lead to an increased risk of neural tube defects (NTDs) [9]. Although both are part of the same biopathway, folate deficiency is much more common and therefore it is of much concern [10]. NTDs are birth defects of the brain, spine, or spinal cord and Spina bifida as well as anencephaly are the most common ones. The spinal column of the fetus does not close completely in spina bifida, whereas in anencephaly, most of the brain and skull do not develop properly due to which babies with anencephaly are either stillborn or may die shortly after birth. The peri-conceptional folate supplementation can reduce the risk of neural tube defects (NTDs) and other congenital abnormalities such as cardiovascular malformations (CVMs), cleft lip and palate [4], urogenital abnormalities, and limb reductions [11]. Supplementation with folic acid reduces the prevalence of NTDs by approximately 70% indicating that 30% of these defects are not folate-dependent and are due to some other reasons, rather than alterations of methylation patterns. Many other genes related to NTDs exist, which may be responsible for folate insensitive NTDs. However, folate deficiency can also occur in people who is suffering from celiac disease that prevents the small intestine from absorbing nutrients from foods (malabsorption syndromes). Deficiency of folic acid can cause a wide range of problems in the human body, which may include tiredness, fatigue, and lethargy, besides muscle weakness and other neurological signs, such as tingling, burning, or peripheral neuropathy leading to numbness. It may also cause psychological problems, such as depression and memory problems, and gastrointestinal symptoms, such as nausea, vomiting, abdominal pain, weight loss and diarrhea, headache and dizziness, and shortness of breath. Anemia, particularly megaloblastic anemia, is often the first sign that there is an underlying folate deficiency, and doctors will usually test for folate deficiencies when they encounter anemia. In pernicious anemia, our immune system attacks healthy cells in our stomach, which prevents the absorption of vitamin B12 from the food we eat, and this is the most common cause of vitamin B12 deficiency. Other factors include the lack of these vitamins in our diet. Besides this, certain medicines, including anticonvulsants and proton pump inhibitors (PPIs), can affect how much of these vitamins our body absorbs both. Vitamin B12 deficiency and folate deficiency are more common in older people. The folates are hydrolyzed to monoglutamate in the gut before absorption by active transport across the intestinal mucosa. Sometimes, passive diffusion also occurs when pharmacological doses of folic acid are consumed. Before it enters the bloodstream, the enzyme dihydrofolate reductase reduces the monoglutamate to tetrahydrofolate [12]. The major folate in plasma is 5-methyl-THF. The activity of dihydrofolate reductase varies among individuals [13]. It is yet not known whether the unmetabolized folic acid has any biological activity or it can be used as a biomarker of folate status [14]. Folate is also synthesized by colonic microbiota and can be absorbed across the colon, although the extent to which colonic folate contributes to folate status is unclear [15]. The folate content of the body is estimated to be around 15–30 mg. Half of this amount is stored in the liver and the rest amount is found in blood and body tissues [16]. Normally, the serum folate concentration is used to assess the folate status of the body. A value higher than 3 ng/mL indicates adequacy [17]. The erythrocyte folate concentration provides a longer-term measure of folate intake; a concentration above 140 ng/mL indicates adequate folate status [13, 17]. A combination of serum or erythrocyte folate concentration can also be utilized to assess folate status. Sometimes plasma homocysteine concentration is also used as a functional indicator of folate status because homocysteine levels rise when the body is unable to convert homocysteine to methionine due to deficiency of 5-methyl tetra hydrafolate [17]. Homocysteine levels, however, are not a highly specific indicator of folate status because they can be influenced by other factors, including kidney

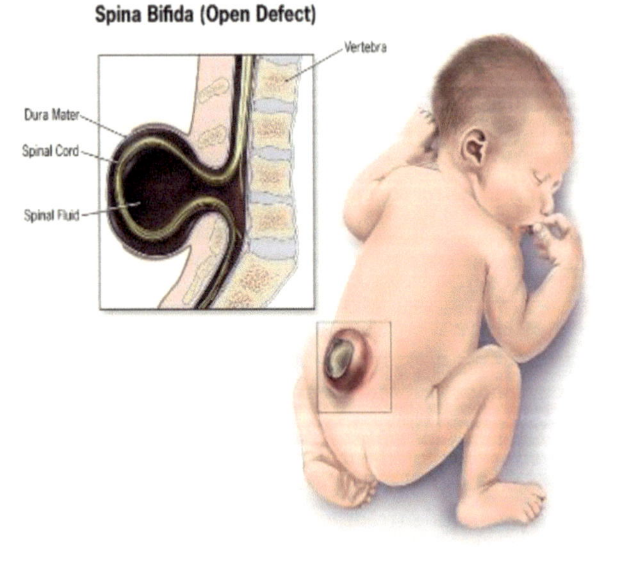

Figure 2.
Neural tube defect—Spina bifida.

dysfunction and deficiencies of vitamin B12 and other micronutrients [17, 18]. The most commonly used cutoff value for elevated homocysteine levels is 16 μmol/L. A homocysteine cutoff of 10 μmol/L has been proposed for assessing folate status in populations (**Figure 2**) [13].

4. Food containing folic acid

Folic acid is not found in natural food sources; however, folate is the natural form of vitamin B9, which is water soluble and naturally found in many foods commonly in dark green and leafy vegetables such as spinach, asparagus, romaine lettuces, broccoli, bok choy, turnip green, beet, dried or fresh beans, and peas (**Table 2**). The main dietary sources of folate are spinach, white beans, asparagus, dark-green leafy vegetables, Brussels sprouts, soybean, orange, and melons [19]. In addition to the aforesaid, beef liver, black-eyed peas, asparagus, lettuce, avocado, broccoli, mustard greens, green peas, kidney beans, canned tomato juice, orange juice, dry-roasted peanuts, fresh orange and grapefruit, papaya, banana, hard-boiled egg, and cantaloupe are also good sources of folate. Other sources of this vitamin include sunflower seeds, avocados, peanuts, orange juice, pineapple juice, cantaloupe, honeydew melon, grapefruit juice, banana, raspberry, papaya, grapefruit, strawberry, corn, and wheat germ. Among animal products, liver (the folate storage organ in mammals) and liver products, whole eggs, and baker's yeast are rich in folates. The major staple crops of the globe *viz.*, rice and maize are low in folates (**Table 3**) [20]. However, pulses and other legumes and green

Sr. No.	Vegetables	Folate content (μg/100 g)
1	Broccoli raw	63
2	Brussels sprout, raw	61
3	Kale, raw	141
4	Collards, raw	129
5	Endive, raw	142
6	Cauliflower green, raw	57
7	Cabbage Chinese (pak-choi), raw	66
8	Cabbage Chinese (pe-tsai), raw	79
9	Cabbage savoy, raw	80
10	Cabbage, raw	43
11	Cauliflower green, raw	57
12	Parsley, fresh	152
13	Spinach, raw	194
14	Peas, edible podded, raw	42
15	Peas green, raw	65
16	Soybean green, raw	165
17	Cowpea (black eyes) immature seeds, raw	168
18	Fava Bean pods, raw	148

Source: USDA Nutrient Database for Standard Reference.

Table 2.
Folate content in raw vegetables.

Sr. No.	Cereals	Folate content (μg/100 g)
1	Rice, brown, long grain, raw	23
2	Rice, white, short grain, raw	06
3	Wheat flour, whole grain, soft wheat	28
4	Wheat flour, whole grain	44
5	Wheat, durum	43
6	Corn grain, yellow	19
Sr. No.	**Pulses**	**Folate content (μg/100 g)**
1	Adzuki bean	622
2	Black bean	444
3	French bean	399
4	Kidney bean red	394
5	Chickpea	557
6	Lentil	479
7	Mung bean	625
8	Peas, split	274
9	Cowpea	659
10	Pigeon pea, immature seeds	173

Source: USDA Nutrient Database for Standard Reference.

Table 3.
Folate content in raw cereal grains and legume seeds.

Food	Serving size	Folic acid/folate per serving* (μg)
Asparagus (cooked)	4 spears	88
Avocados	1 ounce	19
Beans		
Black Beans (cooked from dried)	1 cup	256
Kidney/red beans (canned)	1 cup	131
Lentils (cooked from dried)	1 cup	358
Pinto beans (cooked from dried)	1 cup	294
Chickpeas/Garbanzo beans (canned)	1 cup	161
Chickpeas/garbanzo beans(cooked from dried)	1 cup	282
Blackeye peas (canned)	1 cup	122
Blackeye peas (cooked from dried)	1 cup	358
Bread products (made with enriched flour)		
Bread, white	1 slice	35
Bread, whole wheat	1 slice	14–26
Bread, bagel	1–4 inch bagel	119
Broccoli (cooked)	1 cup	78
Collard greens (cooked)	1 cup	177
Corn on the cob (cooked)	1 ear	35

Food	Serving size	Folic acid/folate per serving* (μg)
Corn (frozen and cooked)	1 cup	51
Okra (cooked)	1 cup	74
Orange	1 orange	39
Orange juice	1 cup	110
Papaya	1 cup	53
Peas		
Green peas, cooked (frozen or canned)	1 cup	75–94
Split peas (cooked from dried)	1 cup	127
Rice, enriched (cooked)	1 cup	195–222
Soybeans (cooked)	1 cup	93
Spinach (raw)	1 cup	58
Spinach (cooked)	1 cup	263

Source: USDA Nutrient Database for Standard Reference, Release 15.

Table 4.
Folate content in cooked food.

leafy vegetables such as spinach, asparagus, lettuce, and Brussels sprouts are rich in folates (**Tables 2** and **3**). It has been reported that folate concentration in rice cultivars ranged from 11.0 to 51 μg/100 g with a mean of 26.0 μg/100 g [20, 21]. Singh [22] have also reported that legumes are a rich source of folates, followed by green vegetables, spices, and cereals. According to USDA [23], the amount of folates varies from 0.1 to 0.5 μg/g in brown rice, which is reduced by 60–67% during milling of rice. The folate content deteriorates on long storage (23%) and also during boiling (48.3%). Folate intake is strongly influenced by various methods of cooking that can degrade the natural forms of the vitamin in foods (**Table 4**). Steaming of spinach or broccoli, in contrast, resulted in no significant decrease in folate content, even for the maximum steaming periods of 4.5 min (spinach) and 15.0 min (broccoli).

5. Folate bioavailability

Folates from natural food sources can enhance the folate status only to a limited extent because of their poor stability while being cooked and also less bioavailability when compared with the synthetic vitamin and folic acid [24]. In addition to the less bioavailability of food folates, the poor stability of folates in foods (particularly green vegetables) while cooking can substantially reduce the amount of vitamin, which is ingested, and this may be an additional factor that limits the ability of food folates from naturally available cooked foods to enhance the folate status. Folate bioavailability from different foods is considered to be dependent on several factors, including the food matrix, the intestinal deconjugation of polyglutamyl folates, the instability of certain labile folates during digestion, and the presence of certain dietary constituents that may enhance folate stability during digestion. However, limited folate bioavailability data are available for vegetables, fruits, cereal products, and fortified foods; hence, it is difficult to evaluate the bioavailability of food folate or whether intervention with food folate improves folate status. The amount of folate that is absorbed and utilized physiologically varies among different food sources and different chemical forms of the vitamin. At least 85% of folic acid is estimated to be bioavailable when taken with food [12, 25], whereas the bioavailability of food folate is commonly estimated at about 50% of folic acid bioavailability [26], but this should be considered as a rough estimate, as data on bioavailability of food folate vary between 30 [24] and 98% [27]. The chemically most stable folate form is synthetic folic acid [28], which is cheap to produce and therefore used for dietary supplements and food fortification. The folic acid consumed as a supplement is highly bioavailable.

It has been reported that the polyglutamyl form of food folates is absorbed in the jejunum as monoglutamyl folate after removal of the polyglutamyl chain by intestinal γ-glutamyl hydrolase [29], which is thereafter reduced and methylated in the enterocyte. The extent of passive diffusion of the reduced and methylated folate across the cell membrane is very limited [30], as it takes place only at high doses. To some extent folate is also absorbed in the colon, and it is suggested that colonic absorption may contribute significantly to total folate absorption [31], but it is still unknown that how relevant this absorption is for maintaining folate status. However, it has been shown for humans [32] that folates synthesized by colon bacteria are bioavailable. Absorbed folate is transported to the liver, which contains about half the body pool of folate [33] and retains 10–20% of absorbed folate due to the first-pass effect [34], while the rest is transported *via* the systemic circulation to body tissues. Some liver folate participates in the enterohepatic circulation and is secreted into bile [35]. However, most biliary folate is reabsorbed, supposedly to moderate between-meal fluctuations in folate supply to cells [36].

National Health and Nutrition Examination Survey data (NHANES 2013–2014) show that the majority population in the United States consume adequate amounts of folate. The average daily intakes of folate from foods range between 417 and 547 μg DFE per day for children between 2 and 19 years of age [37]. However, the mean dietary intakes for males who are 20 years and older are 602 μg DFE and for females, it is 455 μg DFE. It has been reported that although most of the people in the United States consume adequate amounts of folate, there are certain groups such as women of childbearing age and non-Hispanic black women who are still at risk of insufficient folate intakes. It has been further reported that about 35% of adults and 28% of children aged 1–13 years in the United States use supplements containing folic acid [38, 39] to meet their folate requirement. According to estimates of USDA-ARS [37], people aged 2 years and older who consume supplements containing folic acid get a mean of 712 μg DFE from those supplements. Several studies suggest that measurements of folate levels in the erythrocytes further confirm that most people in the United States have adequate folate status. Further there are also some analyses (NHANES 2003–2006), which shows that less than 0.5% of children (aged 1 to 18 years) have deficient folate concentrations in the erythrocytes [40]. Mean concentrations in this age group range from 211 to 294 ng/mL depending on age, dietary habits, and the amount of supplement use. In adults, mean erythrocyte folate concentrations range from 216 to 398 ng/mL, which also indicates the adequate folate status [39].

In contrast to this, there are also reports that some of the population groups are at risk of obtaining excess folic acid, primarily because of the folic acid they obtain from dietary supplements. About 5% of men and women aged between 51 and 70 years and men aged 71 years and older have folic acid intakes exceeding the prescribed upper limit of 1000 μg per day [38]. Furthermore, 30–66% of children aged 1–13 years who take folic acid-containing supplements have intakes of folic acid from both fortified food and dietary supplements exceeding the upper limit of 300–600 μg per day [39]. Almost all children aged 1 to 8 years who consume at least 200 μg/day folic acids from dietary supplements have total intakes that exceed the upper limit [40]. Despite so many reports of excess intakes of folic acid, there is very little information available about the long-term effects of consumption of high folic acid doses in children [14].

6. Folic acid biofortification

Biofortification is a promising and sustainable agriculture-based strategy to minimize Zn and Fe deficiency in dietary food substances [41]. Among the different strategies deployed, the plant breeding approach to develop biofortified crops and agronomic supplementation of micronutrients, such as foliar/soil application along with chemical fertilizers, have received maximum attention [42]. Breeding staple food crops for higher micronutrient contents, where the density of minerals and vitamins in food staples may be increased either through conventional plant breeding or using transgenic techniques. It is recognized as a nutrition-sensitive-agriculture intervention that can reduce vitamin and mineral deficiency [43]. Iron biofortification of beans, cowpea and pearl millet, zinc-biofortification of maize, rice, and wheat, and pro-vitamin A carotenoid-biofortification of cassava, maize, rice, and sweet potato are currently underway and at different stages of development [44, 45]. Results are promising for iron-biofortified crops, as partially iron-biofortified rice has improved the iron stores of reproductive-age women in the Philippines [46], iron-biofortified pearl millet has increased the iron stores and reversed iron deficiency in school children in India [47], and iron-biofortified beans have improved the iron stores in women in Rwanda [48]. The agronomic mode of biofortification includes the application of micronutrient fertilizer directly to the soil and/or foliar application. The agronomic biofortification is most suitable for staple crops with starch as the major component and is mainly practiced on crops such as rice, wheat, maize, sorghum, millet, and sweet potato and also on legumes. The foliar fertilization often results in more uptake of nutrients and ultimately efficient allocation in the edible plant parts [49]. Soil and foliar application combined together gives better results and has been shown as the most effective method for biofortification [42, 50]. Foliar application of micronutrients is generally much more effective in ensuring uptake into the plant because in such cases, immobilization of the nutrients in the soil can be avoided. Alternatively, microorganisms have been bioengineered to overproduce folates. *Bacillus subtilis* was modified at three different levels and an eightfold increase in folate levels was observed [51]. Metabolic engineering has also been developed in *Lactococcus lactis* leading to a more than threefold increase [52].

Several national health authorities have introduced mandatory food fortification with synthetic folic acid, which is considered as convenient fortificant, being cost efficient in production, more stable than natural food folate, and superior in terms of bioavailability and bio-efficacy. It has been reported that the mandatory folic acid fortification in such countries leads to significant increase in both serum and erythrocyte folate concentrations in all sex and age groups. Studies have shown that the mean serum folate concentration increased more than twofold (136%) and the mean erythrocyte folate concentration increased by 57%. The introduction of folic acid-fortified staple foods has effectively decreased the prevalence of NTD in the United States and Canada [53]. It was also observed that fortifying flour with iron and many water-soluble B group vitamins in the United States, Canada, and many other countries has resulted in preventing micronutrient deficiency conditions and is also a very cost-effective prevention of major neural tube defects *viz.*, spina bifida and anencephaly, and also folate deficiency anemia. There are also reports showing that fortification with folic acid has led to a reduction in cases of heart diseases like strokes, which occur due to elevated homocysteine levels. According to the published reports, around 50 countries have implemented the mandatory folic acid fortification program, including the United States and Australia.

In January 1998, the U.S. Food and Drug Administration (FDA) suggested the processing industries to add 140 μg folic acid/100 g to enriched bread, cereals, flours, corn meals, pasta, rice, and other grain products [54] to reduce NTDs. Because cereals and grains are used as staples and are widely consumed, these products have become important supplement of folic acid to the diet. The fortification program increased mean folic acid intakes in the United States by about 190 μg/day [54]. Many other countries, including Costa Rica, Chile, and South Africa, have also established mandatory folic acid fortification programs [55, 56].

7. Analytical methods for the determination of folic acid

Methods reported in the literature for the determination of folic acid include HPLC with different detectors [57], electrophoresis [2], electrochemical methods [58], flow injection analysis [59], and spectrophotometric methods. Recently, some novel spectroscopic methods were reported for routine determination of folic acid. These methods are based on the formation of colored species on binding of folic acid with sodium nitroprusside and ammonia reagent to produce a dark yellow-colored chromogen (λmax at 390). A new, simple, easy, accurate, precise, economic, and sensitive UV spectrophotometric method for the determination of folic acid in commercial tablets has been reported. It was possible to determine the concentration of folic acid in commercial tablets at a λmax of 282.5 nm in a linear range of 1.0–17.5 μg mL^{-1} with an $R2 > 0.9999$ and recovery between 100.6 and 101.1% using a phosphate buffer solution at pH 9.0. De Moura Ribeiro Vinicus et al. [60] reported the spectrophotometric methods for the determination of folic acid in different pharmaceutical formulations, using 0.1 mol L^{-1} NaOH as solvent. This method is simple, selective, and robust.

Author details

Jagdish Singh
ICAR-Indian Institute of Vegetable Research, Varanasi, Uttar Pradesh, India

*Address all correspondence to: jagdish1959@gmail.com

References

[1] Santos LM, Pereira MZ. The effect of folic acid fortification on the reduction of neural tube defects. Biology and Medicine. 2007;**23**(1):17-24

[2] Zhao S, Yuan H, Xie C, Xiao D. Determination of folic acid by capillary electrophoresis with chemiluminescence detection. Journal of Chromatography. A. 2006;**1107**(1):290-293

[3] Crane NT, Wilson DB, Cook DA, Lewis CJ, Yetley EA, Rader JI. Evaluating food fortification options: General principles revisited with folic acid. American Journal of Public Health. 1995;**85**(5):660-666

[4] Czeize AE, Dudas I. Prevention of the first occurrence of neural tube defects by perioconceptional vitamin supplementation. The New England Journal of Medicine. 1992;**327**(226): 1832-1835

[5] Oakley GP, Erickson JD, Adams MJ. Urgent need to increase folic acid consumption. Journal of the American Medical Association. 1995;**274**(21):1717-1718

[6] Aurora-Prado MS, Silva CA, Tavares MF, Altria KD. Determination of folic acid in tablets by microemulsion electrokinetic chromatography. Journal of Chromatography. A. 2004;**1051**(1):291-296

[7] Scott J, R'ebeill'e F, Fletcher J. Folic acid and folate: The feasibility for nutritional enhancement in plant foods. Journal of the Science of Food and Agriculture. 2000;**80**(7):795824

[8] Miller JW. Folic Acid. Encyclopedia of Human Nutrition. 3rd ed2013. eBook ISBN: 9780123848857

[9] Molloy AM, Kirke PN, Troendle JF, Burke H, Sutton M, Brody LC, et al. Maternal vitamin B-12 status and risk of neural tube defects in a population with high neural tube defect prevalence and no folic acid fortification. Pediatrics. 2009;**123**(3):917-923

[10] De-Regil LM, Peña-Rosas JP, Fernández-Gaxiola AC, Rayco-Solon P. Effects and safety of periconceptional oral folate supplementation for preventing birth defects. The Cochrane Database of Systematic Reviews. 2015;**12**:CD007950

[11] Tolarova M. Periconceptional supplementation with vitamins and folic acid to prevent recurrence of cleft lip. Lancet. 1982;**2**(8291):217

[12] Bailey LB. Folate in Health and Disease. 2nd ed. Boca Raton, FL: CRC Press, Taylor and Francis Group; 2010. pp. 473-474

[13] Stover PJ. Folic acid. In: Ross AC, Caballero B, Cousins RJ, Tucker KL, Ziegler TR, editors. Modern Nutrition in Health and Disease. 11th ed. Baltimore, MD: Lippincott Williams & Wilkins; 2012. pp. 358-368

[14] Yetley EA, Pfeiffer CM, Phinney KW, et al. Biomarkers of folate status in NHANES: A roundtable summary. The American Journal of Clinical Nutrition. 2011;**94**:303S-312S

[15] Lakoff A, Fazili Z, Aufreiter S, et al. Folate is absorbed across the human colon: Evidence by using enteric-coated caplets containing 13C-labeled [6S]-5-formyltetrahydrofolate. The American Journal of Clinical Nutrition. 2014;**100**:1278-1286

[16] Bailey LB, Caudill MA. Folate. In: Erdman JW, Macdonald IA, Zeisel SH, editors. Present Knowledge in Nutrition. 10th ed. Washington, DC: Wiley-Blackwell; 2012. pp. 321-342

[17] Bailey LB, Stover PJ, McNulty H, et al. Biomarkers of nutrition for development—Folate review. The Journal of Nutrition. 2015;**145**: 1636S-1680S

[18] Green R. Indicators for assessing folate and vitamin B-12 status and for monitoring the efficacy of intervention strategies. The American Journal of Clinical Nutrition. 2011;**94**:666S-672S

[19] Deconinck E, Crevits S, Baten P, Courselle P, De Beer J. A validated ultra high pressure liquid chromatographic method for qualification and quantification of folic acid in pharmaceutical preparations. Journal of Pharmaceutical and Biomedical Analysis. 2011;**54**(5):995-1000

[20] U.S. Department of Agriculture, Agricultural Research Service. USDA National Nutrient Database for Standard Reference, Release 25, Nutrient Data Laboratory Home Page. 2012. Available from: http://www.ars.usda.gov/ba/bhnrc/ndl.

[21] Ashokkumar K, Sivakumar P, Saradhadevi M. Identification and determination of naturally occurring folates in grains of rice (*Oryza sativa* L.) by UPLC-MS/MS analysis. Natural Product Research. 2017;**23**:1-5

[22] Singh J. Folate content in legumes. Biomedical Journal of Scientific & Technical Research. 2018;**3**(4):3475-3480. DOI: 10.26717/BJSTR.2018.03.000940

[23] U.S. Food and Drug Administration. FDA approves folic acid fortification of corn masa flour. 2016.

[24] Hannon-Fletcher MP, Armstrong NC, Scott JM, Pentieva K, Bradbury I, Ward M, et al. Determining bioavailability of food folates in a controlled intervention study. The American Journal of Clinical Nutrition. 2004;**80**:911-918

[25] Carmel R. Folic acid. In: Shils M, Shike M, Ross A, Caballero B, Cousins RJ, editors. Modern Nutrition in Health and Disease. 11th ed. Baltimore, MD: Lippincott Williams & Wilkins; 2005. pp. 470-481

[26] CDC. Water-Soluble Vitamins & Related Biochemical Compounds. Atlanta, GA, USA: Centers for Disease Control and Prevention; 2009. Available from: http://www.cdc.gov/nutritionreport/part_1.html [Accessed: 13 July 2009]

[27] Brouwer IA, van Dusseldorp M, West C, Meyboom S, Thomas CMG, Duran M, et al. Dietary folate from vegetables and citrus fruit decreases plasma homocysteine concentrations in humans in a dietary controlled trial. The Journal of Nutrition. 1999;**129**: 1135-1139

[28] Gregory JF. Folate. In: Fennema OR, editor. Food Chemistry. New York, NY, USA: Marcel Dekker; 1996. pp. 590-616

[29] Halsted CH. Intestinal absorption of dietary folates. In: Picciano MF, Gregory JF, Stokstad EL, editors. Folic Acid Metabolism in Health and Disease. New York, NY, USA: Wiley-Liss; 1990. pp. 23-46

[30] Zhao R, Matherly LH, Goldman ID. Membrane transporters and folate homeostasis: Intestinal absorption and transport into systemic compartments and tissues. Expert Reviews in Molecular Medicine. 2009;**11**:e4. DOI: 10.1017/S1462399409000969

[31] Aufreiter S, Gregory JF III, Pfeiffer CM, Fazili Z, Kim YI, Marcon N, et al. Folate is absorbed across the colon of adults: Evidence from cecal infusion of 13C-labeled [6*S*]-5-formyltetrahydrofolic acid. The American Journal of Clinical Nutrition. 2009;**90**:116-123

[32] Camilo E, Zimmerman J, Mason JB, Golner B, Russell R, Selhub J, et al.

Folate synthesized by bacteria in the human upper small intestine is assimilated by the host. Gastroenterology. 1996;**110**:991-998

[33] Gregory JF, Williamson J, Liao JF, Bailey LB, Toth JP. Kinetic model of folate metabolism in non pregnant women consuming [H-2(2)] folic acid: Isotopic labeling of urinary folate and the catabolite para-acetamido benzoyl glutamate indicates slow, intake-dependent, turnover of folate pool. The Journal of Nutrition. 1998;**128**: 1896-1906

[34] Gregory JF. The bioavailability of folate. In: Bailey LB, editor. Folate in Health and Disease. New York, NY, USA: Marcel Dekker; 1995. pp. 195-235

[35] Herbert V. Recommended dietary intakes (RDI) of folate in humans. The American Journal of Clinical Nutrition. 1987;**45**:661-670

[36] Lin Y, Dueker SR, Follett JR, Fadel JG, Arjomand A, Schneider PD, et al. Quantitation of in vivo human folate metabolism. The American Journal of Clinical Nutrition. 2004;**80**:680-691

[37] U.S. Department of Agriculture, Agricultural Research Service. What we eat in America, 2013-2014. 2017.

[38] Bailey RL, Dodd KW, Gahche JJ, et al. Total folate and folic acid intake from foods and dietary supplements in the United States: 2003-2006. The American Journal of Clinical Nutrition. 2010a;**91**:231-237

[39] Bailey RL, McDowell MA, Dodd KW, et al. Total folate and folic acid intakes from foods and dietary supplements of US children aged 1-13 y. The American Journal of Clinical Nutrition. 2010b;**92**:353-358

[40] Yeung LF, Cogswell ME, Carriquiry AL, et al. Contributions of enriched cereal-grain products, ready-to-eat cereals, and supplements to folic acid and vitamin B-12 usual intake and folate and vitamin B-12 status in US children: National Health and Nutrition Examination Survey (NHANES), 2003-2006. The American Journal of Clinical Nutrition. 2011;**93**:172-185

[41] Garcia C, Maria N, Juan PPR, Helena P, Luz MDR, Elizabeth CT, et al. Staple crops biofortified with increased micronutrient content: Effects on vitamin and mineral status, as well as health and cognitive function in the general population. Cochrane Database of Systematic Reviews. 2016;**8**. https://doi.org/10.1002/14651858.CD012311

[42] Cakmak I, Pfeiffer WH, McClafferty B. Review: Biofortification of durum wheat with zinc and iron. Cereal Chemistry. 2010;**87**(1):10-20

[43] Ruel MT, Alderman H. Nutrition-sensitive interventions and programmes: How can they help to accelerate progress in improving maternal and child nutrition? Lancet. 2013;**382**(9891):536-551

[44] Bouis HE, Hotz C, McClafferty B, Meenakshi JV, Pfeiffer WH. Biofortification: A new tool to reduce micronutrient malnutrition. Food and Nutrition Bulletin. 2011;**32**(1 Suppl):S31-S40

[45] Saltzman A, Birol B, Bouis HE, Boy E, De Moura FF, Islam Y, et al. Biofortification: Progress toward a more nourishing future. Global Food Security. 2013;**2**:9-17

[46] Haas JD, Beard JL, Murray-Kolb LE, del Mundo AM, Felix A, Gregorio GB. Iron-biofortified rice improves the iron stores of non-anemic Filipino women. The Journal of Nutrition. 2005;**135**: 2823-2830

[47] Finkelstein JL, Mehta S, Udipi SA, Ghugre PS, Luna SV, Wenger MJ, et al. A randomized trial of iron-biofortified

pearl millet in school children in India. The Journal of Nutrition. 2015;**145**(7):1576-1581

[48] Haas JD, Luna SV, Lung'aho MG, Wenger MJ, Murray-Kolb LE, Beebe S, et al. Consuming iron biofortified beans increases iron status in rwandan women after 128 days in a randomized controlled feeding trial. The Journal of Nutrition. 2016;**146**(8):1586-1592

[49] Lawson PG, Daum D, Czaudema R, Meuser H, Harling JW. Soil versus foliar iodine fertilization as a biofortification strategy for field-grown vegetables. Frontiers in Plant Science. 2015;**6**:450

[50] Phattarakul N, Rerkasem B, Li LJ, Wu LH, Zou CQ, Ram H, et al. Biofortification of rice grain with zinc through zinc fertilization in different countries. Plant and Soil. 2012;**361**(1-2): 131-141

[51] Zhu T, Pan Z, Domagalski N, Koepsel R, Ataai MM, Domach MM. Engineering of *Bacillus subtilis* for enhanced total synthesis of folic acid. Applied and Environmental Microbiology. 2005;**71**(11):7122-7129

[52] Sybesma W, Starrenburg M, Kleerebezem M, Mierau I, de Vos WM, Hugenholtz J. Increased production of folate by metabolic engineering of *Lactococcus lactis*. Applied and Environmental Microbiology. 2003;**69**(6):3069-3076

[53] Mosley BS, Cleves MA, Siega-Riz AM, Shaw GM, Canfield MA, Waller DK, et al. Neural tube defects and maternal folate intake among pregnancies conceived after folic acid fortification in the United States. American Journal of Epidemiology. 2009;**169**:9-17

[54] U.S. Food and Drug Administration. Food standards: Amendment of standards of identity for enriched grain products to require addition of folic acid. Federal Register. 1996;**61**: 8781-8797

[55] Centers for Disease Control and Prevention. CDC grand rounds: Additional opportunities to prevent neural tube defects with folic acid fortification. MMWR. Morbidity and Mortality Weekly Report. 2010;**59**: 980-984

[56] Crider KS, Bailey LB, Berry RJ. Folic acid food fortification-its history, effect, concerns, and future directions. Nutrients. 2011;**3**:370-384

[57] Chaudhary A, Wang J, Prabhu S. Development and validation of a high-performance liquid chromatography method for the simultaneous determination of aspirin and folic acid from nano-particulate systems. Biomedical Chromatography. 2010;**24**(9):919-925

[58] Ensafi AA, Karimi-Maleh H. Modified multiwall carbon nanotubes paste electrode as a sensor for simultaneous determination of 6-thioguanine and folic acid using ferrocenedicarboxylic acid as a mediator. Journal of Electroanalytical Chemistry. 2010;**640**(1):75-83

[59] Nie F, He Y, Lu J. An investigation of the chemiluminescence reaction in the sodium hypochlorite-folic acid-emicarbazide hydrochloride system. Microchemical Journal. 2000;**65**(3):319-323

[60] De Moura Ribeiro Vinicus M, Da Silva Melo I, Das Chagas Da Costa Lopes F, Graziella Ciaramella Moita M. Development and validation of a method for the determination of folic acid in different pharmaceutical formulations using derivative spectrophotometry. Brazilian Journal of Pharmaceutical Sciences. 2016:**52**(4). https://doi.org/10.1590/S1984-82502016000400019

Chapter 4

Meat and Vitamin B3: Getting a Grip on Engel's Curve

Adrian C. Williams and Lisa J. Hill

Abstract

We evolved from herbivores to a meat eating "commons" in hunter-gatherer days and then to a non-egalitarian meat power struggle between classes and countries. Egalitarian-ism, trans-egalitarianism and extremes of inequality and hierarchy revolve around the fair-unfair distribution of meat surpluses and ownership of the means of meat production. Poor people on poor diets with too few micronutrients may explain many inequalities of human capital, height and health and divergent development of individuals and nations. Learning from past successes and collapses from switching trophic levels the lesson is that meat moderation toward the top of Engel's curves, not calorie-centrism, is the best recipe for countries and classes. Improved health with longer lives and higher crystallised intelligence comes with an ample supply of micro-nutrients from animal products namely iron, zinc, vitamin A, vitamin B12 and other methyl-donors (such as choline), and nicotinamide (vitamin B3). We concentrate on nicotinamide whose deficits cause the degenerative condition pellagra that manifests as poor emotional and degenerative cognitive states with stunted lives and complex antisocial and dysbiotic effects caused by and causing poverty.

Keywords: Nicotinamide, Vitamin B3, Engel's curve, Poverty, Ageing, Neurodegeneration, Dementia, Cancer, ACE2, Covid-19, Free Energy, Energy gradients

1. Introduction

Seeing world history from a meat perspective was initiated by Hinman and Harris (1939) [1] who believed (as did Homer's heroes and many Gods) meat eating was key to the success and power relationships of nations and to class ascension (as if on the "Great Gatsby Curve"). Later Cokburn (1996) largely concurred as had McCay earlier (1912) [2]. De Castro in his "Geopolitics of Hunger" (1952) [3] was concerned over country and postcode food injustice and argued that Malthus (and Ehrlich 1971), was wrong and cereal dependence with increased fertility was the cause, not the effect, of population booms quoting Doubleday's "True Law of Population" (1842) [4] and significant epidemiological evidence that meat eating reduced fertility. Godwin and Boserup also disagreed with Malthus as they felt population pressure was "the mother of invention" and new agricultural technology as has been true, so far, but has not led to a good diet for the poor ("Golden rice" with extra vitamin A being an exception) and the returns may be diminishing or counter-productive (such as by encouraging population booms) and busts from the neolithic on.

McCarrison (1921) [5] and Boyd-Orr (1936) [6] after observations on differential tribal characteristics (such as the pastoralist Maasai being taller and healthier than cultivators), on different diets in colonial India and Africa, complemented by the discovery of Beri-Beri and Pellagra in the 1940's [7, 8], campaigned for square meals for all. Despite these early initiatives following the science and an analysis of natural experiments, poor nutrition is still the leading cause of death under the age of five as well as causing physical and cognitive stunting and both wasting and obesity. Acute hunger affects 150 million and given little monitoring or screening figures for micronutrient deficiency are likely to be higher than currently recognised, at 2 billion. Geography is critical: if born un-lucky in the lowest 10% of income countries your typical income per head is $3 a day, making meat unaffordable, compared with $3000 plus in the richest countries. Country is now as important as class and that was not as true a century ago [9]. Meat may be the real "stuff" of comparative inequality that could be sorted as a practical non-transcendental move dealing with blighted lives, even if it just a start on fairer initial conditions [10–16]. This is not before time (today even in wealthy Europe 10 cm differences in height exist between the Netherlands and the UK), as even back in 1904 it was realised that poverty and poor diet meant the "tenement" child was stunted and had everlastingly poor odds that meant running the race of life with handicaps [17, 18]. Thomas [17] and Hunter [18] suggested poverty should be seen in clinical, not moral terms, relating to loss of income (including from industrial accidents or illness). Here we use pellagra, an archetypal disease of dietary poverty from a (cotton) market failure, as our generalisable example of such a negative externality easily corrected by social insurance or cash given that Engel's curve predicts that more would get spent on meat [19–22].

Pro-meat views are not exactly the "zeitgeist" of our times as meat is considered to be a major threat to both health and climate but calls to tax it, for instance, could be a serious misstep so an unbiased unpolarised debate is needed [23]. The human right to basic food "entitlements" that enable capabilities, capacities and optimal human capital may need adequate meat and explain the striking benefits of basic income trials [24–28].

2. The demand for meat: predictive pro-active brains

That the demand for meat is high, but elastic, is not surprising as we evolved as an (A (meat) + 2B (vegetable and fruit)), omnivore as have our cuisines and fusion foods. Our closest living relative's (chimpanzees), eat and fish for insects but hunt meat rarely and not fairly. Fairs fare was our social leap in a software with a "Killer App" where needs based and conspicuous sharing of meat obtained as an "affordance" was the social norm (**Figure 1**). More meat in diet, as with the convergent evolution of the Neanderthals, led to bigger brains with nicotinamide as a key ingredient. Meat intake varied by latitude but once obtained there was a sharing culture with shunning of transgressors by the reproductive in-group - even if seeking meat was the usual cause of inter-band warfare and of global diasporas following animal trails [29–31]. Hunter-gatherers had less "stuff" and what they had was often redistributed in feasts and "potlatches": they have been considered an affluent society with leisure and family time for cooperative breeding and teaching [32, 33].

3. Engel's curves: Families, Countries and Classes

The need and umami taste for meat is reflected in Engel's law that he based on family budgets in Belgium (confirmed repeatedly) (**Figure 2**) [34–36]. Modified by Bennett [37], this law states that as families become richer they spend less on starchy food but more on meat. Later attempts to define a poverty food line, where a

Figure 1.
A synopsis of our journey from herbivores to omnivores on a high meat diet and socially compulsory meat sharing to extreme meat inequality. Moderate meat inequality may allow for phenotypic variation in the population with cognitive elites and workers with higher fertility at a cost. Severe meat inequality being the basis of much inequality of opportunity and the diseases of poverty (and affluence) as well as much discrimination and dehumanisation of the poor. Fighting for a fair meat ration has caused much friction and migration in the past and could continue or worsen if corrective action is not taken.

safety net could lie, started with Seebohm Rowntree (1937) [38] and moved on with minimal wages and subsistence rates now $1.9 per day, although this is widely criticised and is not enough to afford any animal products [39, 40]. The nutritional and biochemical advantages of affording meat, other than being a source of calories, revolve around micronutrients not easily obtained from plants such as iron, zinc, Vitamin A and several B vitamins [41–44]. B12 and Choline are important as methyl donors and B3 to the mitochondrial energy supply and as precursor to metabolic nicotinamide-adenine dinucleotide (NAD) - consumer master molecules. NAD is key to brain growth and "human capital" so meat cannot be trivialised as conspicuous consumption as a "Veblen good" (with exceptions), or part of sexual politics (also with exceptions), and our "demonic" past [45] as it is a vital victual.

We may be more like eugenic eusocial insects with diet-induced castes than we like to think [46–48]. In some societies (such as the Taureg), this is explicit with pastoralist

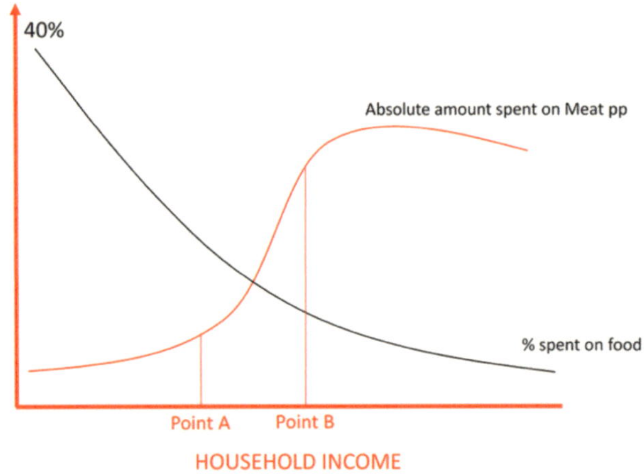

Figure 2.
Engel's law. As incomes rise the proportion spent on food (and rent) falls giving a disposable income for "luxury goods" and education and then services and is the consumer base for modern economies. As incomes rise the absolute spend on meat rises and on starches falls improving human capital and "grey matter" as micronutrient needs are met. There may be increased temptation, compared to the past, to deviate from this law when cheap calories and processed foods, costing 1/5th of vegetable equivalents, are available let alone other commercially available distractions and fashions.

nobles eating the crops of their cultivating slaves [49]. Poor maternal and infant diet may mean that those in poverty are born unequal, making and unmaking minorities, particularly if that is then exacerbated by NAD consuming chemical pollution or excess infections. This is a preferable view point that sets limits that are easily corrected, than the idea of genetically determined human sub-species as first suggested by Linnaeus (1758) [50] and modified by Blumenbach (1795) [51]. Eugenicists and social Darwinists agreed but these views were refuted early by Buffon, Darwin himself and Boas. Boas noted that exposure to new diets changed immigrant people's "fleeting and superficial" differences just as the "Melting Pot" suggested and that hybrids (despite anti-miscegenation laws) were fertile and healthy arguing against separate species - first investigated on Pitcairn Island after the mutiny of the Bounty [52, 53]. Satisfying this need for better diet and meat in new power battles, whose old motives and underhand mechanisms we discuss [54], is important before society can progress and modernise linking meat, disease and demographic transitions, that remain patchy [52–56]. As Darwin noted on his Voyage of the Beagle "***if the misery of our poor be caused not by the laws of nature, but by our institutions, great is our sin***".

4. Supply of meat

Let us first review the supply-side by rough ages of globalisation from pre-history hunting parties to "factory farm" industrialisation in the context of our NAD supply whilst harnessing external sources of energy (originating in NAD dependent photosynthesis) in metabolic network involving the major carbon, hydrogen, phosphorus and nitrogen cycles of life, that we have exploited and disrupted [26, 55–59].

5. Planet of the apes to *Homo sapiens: Meat cornucopia*

The "cradle of mankind" was in the African Rift valley after climate change led to deforestation expanding the savannah grasslands and supporting a population explosion of herbivores. Hominids evolved into this ecological niche and formed prosocial hunting parties aided by new technology from spears to the bow and (poison) arrow and dog domestication. *H. sapiens* emerged with egalitarian "reverse dominance" structures with "reciprocal altruism" to kin and selected non-kin with meat sharing and an extensive cattle vocabulary and spiritual connection with animals and plants [29, 60, 61]. Human phylogeny and ontogeny ("EcoDevo") revolved around cooperative feeding and a shared intentionality over resourcing meat, and over breeding with alloparenting, in three-generation families. Long dependent childhoods allow for gene expression intersecting with diet and epigenetic learning (such as hunting and gathering from tribal encyclopedias), allowing easier adaptation to changing environments [62, 63].

Pit roasting or boiling were the cooking (extra-somatic digestion), methods for meat and drying then salting for storage in newly invented pottery in "delayed return" societies [64]. Hunting and trapping and fishing (some forming the "Salmon nations"), parties crossed the globe, including via Beringia, extirpating animal (such as the woolly mammoth) and fish, sea-mammal and bird species - that alongside climate change led to the demise of megafauna and meat shortages temporarily alleviated by concentrating on smaller game. The interest and obsession about meat and cattle is well preserved early in Cave paintings of Aurochs and Mammoths and sculptures (such as the "Lion man" representing hunting prowess) and other clay artefacts suggesting an early religious element with a veneration of cattle before they were even domesticated – when later some became gods and pastoralism was depicted in rock art, wall paintings and figurines [65].

This was the longest successful subsistence economy in our history and was the environment to which we largely adapted. Extra amylases (to digest starches), and lactase persistence and milk, bread and beer fermentation produce are subsequent genetic and cultural co-evolutions. This period was culturally, with the origins of science and music [66] and metabolically affluent with the sort of joined-up thinking that later influenced the physiocrats [67–69]. The advent of cooking and life of yeasts added to this metabolic thread – now lost and needing a new social contract [70–73]. Populations on the ancestral diet remained low, affecting division of labour, so it may be a "Paleofantasy" that these high meat diets were best, as they must have changed for a reason [74].

6. The (Neolithic) agricultural revolution and cerealization: grass roots

Homo evolved as a food producer and processor first with horticulture and play farming then cultivation of grasses (that we cannot digest easily unlike ruminants), and the co-domestication of cereals [75, 76]. Animal domestication and pastoralism came later in trading or conquest driven relationships that usually involved a degree of specialisation with nomads providing the meat [77–79]. Herdsmen and pastoralists come in various forms, starting around 1000 BC with the humpbacked sanga cattle in Ethiopia and reindeer further north, depending on the seasonal transhumance distances between summer and winter pastures that need to be travelled. Despite managing a distributive and exchange system, specialist pastoralists, such as those with horses and wheels, drove meat inequality hard with property ownership (cattle, rather than land), and controlled meat surpluses [80–83]. Seafood was exploited more variably and often as a "famine food" perhaps as the "protein punch" was low and the effort high or over-fishing is too easy. Some in the "fertile crescent" were better off in "lucky latitudes" across temperate zones with mixed farming roping in animal predomesticates than others in the Americas and Africa (who also had carnivores and parasitic microbes to contend with in the tick and "Tsetse belt".

We can now see various seeds of hierarchy (worse in the hunting season), inequality and how a trans-egalitarianism centred on the meat supply and its ownership might have developed [84, 85]. Sedentism, materialism and social stratification and pathology emerged with a well fed ruling "cognitive" non-food producing class and a less well-fed worker and slave class as a proletariat with, by definition, high fertility [86–88]. Meat became commodified and Engel's curve kicked in and got distorted later with the need for money competing over just about everything that used to be a common good or was a new luxury [89]. Where and when meat was scarce captured proles were used as (as was consecrated sacrificials) meat in the Americas as state-registered cannibalism; less-coercive mechanisms of soft-power elsewhere made sure the rich were easily able to afford meat [90]. A (2nd) cerealization and calorieization event around 1000 AD in Europe specifically separated meat-eaters from grain-eaters with social penalties for transgressors in "Bourdieuian" social enclosures. Another transition phase involved tributary-redistribution by the feudal upper classes putting on regular feast days or by monasteries but then our niche construction comestibles were trusted, perhaps unwisely, to the market place [91–93].

7. Enclosure movements, clearances and conservation

Physical enclosures restricted the commons and hunting and pasture rights for serfs [94–97]. The argument used was that productivity is increased by landowners, by crop rotation improving nitrogen fixation and winter feed (turnips) for animals, compared with the overgrazing or overfishing of the "tragedy of the commons" or "prisoner's dilemma". Such arguments can be true of socialist strategies or privatisation [98] or cooperation as in "Stag Hunt" games and farmer cooperatives. Serfs were denied access to free meat and poachers fined or paid with capital punishment as in the Black act of 1723. "Rule of Capture", or gleaning rarely applied. Thomas' Paine (1796) in his "Agrarian Justice" [99] and the more radical Spence (1829) [100] started a chain of thinking about landowner's debt to others in society and that "homesteading" (as in "go west young man"), or at least compensation for the removal of pastureland should be available to all [101–103]. By forcing workers to work for an income, now usually allotments in cities, the poor were at the mercy of market forces and the price of meat (factory workers often paying three-fifths of their wages in food).

Expropriation of land resources, as in the Scottish Clearances, also damaged local peasant farmers to the benefit of the meat supply to richer cities, becoming yet another basic mechanism of inequality and discrimination with loss of social and human capital for many – who often emigrated [104–106]. Despite higher incomes and more meat per person for a while after the population decline caused by the Black Death, by the 1700s the poor became virtually vegetarian [107–109].

Conservation efforts later had similar effects, even in the American land of abundance, born amongst worries about non-white population increases destroying habitats to the further disadvantage of the poor with less pasture and hunting rights, driving poaching [110, 111]. The Cape is also a good example, as was India, as white hunters emulating aristocrats largely exterminated big game (after ivory and trophies) with improving rifles up to the High Veld and Limpopo river and set up safari reservations whilst denying locals any right to hunting despite their cattle being devastated by Rinderpest creating pellagra outbreaks - eventually the camera replaced the gun [112].

8. The Columbian exchange and aftermath

This early globalisation event is relevant as the Americas were poorly off for animal domesticates (early hunters had killed off both megafauna and horses).

Low meat diets may have contributed to their easy conquest by the Conquistadors' superior technology and cognitive know-how and by infection, as somewhat immunosuppressed [113]. Here is an example of a low meat culture being "laggards" with their late use of the alphabet, wheels and metallurgy. The Spanish brought horses, sheep, and goats and built ranches on an industrial scale with even feral "plagues of sheep" and a glut of cheap meat: whilst the imports of cereals and tubers contributed to "plague of corn" population explosions in the east [114–116]. Appropriation of Indian lands and the extinguishing of Bison and the rise of cattle/cowboy and "red meat republics" led to a high meat diet for white Americans and a poor meat diet for indigenous peoples [117]. The slave trade also led to a poor low (fatty/flap) meat diet for those forcibly imported from Africa and other indentured peoples.

9. Intermission for some early reflections

It cannot be a coincidence that these nutritionally deprived folk were felt to be an "inferior" race although racist, sexist and genetic explanations were invoked even if "Black Spartacus" revolts and humanist ideas gave abolitionists support for the idea we are, indeed, one race. Poor meat, milk and high corn diets were so rife they caused pellagra whose members with poor cognitive and social skills were known as the "Butterfly caste" and considered a degenerate race facing discrimination and the attention of eugenic movements. The eugenicist answer was sterilisation not diet, as discovered by Joseph Goldberger [118]. At times agriculture and maize has been rejected perhaps for this very reason as an "ecology of freedoms" with the ambivalence of "play farming".

This ability to create a man-made economic and dietary environment and then blame the victims or consider them sub-human even in their (spiritual) responses is a fundamental attribution error made easier by physical segregation; even Lombroso, a famous pellagraologist, is best known for proposing (1887) the atavistic "*l'uomo delinquent*" [119–122]. The second amendment did not allow slaves or freedmen to carry arms (except when they ran out of white soldiers), impairing their ability to hunt as did the use of their lands for plantations [123]. Denying such people access to education, health care (believing they have higher pain thresholds or prone to "Cachexia Africana"), or immigration or property rights (losing out on capital gains and heritable wealth increasing their debt) further reinforces stereotyping and ghettos, whilst obscuring the basic dietary cause that is lifelong and transgenerational [124–127].

Perhaps our original social sin is of not sharing, rather than the eating of meat and is the real reason for guilt and blanking over (unfair) access to meat. Displacement to the "narcissism of small differences" with discrimination over surface matters, particularly to people close to us physically or ethnically, is stronger than "clashes of civilisations" or even religions. Such trivia (with no biological basis), may be microagressive "smoke screens" over competition over neighbouring means of meat production (as in the "curse of Cain" the farmer and founder of nations), and is something we need to overcome with a "species think" more empathetic to strangers on lower rungs (with the added benefit of less loneliness and more diverse friends that also drives ill health [128–133]. Hope comes from the relatively recent history of divide and rule racism (after Bacon's rebellion of 1676 that united poor whites and blacks but then "Jim Crow" laws took over for decades until the success of Civil Rights Movements), suggesting these attitudes are not "hard-wired" in our amygdala and pre-frontal/anterior temporal/posterior cingulate cortical areas (that degenerate in pellagra), and that implicit bias can improve once others are seen as part of a coalition for a common cause, such as survival of the species [134–137].

10. Back to colonial times: Space and Race

It is useful here to differentiate between "settler" colonies, the "plantation" and the "conquered" states. "New Europe's" such as in Australasia were similar to the Americas in that, at the expense of genocides/holocausts of indigenous populations, they created a meat surplus to export to the European core [138, 139]. Plantation colony peoples (other than the "planters"), suffered from a deterioration in their diet as land and labour were swept up in cash crop farming. Conquered states, such as India, had peasants taxed (providing the European centre with "free lunches"), in a "beggar thy neighbour" policy reducing incomes and impoverishing an already vegetarian diet - with no mercy given in the book-end Bengal famines [140–144]. Repercussions of these policies often born during the cotton empires continue to shape our world with little sight of reparations and restitution to those worst affected [145–148]. For instance, it is well established that dietary intervention during pregnancy improves birth weight and development and academic achievement and blocks the programming of future ailments (including the metabolic syndrome) as first noted in the Dutch "Hunger Winter" of 1944 [149]. So it is important that "battle fatigue" does not set in to this opportunity to correct constructively a past injustice.

More positively imperialism did, like the Columbian Exchange, lead to significant transfers of plants with nutritional or economic - such as rubber) or medicinal value (such as quinine) largely by the Kew garden network instigated by botanists Banks and Hooker) - and animal species, some more productive such as the Hereford and Aberdeen Angus cattle (that largely replaced the shorthorn and Texas longhorn in all but the most arid areas) [150–152].

11. 1850s meat transition in Western Europe

After the 1700s when only the rich ate much meat a meat transition started in Britain ("Rosbifs" as depicted by Hogarth) with its "Hungry Empire" [153]. These times had started with well documented ill health, height and opiate and alcohol issues and (Chartist) hunger-revolts and "knife and fork" "milk famine" issues that spawned the luddites and trades unions as incomes were falling (some families living on boiled nettles). The new Poor Laws were harsh and the country flirted with revolution as happened across Europe [154]. Meat intake then doubled [155, 156]. Innovative agricultural practices and stock breeding increased production that just about kept up with population growth. The important increase came from imports from the "settler" colonies, and "Atlantic" cod, aided by advances in steam transportation and refrigeration (frozen meat imports to Britain were 200,000 tons in 1900 and 350,000 tons a decade later), with price drops that, for a change, benefited the urban poor, all paid for by profits from the cotton trade and other plantation income [157]. Britain accounted for a huge proportion of the transcontinental and transequatorial meat trade [158]. Repeal of the Corn Laws and repeal of excises on meat also helped the working class [23]. This was a period of significant food price deflation with cheap meat and vegetables attributed to increased productivity from technological improvements (including the cotton gin market but the market collapsed triggering the pellagra outbreak in the Americas) [159]. The influence of luminaries such as Thomas Paine's "Rights of Man" (1791) [160] and Robert Owen (1813) [161] "*man's in-humanity to man*" helped let alone Marx and Engels (Condition of the Working Class in England) and the Quaker movement. All efforts contributed to better diet and better (brain) health and longevity.

Infectious diseases particularly Tuberculosis virtually disappeared even if replaced by autoimmune diseases whilst fertility declined completing the first modern disease and demographic transitions [162, 163].

12. More recent times: place of markets

These have seen supply side "Green" revolutions using artificial fertilisers and hybrid or genetically modified seeds with higher yields and higher micronutrient content [164, 165]. This has resulted in continued "Calorieization" and "Cerealization" for the poor but "Meatification" for the rich and aspirant middle-classes. Grain Aid and subsidies concentrate on cereals (some used to feed animals). Rich countries such as the USA and Russia (meat and wheat diplomacy) have turned into major exporters of food with less produced locally putting poor nations at the mercy of international price hikes – as happened with the price hikes of 2008 after the economic meltdown bringing the poorest to the brink of famine [116, 166]. This reversal has paradoxes given Africa has enormous land banks and plenty of sun yet its agriculture would be familiar to those alive at the dawn of Christianity with a lack of irrigation - leaving it ill prepared for already baffling rainfall conditions and friction with nomads as the Sahara marches south. Russia, by contrast, may gain from melting permafrost. This is harsh as Africa contribute almost nothing to climate change even if there is now an opportunity, if supported, to leap frog with clean and electric technology. Rich countries in the Global North, who have greater access to the latest technology and economies of scale (favouring monocrops), and have political power through mega-merged corporations, including meat processors, have by manipulating tariffs and trade rules supported their own farmers with subsidies (such as in the USA and EU). World Bank rules insist on neoliberal free-trade policies for others earlier in development even though this distorts the market place to the benefit of the north.

Pushback on this faultline of neoliberal policies and the "dominium" of neo-colonialism, being grimmer than the "imperium," was helped by the oil embargo of 1973–4 and the formation of the G-77 as a global south poor nations trade union, achieving some potency including over food supplies [167–169]. However these policies shut out many small farmers (many of them women), from markets who then become impoverished themselves on poor diets [170–172]. This all inhibits local development that needs early protection from cheap imports unless there is a local competitive advantage which there is in the tropics as in "Banana republics" but this is not true of staples or cattle that prefer temperate zones. These 'meatonomics' with subsidiaries cause meat deflation in the already rich world aided by ample oil induced efficiency gains.

13. Pastoralism – mixed farming

Pastoralists in Africa are threatened despite producing a high proportion of meat and milk and using geographically isolated range-land unsuitable for crops as common pastureland and water resources get enclosed by developers and transhumance routes (necessary for the best seasonal grazing and to avoid the tsetse fly), blocked so that they get marginalised and cannot afford their own produce and do not get integrated with mixed farming and agro-pastoralism that is not the simplest of evolutionary progressions. There is a false assumption that pastoralists damage ecosystems and equilibria through over-grazing and exceeding the "carrying capacity" of the land. More thought needs to go in to this if meat production is to be increased including optimising vaccination programs [173, 174].

14. Reap and sow

The vast grain export as Aid helped farmers in the USA dispose of their surpluses helping short term but did not encourage local development. Some interventions back-fired badly such as peanut farming in Tanzania, apple farming in Nepal and the exacerbation of a food crisis in Malawi [175]. Green developments can be criticised for fueling population explosions rather than credited for alleviating hunger. Import substitution encourages empty calorization of locally processed foods and drinks and is responsible for the "commercial determinants" of ill health and the triple burden of combining diseases of poverty with obesity rather than providing a balanced omnivorous diet [176–179].

The modern period has, despite the above, seen some telling upswings for the majority (such as 1945–1975), that relate to more egalitarian diets with more milk and meat after World War 2 rationing. A 30 years "Trente Glorieuses", helped by the rise of welfare states, lasted longer than relief from the exigencies of war would explain. Divergences related to more neoliberal policies and less attention to or sympathy for welfare and the incomplete success of civil rights, consumer or cultural power (music, sport) suggest something more fundamental [180–182]. Meat transitions are nevertheless in progress as in China (the high price however favouring the upper and urban middle classes), but are only slowly occurring in India and only arguably in sub-Saharan Africa.

15. Meat downsides

Downsides of meat production and consumption given the "Long Shadow" of the livestock revolution and ecological "Hoof-prints" are well rehearsed as are health consequences of meat gluttony from gout onward to (bowel) cancer [183–185]. The knives are out for "hamburger" carnivores and this is not surprising given meat's green-house gas effects (CO_2, CH_4 and N_2O [186]), let alone animal rights concerns and risk of food poisoning (e.g. E.coli, Salmonella and Campylobacter), and emergent zoonoses, including COVID-19).

Agri-businesses encroach on land over and above the pastureland that could not be used for crops, that is grabbed and de-forested with little attention to improving the soil long-term leading to desertification -and uses gigantic quantities of fossil fuels as "food-miles" and for artificial fertilisers with consequent loss of biodiversity and insect apocalypses, from pesticide use [187, 188]. Meat markets and farms are under intense pressure, by the poor and the rich (with their exotic tastes), to produce in close ecological proximities often linked to de-forestation increasing risk of (bat) coronaviruses crossing species boundaries.

Rich Americans eat their body weight or more in meat per annum [189]. The poor in sub-Saharan Africa and other places in the "Dickensian" Global South eat negligible amounts. In middle- and low-income countries there is a marked class divide with the rich eating a great deal. Even in the USA or UK, the very poor can live in food deserts and metabolic ghettoes reliant on food banks when austerity is policy [190]. In 1962 the average for the Chinese was 6Kg of meat pa but that figure is now 60Kg and rising toward the American average of 120Kg [191].

Ten calories of animal feed (grain, soybean and fishmeal more than grass), produce one calorie of meat and require enormous quantities of land (deforestation), water, oil, artificial fertilisers (fixing nitrogen using fossil fuels and allowing eutrophication), antibiotics (enabling evolution of superbugs and not even given for therapeutic purposes), and pesticides ("silent springs") [192]. Some relief comes from less use of ruminants and more of mono-gastric chickens and pigs (badly affected by pandemic induced culls) and plant-based green-marketed foods and drinks as meat and dairy substitutes.

The case against meat is, as a result of the above, strong as a major contributor (30–40%) to greenhouse gas emissions (methane and N_2O ' tipping points' as they are potent but have shorter half-lives than CO_2 but do not have its fertiliser effect), and climate change (1.4%F since 1880) [193]. Meat intake is currently running at 300 million metric tonnes having been 7 million in 1960 and could rise by another 75% by 2050: artificial meats are unlikely to be the whole answer to curb this hunger but other mitigations such as less beef and more insects in diet are possible [194].

16. Vegetarianism

Widespread vegetarianism is usually proposed by those in the rich world who have all micronutrients available from other, often supplemented, foods not available to poor economic vegetarians [195, 196]. Embarrassment over meat eating and distancing from the sight of slaughter has a long history dating back to Pythagoras and tied up with concerns about man's place in nature "***Every moving thing that liveth shall be meat for you***" (Genesis ix.2–3), versus "***Take not away the life you cannot give: For all things have a right to live***" (Dryden's Ovid),- up to 17th and 18th C tracts and modern animal rights and animal sentience campaigns and research [197, 198]. There is resistance to vegetarianism culturally, despite "Veganuary," and converting to some plant-based diets that are heavily processed or engineered corn now 'super-sweet' can increase emissions and may have health risks or be too expensive [199–203]. Our new found feelings about stewardship of animals should extend to different geographies as "out of sight out of mind" may shield us emotionally from slaughter houses but should not work for other human-beings on poor monophagic vegetarian diets [204].

17. Omnivore's dilemma or dialectic?

Thus, we have contradictory views on meat – is it our maker or breaker? The argument against meat is aimed at those who eat and waste a lot but relatively little discourse is about the needs of the poor and health inequalities and their "killing fields" [205–208]. Perhaps it is less of a dilemma if it is the high variances of meat intake that are the real problem alongside the replacement of vegetables with processed cereals and their products (such as high fructose corn syrup) and that this basic inequality affecting "food for thought" is driving many of the more obvious "Syndemic" problems from zoonoses to climate change to risk of wars and much ill health at the lower end of the income scales [209]. Let us look again at Engel's Law [210].

18. Engel's Law: Basic niche construction

In 2011 in Norway (GDP per person $62000) food budget share was 10% whereas in Chad (GDP per person $2000) the share was 52%. Poor households have to move up the income scale before the spend on starchy foods falls and the amount spent on the more elastic need for meat rises. It then can fall as the rich become "Flexitarian" in their eating habits (as suggested in Britain by George Cheyne (1733) and by Hippocrates with his "Regimens" much earlier [211].

An unfortunate modern twist, whether poor in America or Africa, is to achieve food security by spending on high fat and high sugar processed foods (or materialistic goods from cars to phones), rather than low density fruit and vegetables that used to supply some micronutrients (such as folate and flavonoids let alone psychoactive com-pounds in Victorian times when they were cheap) even when meat was unaffordable. This plays into the commercial determinants of disease in "fast food swamps" – with individuals then stigmatised for their life-style rather than the correct label of poor life circumstances [212, 213].

States have traditionally taken some interest up to point 1 providing grain but not point 2 as meat is often felt to be a hedonic "luxury" rather than having a health utility. (Henri IV was an exception in his push for "a chicken in every pot" and Papin's "New Digester" pressure cooker (1679) helped poor to cook cheap animal products). Inflation is a painful tax on low to moderate income house-holds as are many austerity measures or actual taxes on food and drink – as is sometimes proposed for meat. Many rulers, whether fascist or socialist, have even put conditions on supplying grain to useful workers and starved the "undeserving poor" or dissidents. Exceptions included the Kennedy's who pointed out "***the obscenity of conspicuous wealth amongst public squalor***" backed up by some intellectuals in a tradition from Aristotle, and glimpsed by Mill, on emphasising a basic need for a good diet before other human attributes can really take off [214–218].

States and markets often ensure the extra meat goes to the urban middle classes (or middling diets) turning Engel's Law into a political mechanism. Egypt, for example, in the 1970s imported grain for animal feed so that the richest 25% ate four times as much meat as the poorest 25% (and had form as their rulers suppressed wages after the Black Death such that there was no survivor benefit for workers diets, as was seen in most of Europe) [219].

Other societies channel meat to the rich by restricting access to wildlife that normally favour the poor but then becomes an exotic commodity food for the rich. In Samoa hunting pigeons is comparable to English aristocrats hunting deer and as in China and earlier in the West wild delicacies from the land or the high seas were only inhibited when enforced by much stricter conservation laws [220, 221].

19. Preston's curvilinear curves

Meat is an important component of any "unified growth theory" and is an achievable developmental goal. Prestons' curves suggest that some element of modernity with a low ceiling effect after which there are diminishing returns can improve longevity, health and happiness as measured, for instance, by Happy Planet indices (**Figure 3**) [222]. Low-income poverty traps and middle-income traps (the "Argentina paradox") with economic divergences may be because Engel's curves were not satisfied for enough of the people who then can not obtain college degrees [40]. Point 2 on Engel's curve should be the target to avoid the "hidden hunger" of micronutrient deficiencies [223]. As Kristof has written "***Future generations will be baffled at our heartlessness and our indifference to suffering in impoverished countries***" [224].

20. Wealth and health of nations: Lands of milk, meat and honey

Until a reasonable proportion of the population have climbed Engel's curve either because the cost of grain and meat has fallen (due to increased productivity

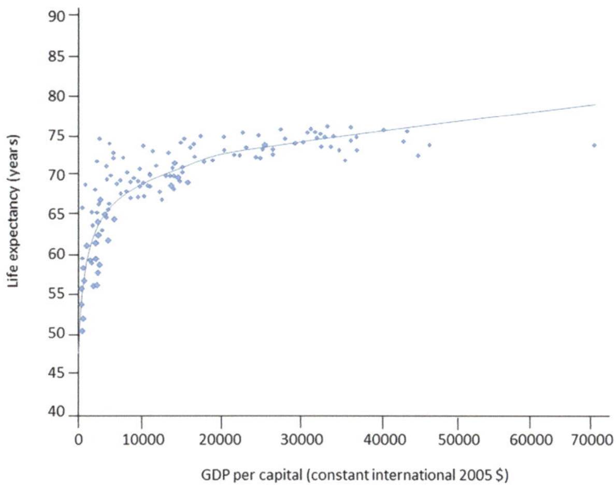

Figure 3.
Preston's curves show the importance of income inequality to longevity and health and happiness and chances of rising up the social elevator. These curves rapidly flatten with a low ceiling effect. Better hygiene and education or medical innovations have been considered causative agents but better diets are the more likely factor. The ceiling effect would accord with Engel's curves and would not be that hard to achieve at a low cost and with more sharing and less waste.

in the agricultural sector or because that country can afford to import more) consumers do not have enough discretionary income and the motivation or "marginal propensity to consume." [214, 225]. Living from paycheck to paycheck with debt and little capital does not allow spend on luxury goods (that include animal secondary products) or when that is satiated services – in other words the engine of want with a modern economy and a market that appreciates new ideas, technology and institutional governance [226–228]. The rise of consumerism, such as clothes, may however, as first noted by Marshall (1890) [229] and Veblen (1899) [230], be a chief driver of climate change and is fuelled by inequality and status anxiety that might be constrained by more equality. More educated better fed "sympathetic consumers" date back to abolitionists campaigns to reduce consumption of sugar and rum and is now evident in "Fair Trade" movements aware of invisible labourers and animal rights would be the aim resuscitating environmental Kuznets curves claims [231]. Power disparities should lessen with less income inequality allowing all to live in cleaner places and move up on the Great Gatsby curve by allowing all to "Green-up" as environmentalists [232, 233].

21. Declines and falls: Milk and meat famines

This underlying process for the wealth and health of nations has in the past led to the intellectual rise of hegemonies in the positive phase – such as the classical "Axial" age that independently at 4 different Eurasian sites in the "lucky latitudes" for domesticates fostered philosophical and scientific advances [234] and falls of Empires as in Ancient Greece and Rome and Mesoamerica from environ-mental "ecocides" with damage to the soil and to diet [235–238]. Easter and other Island's "parable" collapses show the close relationship between diet, meat and ecology – eating all the birds led to eating all rats and "slash and burn" deforestation and then on a plant-based diet (as in "peak corn" in Mayan and elsewhere in post-Columbian times), boom bust demographics [239, 240]. Success or collapse can also be seen in terms of ecological succession and transfers between "r" selection and "K" selection for lower populations of higher quality that has to be driven by omnivorous diets avoiding "catabolic" and negative phases of the cycle that often follow peak inequality in states [241, 242]. Current failures to help a levelling up of under developed countries ignore the role of building human capital based first and foremost on a good diet and was not true of the Tiger economies [243–247].

22. National and personal exceptionalism: myths

The dietary "lurking variable" impact on demographic and disease transitions may have been under-played and the more superficial correlates with educational level or hygiene overplayed [248]. Education is dependent on cognitive skills (of both teachers and pupils), so a better diet must help: as well as fueling innovation. Patents correlate with their owner's parental income but not with educational opportunities so early diet could be crucial [249]. Good diet must surely be a factor in the "Flynn" effect where improvements in performance on IQ tests improves fast with improved income so cannot be genetic or due to discrimination [250, 251]. Diet may not be any more deterministic than genes but may have more predictive power over success and gene expression by diet is easier to alter than genomes.

The "Great Enrichment" of the globe that began in the Netherlands ("Embarrassment of Riches") then England is not easily explained by classical or neoclassical economics but had ideas from the well-fed "Bourgeoise" as their real "steam" power as a greater mass of people "could have a go" at innovation and invention - with better institutions and capital accumulation as secondary reinforcements to avoid intellectual "enclosures" from patents and monopolies [252, 253]. Fogel championed diet as being the prime mover with a well- nourished well paid workforce being a pre-condition for development [254–257] and has, at times, been the policy of the United Nations [258] as advised by others [259, 260]. Mechanisms that go beyond the need for calories with micronutrients create virtuous cycles: lessons on vicious cycles come from pellagra [261, 262] - affected "cotton" states still scored the lowest for social and economic well-being over 50 years later and the highest for incarceration [263].

23. Less infection, lower fertility, autoimmunity, longer lives

Better diet improves host resistance as even if not catching an infection people are less likely to die - TB and measles are good examples - as well as increasing crystallised intelligence from living longer. Not catching TB or other "Old Friends" because of better diet may contribute at least as much as better hygiene to the switch between chronic infections and auto-immune diseases so characteristic of modern states. Furthermore fertility falls, but quality of offspring improves; this being a non-coercive form of population control and age structures from dividend to drag that is a mixed blessing in some economies with see-saws between birth control and infertility clinics, but an advantage to others, many in the south such as in Nigeria where population booms are dramatic and could eclipse the population of the United States, so once again meat moderation may be best to avoid "arcs of instability" [264, 265].

24. Clean, safe meat and justice

Reducing the variances in meat (and milk, fish and shrimp) intake perhaps to a mean of around 30–50 Kg per person per annum may be the way forward through

DOI: http://dx.doi.org/10.5772/intechopen.100056

more sharing and less waste and better husbandry but is still, in essence, a power battle [266]. This meat pathway should be made cleaner and use emerging meat technologies as they become available and affordable. Risk of emergent zoonoses should be reduced. Only then would the lottery of one's life chances so related to geography, including within states, and discrimination and snobbery often based on educational achievement even if fronted by the identity politics of the day, such as skin colour reduce or at the least enter a new chapter [267, 268].

Redistributive food justice as in school meals programmes (as advocated by the Black Panthers and opposed by the FBI who understood the power of food), would deliver more "bang for the buck" than high military and security budgets or even the high healthcare costs for those in their last few years [269]. Free school meals remains a marker of deprivation in many rich countries that correlates intellectual attainment and college entry: only when a strong dietary base is satisfied would a meritocracy make sense [270, 271]. A backlash would doubtless need to be overcome as previously disadvantaged people become more assertive as seen with the Wilmington insurrection [272] (1898) or "Tulsa" riots on "Black Wall Street" (no food desert with 38 grocery stores), who stopped seeing themselves as second class citizens to the irritation of societies status quo ante [273].

25. Metabolism and free energy scales: Prometheus unbound

As Stephen J. Gould said in 1981 ***"Few tragedies or injustices can be more extensive than the stunting of life by a limit imposed from without, but falsely identified as lying within."*** Gould pointed out that intelligence like height could have high heritability within a family yet be very sensitive to poor diet in the population. Our propensity to virtue or violence may also have a lot to do with whether we are toward the top of Engel's curves and full of energy and not too worried about being displaced [274]. A metabolic and more classic liberal approach, such as of Mill and Smith, that encompasses ethical commitment to others welfare is needed [275–277]. Ultimately this may all be about flux (as noted by Heraclitus) and thermodynamics turned on it's head by use of oil. Our predicted needs are for NAD(H)-based internal (and external) energy gradients. Josiah Gibbs (1878) and followers understood that

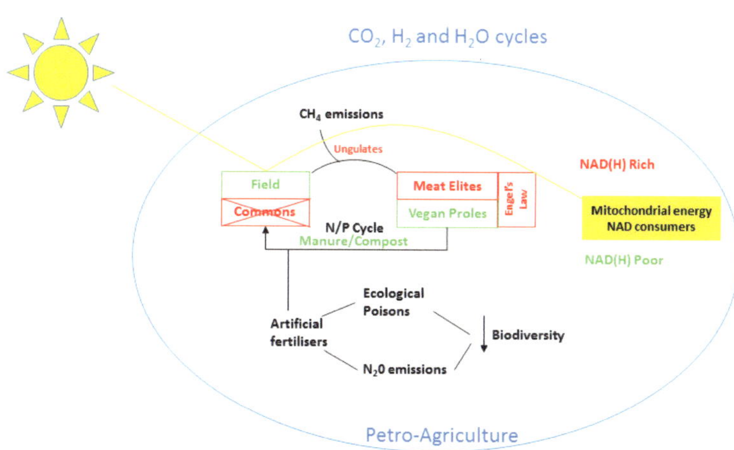

Figure 4.
Metabolic health and metabolic ghettoes and rifts seen from a NAD(H) perspective. Poverty may one day be defined in biochemical not sociological terms, as NAD deficient. NAD deficiency then exacerbates infections, chemical pollution, stress and poor education to which the poor are more exposed, particularly when residentially segregated.

chemiosmotic proton motive forces in mitochondria and chloroplasts were key as free energy usage based on Hydrogen ((H) carried by NAD and stored as ATP) connects our brains to the great cycles of life on a symbiotic planet connecting photosynthesis to respiration and combustion (**Figure 4**) [278–281].

26. Metabolic rifts

Local farming "100-Mile Diets" (as was normal until recently) and less deforestation or loss of mangrove swamps and less mono-cropping and better ground cover (absorbing CO2) and care of the soil reducing desertification and dustbowls with less artificial fertilisers and lower transportation needs will help. Humus and the soil are both chemical and biological systems (Darwin's worms (1881) and their microbiome) with complex interactions. Losing the "commons" pastureland caused issues from the original metabolic rift between humans, animals and nature, as proposed by Rousseau and Marx, such that manure or compost does not get returned to the soil through to modern views on the circular biomimicking economy [68, 238, 282, 283]. Even losing the right to produce your own natural greenhouse gas emissions is part of a multiple "Enclosure" movement making everything a commodity that is entrusted and traded for a profit in the market but not returning proceeds to the underpaid whose diet then loses out [284, 285]. A symptom of capitalism is that food chain workers on the land are in the frontline at food banks. Significant "*Schumacherian Small is Beautiful*" locavore counter-movements emanate from community peasant farmers, such as "La Via Campesina", "Seikatsu" (allotments and "Dig for Victory" in WW2 that included chickens and pigs), with experiments in countries such as Cuba and slow food movements although they rarely major on meat and milk [286, 287]. Gardeners and horticulturalists and not only industrial farmers, have much to learn about sustainability tailored to the local landscape - desperately poor farmers will mine their land for survival (and pastoralists invented zero-grazing stall fed cattle), as often as capitalists in tendencies that date back to the bronze age as first noted by Xenophon in the fourth century BC [288].

A return to universal common arcadian pastureland is too big a Utopian ask but the idea of "Sitopias" with a "Humanomics" and restraints on the "invisible hand" with a "doughnut" economy allowing meat for all in a resilient panarchy - meat for the rich is subsidised - is on the cards and is not a zero-sum game reducing many of the current risks to ourselves and to the planet [289, 290].

27. Mayday: wars and pandemics?

"If a free society cannot help the many who are poor, it cannot save the few who are rich" was the socially sustainable message from president J.F. Kennedy's inaugural back in 1961. The alternative of an increasingly unequal world and further geographies of poverty with new class structures and aggrandizers as global warning damages many agricultural systems would create a vicious cycle of biodiversity loss, even our "good-selves" [291–293]. Stressing of the meat supply risks emergent zoonoses - and war [294–296]. Dyer (2021) writing on wars from resource stress points out that primate and human rustling and outright wars, from the fierce Yanomamo on, were often over meat [297]. Livestock raiding may be a rite of passage for some but easily escalates to predatory raiding and has a lethal synergy with oil interests that can lead to geopolitical conflicts.

Revolutions and wars were often triggered by an excess of an elite class colluding with the hungry poor. Aggressors cottoned on to the value of starving their opponents in the American Civil war and World Wars - whose original purpose (like all empires as noted by Sophocles) was to gain pastureland ("Lebenstraum"). (Napoleon's earlier serial mistakes had been sending grain to Britain thinking that the gold spent would be more destructive, then ignoring his own supply lines in Russia and again in the Haitian revolution of the enslaved [123]). Civil wars are vicious often starving opponents as is being seen now in South African insurrections that bring access to food to its knees with child support coming nowhere close to satisfying – and tragically not representing the ideals of Mandela [298]. Despite somewhat different histories this Anglo-Dutch colony (note the extraordinary cattle-killing episode of 1857) and the American issues south of the Mason-Dixon line, including the Caribbean and down to the Rio de la Plata, are the hard core of white supremacy and both being prone to pellagra are key to our argument – as is also supported by the Australian experience [299].

Worse could come with a dystopia of climate change with droughts or floods affecting global food production first in the southern tropics and subtropics (often affecting pastoralist first who have to move), and triggering forcible attempts to break down walls (the fall of the Berlin Wall was triggered by meat queues and "solidarity" revolts), constructed as enclosures of rich gated communities protecting their own food supplies, is not too hard to imagine [300]. The end of the Cold War changed the political worry to migration and "the Coming Anarchy" from movements from low meat zones (as has been the case for all major human migrations from "Out of Africa" on) with terrorists joining up for remuneration as much as religious fervour to feed their families [40]. Water shortages have led to migrations but cooperation between states have often mitigated friction so far but remains a danger - the same might happen if meat was equilibrated increasing resilience [301, 302]. Wars and even "wars" against pathogens, often blamed on foreigners fueling xenophobia, is not an answer [303]. Relying on vaccines does not address the underlying driver but a more ecological and veterinary "One Health" approach with an emancipated public health concentrating on natural resistance from a good diet and safer sources avoids "Red Queen" evolutionary dynamics - or microbes will indeed have the last say [304, 305].

28. Conclusion

What we have tried to demonstrate is that, despite the importance of meat in our evolution and sharing of meat during our long less hierarchal hunter-gatherer phase, since recorded history began getting a grip on Engel's curve has been rigged for any but the rich and this mismatch has caused much inequality, disease, poor demographics and friction. Border wars and pandemics [306] would be a tragic own goal for rich and poor alike, given the political and societal impacts, if we cannot develop a new concept of the world from "emancipatory catastrophes" and cafeterias for all, not banquets for the few [40, 307]. These natural experiments and lived experiences aggregate evidence and by appreciating our "Bayesian" priors and precarity of the poor (all precarious), we can plan better for the future. Covid gives a current example of the poor being put at risk of late onset ailments from commercial 'bliss points' that then become risk factors as well as the hunt for meat leading to unsafe farming practices allowing microbes to jump species (as also happened with the bovine and human prion disease epidemic). We have been mean, sometimes inadvertently, about recreating our ancestral meat commons and stunting others lives in order to be in charge then racialising conduct and cognition. Not thinking like a "species being", but as classes or nations, puts our own species under threat let alone endangering many other species with extinction [308]. As Lewis Carrol said (in Through the Looking Glass, 1872) ***"the question is, said Humpty Dumpty, who***

is to be master- that's all". Being more generous is to be more masterful over the longer term and more "Nordic" [309] and as with ancient potlatches would avoid the dangers of surplus wealth in general, let alone the specific dangers of meat and nicotinamide overload, as money has to be spent somehow often propping up the "rentier" class or in dangerous arms races or housing or stock market bubbles bailed out by governments at the expense of the poor, as pointed out by Bataille (1949) [218, 310]. Only a proportion of capital is spent in useful reinvestment, research and development but is hoarded by the 1% [311] or as Galbraith said earlier ***"beyond doubt wealth is the relentless enemy of understanding"*** echoing long held biblical concerns about 'Mammon' and 'Paradise Lost'. A more anthropological vision of the world that questions the least questioned assumptions, as Paul Broca (a 19thC neurologist) suggested, would accept the lifetime importance of a good diet, with nicotinamide pathways as the "secret sauce" rather than carrying on with the disadvantaged "left behind" being subject to the most unlikely explanations for their plight and discrimination and snobbery from the elites - that wastes so much human talent and could cancel race matters that already are not an issue amongst well fed well educated elites [277, 312].

Acknowledgements

We would like to thank Queen Elizabeth Hospital Birmingham Charity for funding this study.

Conflict of interest

The author(s) declared no potential conflicts of interest with respect to the research, authorship, and/or publication of this article.

Author details

Adrian C. Williams[1*] and Lisa J. Hill[2]

1 Department of Neurology, University Hospitals Birmingham NHS Foundation Trust, Birmingham, UK

2 School of Biomedical Sciences, Institute of Clinical Sciences, University of Birmingham, UK

*Address all correspondence to: adrian.williams1949@gmail.com

References

[1] Hinman RB, Harris RB. The Story of Meat: Swift; 1942.

[2] Cokburn A. A short, meat-oriented history of the world. From Eden to the Mattole. New Left Review. 1996:16-42.

[3] De Castro J. The Geography of Hunger: Little, Brown; 1952.

[4] Doubleday T. The True Law of Population Shewn to be Connected with the Food of the People: Simpkin, Marshall; 1842.

[5] Carpenter KJ. Beriberi, White Rice, and Vitamin B: A Disease, a Cause, and a Cure: University of California Press; 2000.

[6] Boyd-Orr JBOB, Rowett Research Institute A, Scotland, Institute RR, Committee MS. Food, Health and Income: Report on a Survey of Adequacy of Diet in Relation to Income: Macmillan; 1936.

[7] Kiple KF, Ornelas K. The Cambridge world history of food: Cambridge University Press; 2000.

[8] Toussaint-Samat M. Histoire de la cuisine bourgeoise du Moyen Age à nos jours: Albin Michel; 2001.

[9] Milanovic B. Worlds Apart: Measuring International and Global Inequality: Princeton University Press; 2011.

[10] Sen AK. The idea of justice: Harvard University Press; 2009.

[11] Nussbaum MC. Creating Capabilities: The Human Development Approach: Harvard University Press; 2013.

[12] Atkinson AB. Inequality: Harvard University Press; 2015.

[13] Marmot M. The Health Gap: The Challenge of an Unequal World: Bloomsbury Publishing; 2015.

[14] Milanovic B. Global Inequality: A New Approach for the Age of Globalization: Harvard University Press; 2016.

[15] Wilkinson R, Pickett K. The Inner Level: How More Equal Societies Reduce Stress, Restore Sanity and Improve Everyone's Well-Being: Penguin Publishing Group; 2020.

[16] Savage M. The Return of Inequality: Harvard University Press; 2021.

[17] Thomas K, Bremmer J, Roodenberg H. Cultural History of Gesture. Ithaca, New York: Cornell University Press; 1992.

[18] Hunter R. Social conscience in the progressive Era: New York: Harper Torchbooks (1965); 1904.

[19] Rubinow IM. Social insurance: Holt; 1913.

[20] Pigou A. The Economics of Welfare. Macmillan. New York. 1920.

[21] Tobin DM, Vary JC, Ray JP, Walsh GS, Dunstan SJ, Bang ND. The lta4h locus modulates susceptibility to mycobacterial infection in zebrafish and humans. Cell. 2010;140.

[22] Wood EM. The Origin of Capitalism: A Longer View: Verso; 2017.

[23] Keen M, Slemrod J. Rebellion, Rascals, and Revenue: Tax Follies and Wisdom Through the Ages: Princeton University Press; 2021.

[24] Drèze J, Sen A, Press OU. Hunger and Public Action: Clarendon Press; 1989.

[25] Vernon J. Hunger: A Modern History: Harvard University Press; 2009.

[26] Pimentel D, Marcia H. Pimentel MS. Food, Energy, and Society, Third Edition: CRC Press; 2007.

[27] Rieff D. The Reproach of Hunger: Food, Justice, and Money in the Twenty-First Century: Simon & Schuster; 2015.

[28] Clapp J, Moseley WG. This food crisis is different: COVID-19 and the fragility of the neoliberal food security order. The Journal of Peasant Studies. 2020;47(7):1393-1417.

[29] Stanford CB, Bunn HT. Meat-eating and human evolution: Oxford University Press; 2001.

[30] Teaford MF, Ungar PS. Diet and the evolution of the earliest human ancestors. Proceedings of the National Academy of Sciences. 2000;97(25): 13506-13511.

[31] cocks d. Global Overshoot: Contemplating the World's Converging Problems: Springer New York; 2013.

[32] Lee RB, DeVore I, Research W-GFfA. Man the Hunter: Aldine; 1968.

[33] Sahlins MD. Stone Age Economics. Chicago: Aldine; 1972.

[34] Engel E. Die vorherrschenden Gewerbszweige in den Gerichtsamtern mit Beziehung auf die Productions-und Consumtionsverhaltnisse des Konigreichs Sachsen. Zeitschrift des Statistischen Bureaus. 1857;3.

[35] Prais SJ, Houthakker HS. The Analysis of Family Budgets: Cambridge University Press; 1971.

[36] Davis GC, Serrano EL. Food and Nutrition Economics: Fundamentals for Health Sciences: Oxford University Press; 2016.

[37] Bennett MK. International contrasts in food consumption. Geographical Review. 1941;31(3):365-376.

[38] Rowntree BS. The human needs of labour. The human needs of labour. 1937.

[39] Waltman JL. Minimum wage policy in Great Britain and the United States: Algora Publishing; 2008.

[40] Brown G. Seven Ways to Change the World: How To Fix The Most Pressing Problems We Face: Simon & Schuster UK; 2021.

[41] Crichton R. Iron Metabolism: From Molecular Mechanisms to Clinical Consequences: Wiley; 2016.

[42] Prado EL, Abbeddou S, Yakes Jimenez E, Somé JW, Ouédraogo ZP, Vosti SA, Dewey KG, Brown KH, Hess SY, Ouédraogo J-B. Lipid-based nutrient supplements plus malaria and diarrhea treatment increase infant development scores in a cluster-randomized trial in Burkina Faso. The Journal of nutrition. 2015;146(4): 814-822.

[43] Bailey RL, West Jr KP, Black RE. The epidemiology of global micronutrient deficiencies. Annals of Nutrition and Metabolism. 2015;66(Suppl. 2):22-33.

[44] Gille D, Schmid A. Vitamin B12 in meat and dairy products. Nutrition reviews. 2015;73(2):106-115.

[45] Millett K, MacKinnon C, Mead R. Sexual Politics: Columbia University Press; 2016.

[46] Mandeville B. The Grumbling Hive: or, knaves Turn'd Honest: Printed in the year; 1705.

[47] Huxley A, Calcio A. Brave New World: CreateSpace Independent Publishing Platform; 2014.

[48] Wilson EO. The Social Conquest of Earth: Liveright Publishing Corporation; 2013.

[49] Ehret C. The Civilizations of Africa: A History to 1800: James Currey; 2002.

[50] Davenport J. Synopsis of Biological Data on the Lumpsucker, *Cyclopterus Lumpus* (Linnaeus, 1758): Food and Agriculture Organization of the United Nations; 1985.

[51] Blumenbach JF. On the Natural Varieties of Mankind: De Generis Humani Varietate Nativa: Bergman Publishers; 1969.

[52] Benton-Cohen K. Inventing the Immigration Problem: The Dillingham Commission and Its Legacy: Harvard University Press; 2018.

[53] Shah S. The Next Great Migration: The Beauty and Terror of Life on the Move: Bloomsbury Publishing; 2021.

[54] Reskin BF. Getting it right: Sex and race inequality in work organizations. Annual Review of Sociology. 2000;26(1):707-709.

[55] Ruddiman WF. Plows, plagues, and petroleum: Princeton University Press; 2010.

[56] McMichael P. Development and social change: A global perspective: A global perspective. Thousand Oaks, US: Pine Forge; 2008.

[57] Malm A. Fossil capital: The rise of steam power and the roots of global warming: Verso Books; 2016.

[58] Sachs JD. The Ages of Globalization: Geography, Technology, and Institutions: Columbia University Press; 2020.

[59] Smil V. Energy and Civilization: A History: MIT Press; 2017.

[60] Evans-Pritchard EE. The nuer: Oxford Clarendon; 1940.

[61] Waal FB. Our inner ape: The best and worst of human nature: Granta; 2006.

[62] West-Eberhard MJ. Developmental plasticity and evolution: Oxford University Press; 2003.

[63] Tomasello M. Becoming human: Harvard University Press; 2019.

[64] Price B. Fifty Foods that Changed the Course of History: Apple; 2014.

[65] Bahn PG. de la source Images of the Ice Age: distributeur Oxford University Press; 2017.

[66] Spitzer M. The Musical Human: A History of Life on Earth-A Radio 4 Book of the Week: Bloomsbury Publishing; 2021.

[67] Vardi L. The Physiocrats and the World of the Enlightenment: Cambridge University Press; 2012.

[68] Carney M. Value(s): Building a Better World For All: HarperCollins Publishers; 2021.

[69] Brillat-Savarin JA. The Physiology of Taste: Or Meditations on Transcendental Gastronomy, trans. MFK Fisher, New York, Everyman's Library. 1825.

[70] Wrangham RW, Carmody RN. Human adaptation to the control of fire. Evolutionary Anthropology. 2010.

[71] Mazzucato M. Mission economy: A moonshot guide to changing capitalism: Penguin UK; 2021.

[72] Shafik M. What we owe each other: Princeton University Press; 2021.

[73] McCloskey DN. Bourgeois equality: University of Chicago Press; 2021.

[74] Zuk M. Paleofantasy: What evolution really tells us about sex, diet, and how we live: WW Norton & Company; 2013.

[75] Scott JC. Yale Agrarian Studies Series James C. Scott, Series Editor. Against the Grain: Yale University Press; 2017. p. xviii-xxii.

[76] Wells *S. Pandora*'s Seed: The Unforeseen Cost of Civilization: Random House; 2010.

[77] Bellwood P. First migrants: ancient migration in global perspective: John Wiley & Sons; 2014.

[78] Reich D. Who We Are and How We Got Here: Ancient DNA and the New Science of the Human Past: Oxford University Press; 2018.

[79] Favereau M. The Horde: How the Mongols Changed the World: Belknap Press; 2021.

[80] Khazanov AM. Nomads and the outside world: University of Wisconsin Press; 1994.

[81] Anthony DW. The horse, the wheel, and language: Princeton University Press; 2010.

[82] Zeder MA. The domestication of animals. Journal of anthropological research. 2012;68(2):161-190.

[83] Lucassen J. The Story of Work: A New History of Humankind: Yale University Press; 2021.

[84] Flannery K, Marcus J. The Creation of Inequality. Cambridge: Harvard University Press. 630 p. 2012.

[85] Price TD, Feinman GM. Pathways to power. New perspectives on the emergence of social inequality. 2010.

[86] Sjoberg G. Contradictory functional requirements and social systems. Journal of Conflict Resolution. 1960;4(2):198-208.

[87] Chandler T, Fox G. 3000 years of urban growth: Elsevier; 2013.

[88] Goodman DC, Chant C. European cities & technology: Industrial to post-industrial city: Psychology Press; 1999.

[89] Sandel MJ. What Money Can't Buy: The Moral Limits of Markets: Farrar, Straus and Giroux; 2012.

[90] Mann M. The sources of social power: volume 1, a history of power from the beginning to AD 1760: Cambridge university press; 2012.

[91] Scheidel W. The Great Leveler: Violence and the History of Inequality from the Stone Age to the Twenty-First Century: Princeton University Press; 2018.

[92] Grenfell MJ. Pierre Bourdieu: key concepts: Routledge; 2014.

[93] Lucassen J. The Story of Work. The Story of Work: Yale University Press.

[94] Bollier D, Helfrich S. The wealth of the commons: A world beyond market and state: Levellers Press; 2014.

[95] Grancini G, Roldán-Carmona C, Zimmermann I, Mosconi E, Lee X, Martineau D, Narbey S, Oswald F, De Angelis F, Graetzel M. One-Year stable perovskite solar cells by 2D/3D interface engineering. Nature Communications. 2017;8(1):1-8.

[96] Vivero-Pol JL, Ferrando T, De Schutter O, Mattei U. Routledge handbook of food as a commons: Routledge; 2018.

[97] Shiva V. Reclaiming the Commons: Biodiversity, Indigenous Knowledge, and the Rights of Mother Earth: Synergetic Press; 2020.

DOI: http://dx.doi.org/10.5772/intechopen.100056

[98] Winchester S. Land: the Ownership of Everywhere: HarperCollins Publishers Limited; 2021.

[99] Paine T. Agrarian justice (1797). The Origins of Universal Grants: Springer; 2004. p. 3-16.

[100] Spence W. Remarks on the Education and Amusements of the Lower Classes. Gard Mag. 1829;5:126.

[101] Bremner RH. The discovery of poverty in the United States: Transaction publishers; 1992.

[102] Bronstein JL. Land reform and working-class experience in Britain and the United States, 1800-1862: Stanford University Press; 1999.

[103] Kończal K. Mnemonic Populism: The Polish Holocaust Law and its Afterlife. European Review. 2021;29(4):457-469.

[104] Tilly C. Durable inequality: University of California Press; 1998.

[105] Massey DS. Categorically unequal: The American stratification system: Russell Sage Foundation; 2007.

[106] Sunderland D. Social Capital, Trust and the Industrial Revolution: 1780 1880: Routledge; 2007.

[107] Berry BJ. Divergent Paths in Twentieth-Century Urbanisation. The Human Consequences of Urbanisation: Springer; 1973. p. 164-181.

[108] Dyer G. The Shortest History of War: Black Incorporated; 2021.

[109] Thomas K. Man and the natural world: Changing attitudes in England 1500-1800: Penguin UK; 1991.

[110] Thompson A, Ward S, MacKenzie JM. British culture and the end of empire: Manchester University Press; 2001.

[111] Robertson T. The Malthusian Moment: Rutgers University Press; 2012.

[112] Dalziel N, Mackenzie J. The Penguin Historical Atlas of the British Empire: Penguin; 2006.

[113] Diamond JM. Guns, Germs and Steel: A Short History of Everybody for the Last 13,000 Years: Vintage; 1998.

[114] Crosby AW. Ecological Imperialism: The Biological Expansion of Europe, 900-1900: Cambridge University Press; 1993.

[115] Collingham EM. Imperial Bodies: The Physical Experience of the Raj, C.1800-1947: Wiley; 2001.

[116] Fraser EDG, Rimas A. Empires of Food: Feast, Famine and the Rise and Fall of Civilizations: Arrow Books; 2011.

[117] Specht J. Red Meat Republic: Princeton University Press; 2019.

[118] Etheridge EW. The Butterfly Caste: A Social History of Pellagra in the South: Greenwood Publishing Company; 1972.

[119] Ross L, Greene D, House P. The "false consensus effect": An egocentric bias in social perception and attribution processes. Journal of experimental social psychology. 1977;13(3):279-301.

[120] Black E. War Against the Weak: Eugenics and America's Campaign to Create a Master Race, Expanded Edition: Dialog Press; 2012.

[121] Harris LT, Fiske ST. Dehumanizing the lowest of the low: Neuroimaging responses to extreme out-groups. Psychological Science. 2006;17(10): 847-853.

[122] Sussman RW. The myth of Race: Harvard University Press; 2014.

[123] Anderson C. The Second: Race and Guns in a Fatally Unequal America: Bloomsbury Publishing; 2021.

[124] Schultz TW. The economic value of education: Columbia University Press; 1963.

[125] Wirth L. The Ghetto: Transaction; 1998.

[126] Kozol J. Still separate, still unequal. Harper's Magazine. 2005;9:41-55.

[127] Zolberg AR, Zolberg AR. A Nation by Design: immigration policy in the fashioning of America: Harvard University Press; 2009.

[128] Freud S, McLintock D. Civilization and Its Discontents: Penguin Adult; 2004.

[129] Wilmer F. The Social Construction of Man, the State and War: Identity, Conflict, and Violence in Former Yugoslavia: Taylor & Francis; 2004.

[130] Radtke K. Seek You: A Journey Through American Loneliness: Pantheon Books; 2021.

[131] Dunbar R. Friends: Understanding the Power of Our Most Important Relationships: Little, Brown Book Group Limited; 2022.

[132] Keohane J. The Power of Strangers: The Benefits of Connecting in a Suspicious World: Penguin Books Limited; 2021.

[133] Sue DW, Spanierman L. Microaggressions in Everyday Life: Wiley; 2020.

[134] Sowell T. Discrimination and Disparities: Basic Books; 2019.

[135] Morris A. The Scholar Denied: W. E. B. Du Bois and the Birth of Modern Sociology: University of California Press; 2017.

[136] Agarwal P. Sway: Unravelling Unconscious Bias: Bloomsbury Publishing; 2020.

[137] Charlesworth TE, Banaji MR. Patterns of implicit and explicit attitudes: I. Long-term change and stability from 2007 to 2016. Psychological Science. 2019;30(2): 174-192.

[138] Davis M. Late Victorian holocausts: El Niño famines and the making of the third world: Verso Books; 2002.

[139] Madley B. An American Genocide: Yale University Press; 2016.

[140] Rodney W. How Europe Underdeveloped Africa: Bogle-L'Ouverture Publications; 1972.

[141] Findlay R. 'Rourke, HK (2007) Power and Plenty: Trade, War, and the World Economy in the Second Millennium. Princeton: Princeton University Press.

[142] Hickel J. The Divide: A Brief Guide to Global Inequality and Its Solutions: William Heinemann; 2017.

[143] Patnaik U, Patnaik P. Capital and Imperialism: Theory, History, and the Present: Monthly Review Press; 2021.

[144] Patel P. The Tyranny of Nations: How the Last 500 Years Shaped Today's Global Economy: Bifocal; 2021.

[145] Beckert S. Empire of Cotton: A Global History: Knopf Doubleday Publishing Group; 2014.

[146] Bedat M. Unraveled: The Life and Death of a Garment: Penguin Publishing Group; 2021.

[147] Andrews K. The New Age of Empire: How Racism and Colonialism Still Rule the World: Penguin UK; 2021.

[148] Sanghera S. Empireland: How Imperialism Has Shaped Modern Britain: Penguin Books Limited; 2021.

[149] Aizer A, Currie J. The intergenerational transmission of

inequality: maternal disadvantage and health at birth. Science. 2014;344(6186):856-861.

[150] MacKenzie JM. Imperialism and the natural world: Manchester University Press Manchester; 1990.

[151] McCracken DP. Gardens of empire: Leicester University Press; 1997.

[152] Drayton R. Nature's Government: Science, Imperial Britain, and the 'Improvement'of the World. New Haven and London: Yale University Press, 2000.

[153] De Vriese D. Lizzie Collingham, The Hungry Empire. How Britain's Quest for Food Shaped the Modern World. TSEG-The Low Countries Journal of Social and Economic History. 2018;15(1):115-117.

[154] Cole GDH, Postgate RW. common people, 1746-1938. 1938.

[155] Perren R. meat trade in Britain, 1840-1914: Routledge and Kegan Paul; 1978.

[156] Rixson D. history of meat trading: Nottingham University Press; 2000.

[157] Parker P, Books CC. History of World Trade in Maps: Collins; 2021.

[158] Cronon W. Nature's Metropolis: Chicago and the Great West: W. W. Norton; 1992.

[159] Landes DS. The Unbound Prometheus: Technological Change and Industrial Development in Western Europe from 1750 to the Present: Cambridge University Press; 2003.

[160] Paine T. The Rights of Man: Start Publishing LLC; 2013.

[161] Sewell R. In the Cause of Labour – A History of British Trade Unionism: Wellred Publications; 2003.

[162] McKeown T. The origins of human diseasecontinued: Basil Blackwell; 1988.

[163] Rook GA, Lowry CA. The hygiene hypothesis and affective and anxiety disorders. The hygiene hypothesis and Darwinian Medicine: Springer; 2009. p. 189-220.

[164] McMichael P. Development and Social Change: A Global Perspective: Pine Forge Press; 2008.

[165] Van Bavel BJ, van Bavel B. The invisible hand?: how market economies have emerged and declined since AD 500: Oxford University Press; 2016.

[166] Slobodian Q. Globalists: The End of Empire and the Birth of Neoliberalism: Harvard University Press; 2020.

[167] Nkrumah K. Neo-colonialism: The Last Stage of Imperialism: International Publishers; 1965.

[168] Haq MU. The poverty curtain; choices for the third world. 1976.

[169] Lang A. World Trade Law after Neoliberalism: Reimagining the Global Economic Order: OUP Oxford; 2013.

[170] Knuth L, Food, Nations AOotU. The Right to Adequate Food and Indigenous Peoples: How Can the Right to Food Benefit Indigenous Peoples?: FAO; 2009.

[171] Norheim OF, Asada Y. The ideal of equal health revisited: definitions and measures of inequity in health should be better integrated with theories of distributive justice. International Journal for Equity in Health. 2009;8(1):1-9.

[172] De Schutter O. The political economy of food systems reform. European Review of Agricultural Economics. 2017;44(4):705-731.

[173] Scoones I, Wolmer W. Pathways of change in Africa: crops, livestock & livelihoods in Mali, Ethiopia & Zimbabwe: James Currey Ltd; 2002.

[174] Homewood K. Ecology of African Pastoralist Societies: James Currey; 2008.

[175] Harris J, Sen A, Wise T, Gallagher K, Goodwin NR. A Survey of Sustainable Development: Social And Economic Dimensions: Island Press; 2013.

[176] Nestle M, Pollan M. Food Politics: How the Food Industry Influences Nutrition and Health: University of California Press; 2013.

[177] Patel R. Stuffed and Starved: The hidden battle for the food system. Melville House Publishing: Brooklyn; 2007.

[178] Carolan MS. Reclaiming food security: Routledge; 2013.

[179] Caradonna JL. Sustainability: A history: Oxford University Press; 2014.

[180] Maddison A. The world economy: Developmental centre studies. 2001.

[181] Pomeranz K. The Great Divergence: China, Europe, and the Making of the Modern World Economy: Princeton University Press; 2000.

[182] Putnam RD, Garrett SR. The Upswing: How We Came Together a Century Ago and how We Can Do it Again: Swift Press; 2020.

[183] Rifkin J. Beyond beef: the rise and fall of the cattle culture. Ringwood. Victoria: Penguin Books. 1992.

[184] Marcus E. Meat Market: Animals, Ethics, & Money: Brio Press; 2005.

[185] Mance H. How to Love Animals: In a Human-Shaped World: Penguin Publishing Group; 2021.

[186] Lamb HH. Climate, History and the Modern World: Routledge; 1995.

[187] Pearce F. A Trillion Trees: How We Can Reforest Our World: Granta Books; 2021.

[188] Goulson D. Silent Earth: Averting the Insect Apocalypse: Random House; 2021.

[189] Emel J, Neo H. Political Ecologies of Meat: Routledge; 2015.

[190] Harrington M. The Other America: Scribner; 1997.

[191] Clements J. The Emperor's Feast:'A tasty portrait of a nation'–Sunday Telegraph: Hodder & Stoughton; 2021.

[192] Weis T. The Ecological Hoofprint: The Global Burden of Industrial Livestock: Bloomsbury Publishing; 2013.

[193] Volk T. CO2 Rising: The World's Greatest Environmental Challenge: MIT Press; 2010.

[194] Mellon J. Moo's Law: An Investor's Guide to the New Agrarian Revolution: Harriman House Limited; 2020.

[195] Keith L. The Vegetarian Myth: Food, Justice, and Sustainability: PM Press; 2009.

[196] Stuart T. The Bloodless Revolution: A Cultural History of Vegetarianism from 1600 to Modern Times: W.W. Norton & Company; 2008.

[197] Singer P. The Most Good You Can Do: How Effective Altruism is Changing Ideas about Living Ethically: Yale University Press; 2015.

[198] Godfrey-Smith P. Other Minds: The Octopus and the Evolution of Intelligent Life: HarperCollins Publishers; 2017.

[199] Pollan C, Pollan L, Pollan D, Pollan T, Pollan M. The Pollan Family

Table: The Best Recipes and Kitchen Wisdom for Delicious, Healthy Family Meals: Scribner; 2016.

[200] Vieux F, Darmon N, Touazi D, Soler LG. Greenhouse gas emissions of self-selected individual diets in France: changing the diet structure or consuming less? Ecological Economics. 2012;75:91-101.

[201] Macdiarmid JI, Kyle J, Horgan GW, Loe J, Fyfe C, Johnstone A, McNeill G. Sustainable diets for the future: can we contribute to reducing greenhouse gas emissions by eating a healthy diet? The American journal of clinical nutrition. 2012;96(3):632-639.

[202] Markowski KL, Roxburgh S. "If I became a vegan, my family and friends would hate me:" Anticipating vegan stigma as a barrier to plant-based diets. Appetite. 2019;135:1-9.

[203] Leese EA, Charalambides EJ. Think Like a Vegan: What everyone can learn from vegan ethics: Unbound; 2021.

[204] Davies W. Nervous States: How Feeling Took Over the World: Random House; 2018.

[205] Frankfurt HG. On inequality: Princeton University Press; 2015.

[206] Lister R. Poverty: Wiley; 2021.

[207] Bartley M. Health inequality: an introduction to concepts, theories and methods: John Wiley & Sons; 2016.

[208] Therborn G. Inequality And World-Political Landscapes. New Left Review. 2021(129):5-26.

[209] Ostry JD, Loungani P, Berg A, Ostry JD. Confronting Inequality: Columbia University Press; 2019.

[210] Aghion P, Antonin C, Bunel S. The Power of Creative Destruction: Harvard University Press; 2021.

[211] Jones A. One: Pot, Pan, Planet: a Greener Way to Cook for You, Your Family and the Planet: HarperCollins Publishers Limited; 2021.

[212] Drewnowski A, Eichelsdoerfer P. Can low-income Americans afford a healthy diet? Nutrition today. 2010;44(6):246.

[213] Allen TN. Food Inequalities: ABC-CLIO; 2021.

[214] Maslow AH. Motivation And Personality: Harper & Row; 1981.

[215] Schlesinger ER, Overton DE, Chase HC, Cantwell KT. Newburgh-Kingston caries fluorine study. 13. Pediatric findings after ten years. Journal of the American Dental Association. 1956;52:296-306.

[216] Arendt H. The human condition: University of Chicago Press; 2013.

[217] Daly HE. Steady-state economics: with new essays: Island press; 1991.

[218] Galbraith JK. The Affluent Society: Houghton Mifflin; 1998.

[219] Mitchell T. Colonising Egypt: University of California Press; 1991.

[220] MacKinnon JB. The Day the World Stops Shopping: How ending consumerism gives us a better life and a greener world: Random House; 2021.

[221] Stirnemann R, Stirnemann I, Abbot D, Biggs D, Heinsohn R. Interactive impacts of by-catch take and elite consumption of illegal wildlife. Biodiversity and conservation. 2018;27(4):931-946.

[222] Masters WA, Rosenblum NZ, Alemu RG. Agricultural Transformation, Nutrition Transition and Food Policy in Africa: Preston

Curves Reveal New Stylised Facts. The Journal of Development Studies. 2018;54(5):788-802.

[223] Mies M, Bennholdt-Thomsen V. Subsistence and agriculture. éd), Subsistence perspective, London & New York: Zed Books. 1999:65-108.

[224] Kristof ND, WuDunn S. Tightrope: Americans Reaching for Hope: Knopf Doubleday Publishing Group; 2020.

[225] Csikszentmihalyi M, Csikszentmihalyi IS. Optimal Experience: Psychological Studies of Flow in Consciousness: Cambridge University Press; 1992.

[226] Baumol WJ, Bowen WG. On the performing arts: The anatomy of their economic problems. The American Economic Review. 1965;55(1/2):495-502.

[227] Boppart T. Structural change and the Kaldor facts in a growth model with relative price effects and non-Gorman preferences. Econometrica. 2014;82(6):2167-2196.

[228] Acemoglu D. Introduction to Modern Economic Growth: Princeton University Press; 2009.

[229] Marshall A. Principles of Economics: Macmillan; 1925.

[230] Veblen T, Howells WD. The Theory of the Leisure Class. 1899: A. M. Kelley, bookseller; 1965.

[231] Kuznets S. Economic growth and income inequality. The American Economic Review. 1955;45(1):1-28.

[232] Jones V. The Green Collar Economy: How One Solution Can Fix Our Two Biggest Problems: HarperCollins; 2009.

[233] Boyce J. Economics for People and the Planet: Inequality in the Era of Climate Change: Anthem Press; 2019.

[234] Jaspers K. The Origin and Goal of History (Routledge Revivals): Routledge; 2014.

[235] Simkhovitch VG. Hay and history. Political Science Quarterly. 1913;28(3):385-403.

[236] Hyams E. Soil and Civilisation: Thames & Hudson; 1952.

[237] Ohlson K. The Soil Will Save Us: How Scientists, Farmers, and Foodies Are Healing the Soil to Save the Planet: Rodale Books; 2014.

[238] Montgomery DR. Dirt: The erosion of civilizations: Univ of California Press; 2012.

[239] Ponting C. A Green History of the World: The Environment and the Collapse of Great Civilizations: Penguin Books; 1993.

[240] Williams M. Deforesting the Earth: From Prehistory to Global Crisis: University of Chicago Press; 2003.

[241] Odum HT. Environment, power, and society for the twenty-first century: the hierarchy of energy: Columbia University Press; 2007.

[242] Greer JM. The Long Descent: A User's Guide to the End of the Industrial Age: Founders House Pub Llc; 2019.

[243] Tainter J. The collapse of complex societies: Cambridge university press; 1988.

[244] O'Sullivan M. The levelling: What's next after globalization: Hachette UK; 2019.

[245] Beckford GL. Persistent Poverty: Underdevelopment in Plantation Economies of the Third World: Oxford University Press; 1972.

[246] Folbre N. The Invisible Heart: Economics and Family Values: New Press; 2001.

[247] Becker GS. Human capital revisited. Human Capital: A Theoretical and Empirical Analysis with Special Reference to Education, Third Edition: The university of Chicago press; 1994. p. 15-28.

[248] Goodhart C, Pradhan M. The Great Demographic Reversal: Ageing Societies, Waning Inequality, and an Inflation Revival: Springer International Publishing; 2020.

[249] Aghion P, Antonin C, Bunel S. The Power of Creative Destruction: Economic Upheaval and the Wealth of Nations: Harvard University Press; 2021.

[250] Gould S. The mismeasure of man WW Norton. New York. 1981.

[251] Flynn JR. What Is Intelligence?: Beyond the Flynn Effect. Cambridge: Cambridge University Press; 2009.

[252] McCloskey DN. Bettering Humanomics: A New, and Old, Approach to Economic Science: University of Chicago Press; 2021.

[253] Breznitz D. Innovation in real places: Strategies for prosperity in an unforgiving world: Oxford University Press, USA; 2021.

[254] Fogel RW. The Escape from Hunger and Premature Death, 1700-2100: Europe, America, and the Third World: Cambridge University Press; 2004.

[255] Alderman H, Behrman JR, Hoddinott J. Economic and nutritional analyses offer substantial synergies for understanding human nutrition. The Journal of nutrition. 2007;137(3):537-544.

[256] Allen RC. The British industrial revolution in global perspective: Cambridge University Press; 2009.

[257] Floud R, Floud PR, Gregory A, Wachter K, De Vries J, Johnson P, Smith R, Wrightson K. Height, Health and History: Nutritional Status in the United Kingdom, 1750-1980: Cambridge University Press; 1990.

[258] FAO (Food and Agriculture Organization). FAOSTAT Agriculture Database 2002 [Available from: http://apps.fao.org.

[259] Pinstrup-Andersen P, Watson DD, Frandsen SE, von Braun J, Kuyvenhoven A. Food Policy for Developing Countries: The Role of Government in Global, National, and Local Food Systems: Cornell University Press; 2011.

[260] Lang T, Heasman M. Food Wars: The Global Battle for Mouths, Minds and Markets: Routledge; 2015.

[261] Williams AC, Dunbar RI. Big brains, meat, tuberculosis, and the nicotinamide switches: co-evolutionary relationships with modern repercussions? International Journal of Tryptophan Research. 2013;6:IJTR. S12838.

[262] Rosales FJ, Reznick JS, Zeisel SH. Understanding the role of nutrition in the brain and behavioral development of toddlers and preschool children: identifying and addressing methodological barriers. Nutritional neuroscience. 2009;12(5):190-202.

[263] Smith D. The geography of social well-being in the US New York. McGraw-Hill; 1973.

[264] MacNamara T. Birth Control and American Modernity: A History of Popular Ideas: Cambridge University Press; 2018.

[265] Williams AC, Hill LJ. Nicotinamide as Independent Variable for Intelligence, Fertility, and Health: Origin of Human Creative Explosions? International Journal of Tryptophan Research. 2019;12:1178646919855944.

[266] Prato S. The struggle for equity: Rights, food sovereignty and the rethinking of modernity. Development. 2014;57(3):311-319.

[267] Alkon AH, Agyeman J. Cultivating the fertile field of food justice. Cultivating food justice: Race, class, and sustainability. 2011:331-347.

[268] Coates BE, Johnston RJ, Knox PL, Geografie S. Geography and inequality: Oxford University Press Oxford; 1977.

[269] Levine S. School lunch politics: Princeton University Press; 2011.

[270] Sandel MJ, Milijić P. Tyranny of Merit: What's Become of the Common Good? Review Of Economics And Economic Methodology.109.

[271] Wooldridge A. The Aristocracy of Talent: How Meritocracy Made the Modern World: Penguin Books Limited; 2021.

[272] Zucchino D. Wilmington's Lie: The Murderous Coup of 1898 and the Rise of White Supremacy: Grove Atlantic; 2021.

[273] Ellsworth S. The Ground Breaking: The Tulsa Race Massacre and an American City's Search for Justice: Icon Books; 2021.

[274] Pinker S. The better angels of our nature: A history of violence and humanity: Penguin Books London; 2012.

[275] Amadae SM. Prisoners of reason: Game theory and neoliberal political economy: Cambridge University Press; 2016.

[276] Smith VL, Wilson BJ. Humanomics: Moral sentiments and the wealth of nations for the twenty-first century: Cambridge University Press; 2019.

[277] Tett G. Anthro-Vision: A New Way to See in Business and Life: Avid Reader Press / Simon & Schuster; 2021.

[278] Margulis L. The Symbiotic Planet: A New Look At Evolution: Orion; 2013.

[279] Williams AC, Ramsden DB. Pellagra: A clue as to why energy failure causes diseases? Med Hypotheses. 2007;69(3):618-628.

[280] Carroll SM. The big picture: on the origins of life, meaning, and the universe itself: Penguin; 2017.

[281] Rhodes R. Energy: A Human History: Simon & Schuster; 2019.

[282] Foster JB, Clark B, York R. The Ecological Rift: Capitalism's War on the Earth: Monthly Review Press; 2011.

[283] Raworth K. Doughnut Economics: Seven Ways to Think Like a 21st Century Economist: Penguin Random House; 2018.

[284] Leopold A. A Sand County Almanac: Random House Publishing Group; 1984.

[285] MacGillis A. Fulfillment: winning and losing in one-click America: Scribe Publications Pty Limited; 2021.

[286] Feenstra RC, Hanson GH. Foreign direct investment and relative wages: Evidence from Mexico's maquiladoras. Journal of international economics. 1997;42(3-4):371-393.

[287] O'Kane G. A moveable feast: Exploring barriers and enablers to food citizenship. Appetite. 2016;105: 674-687.

[288] White G, Secord A. The Natural History of Selborne: OUP Oxford; 2013.

[289] Ostrom E. Governing the commons: The evolution of institutions for collective action: Cambridge university press; 1990.

[290] Tilman D, Clark M. Global diets link environmental sustainability and

human health. Nature. 2014;515(7528):518-522.

[291] Brooks D. Bobos in Paradise: The New Upper Class and How They Got There: Simon & Schuster; 2010.

[292] Sennett R. Respect: the formation of character in an age of inequality (London, Allen Lane/Penguin). 2003.

[293] Crossley S. In Their Place: The Imagined Geographies of Poverty: Pluto Press; 2017.

[294] Pettifor A. The Case for the Green New Deal: Verso Books; 2020.

[295] Freeman J. Weight. The Virginia Quarterly Review. 2020;96(1):62-67.

[296] Wallace R. Dead Epidemiologists: On the Origins of COVID-19: Monthly Review Press; 2020.

[297] MacMillan M. War: How Conflict Shaped Us: Random House Publishing Group; 2020.

[298] Johnson RW. How Long Will South Africa Survive?: Hurst & Company; 2015.

[299] Fredrickson GM. White Supremacy: A Comparative Study of American and South African History: Oxford University Press, USA; 1981.

[300] Wapner J. Wall Disease: The Psychological Toll of Living Up Against a Border: The Experiment; 2020.

[301] Crawford MH, Campbell BC. Causes and consequences of human migration: An evolutionary perspective: Cambridge University Press; 2012.

[302] Mavroudi E, Nagel C. Global migration: Patterns, processes, and politics: Routledge; 2016.

[303] Kraut AM. Goldberger's War: The Life and Work of a Public Health Crusader: Farrar, Straus and Giroux; 2021.

[304] Caduff C. The Pandemic Perhaps: Dramatic Events in a Public Culture of Danger: University of California Press; 2015.

[305] Morens DM, Fauci AS. Emerging pandemic diseases: how we got to COVID-19. Cell. 2020.

[306] Furtado P. Plague, Pestilence and Pandemic: Voices from History: Thames & Hudson; 2021.

[307] Beck U. The metamorphosis of the world: How climate change is transforming our concept of the world: John Wiley & Sons; 2016.

[308] Kolbert E. The Sixth Extinction: An Unnatural History: Bloomsbury; 2014.

[309] Partanen A. The Nordic Theory of Everything: In Search of a Better Life: Duckworth Books; 2017.

[310] Graeber D. Toward An Anthropological Theory of Value: The False Coin of Our Own Dreams: Palgrave Macmillan US; 2002.

[311] Stiglitz J. Of the 1%, by the 1%, for the 1%. Vanity fair. 2011;11(64):156-111.

[312] Skelton D. The New Snobbery: Biteback Publishing; 2021.

Chapter 5

Deficiency of Vitamin B-Complex and Its Relation with Body Disorders

Muhammad Farhan Sarwar, Muhammad Haroon Sarwar and Muhammad Sarwar

Abstract

Vitamins B denote to some diverse kinds of vitamins which collectively, are recognized as B-complex vitamin. At hand are eight types of vitamins in vitamin B complex; thiamine (B1), riboflavin (B2), niacin (B3), pantothenic acid (B5), pyridoxine (B6), biotin (B7), folate (B9) also known as folic acid and cobalamin (B12). B vitamins have a direct impact on body energy levels, brain function and cell metabolism. There is a roundup of four top causes of vitamin B deficiency; a non-balanced diet, excessive alcohol consumption, various medications and gut malabsorption conditions. Deficiencies in these B vitamins can lead to a number of different symptoms like paresthesias, peripheral neuropathy, psychosis and heart attack and stroke over time if the deficiency is not reversed. Vitamins are found in highest abundance in meat, eggs and dairy or milk products such as butter, yogurt and cheese produced from milk of mammals usually buffaloes, cattle, goats, sheep and camels. Most people can get many nutrients they need, including B vitamins, by eating a varied diet of lean meats, grains, fruits and vegetables. This chapter provides an affluent of the most common types of vitamins B, including why body needs these, their deficiency symptoms and which foods contain them.

Keywords: B vitamins, B-complex vitamin, disorder, neuropathy, biochemical action

1. Introduction

Within little earnings nations, insufficient quantities of foodstuff (resulting situations like kid malnourishment and undersized development) and limited variety of foodstuff (resulting deficiency of vigorous micronutrients like vitamins, minerals or trace elements) remain to be urgency healthiness complications. Undernourishment entirely in its practices rises the threat of illness and premature expiry. Almost millions of persons in the biosphere do not have sufficient diet to consume. Undernourishment disturbs entirely age crowds, however it is particularly common amongst poor persons and those with insufficient entrance to fitness teaching, good sanitation and clean water. Maximum of the malnutrition-related nervous complaints are escapable [1–3].

The body requires nutrients such as protein, fats and sugars to build tissues and fuel biological processes, but even when calories are plentiful, there are some vital nutrients that when missing, would cause catastrophic illness and death. One of the first of these deficiency syndromes to be identified is scurvy, due to a deficiency of vitamin C that led to the discovery of the B vitamins, which is a group of water-soluble chemicals working with enzymes to support a wide range of functions in the body [4].

The B vitamins are a group of eight nutrients, each with unique roles in keeping the body healthy. Though these vitamins share similar names (B1, B2, B3, etc.), they are chemically distinct compounds that often coexist in the same foods. Generally, dietary supplements containing all eight are referred to as a vitamin B complex. Each B vitamin is either a cofactor (normally a coenzyme) for key metabolic processes otherwise a precursor needed to make one. They are especially important for maintaining cell health and keeping energized with a unique function in the body. There are eight kinds of vitamins in the vitamin B complex: thiamine (B1), riboflavin (B2), niacin (B3), pantothenic acid (B5), pyridoxine (B6), biotin (B7), folate (B9, also known as folic acid) and cobalamin (B12) [5].

All B vitamins play crucial roles as coenzymes for enzymatic reactions in different biological systems. The eight B vitamins form a group of chemically very heterogeneous essential substances, which have a wide variety of functions in the human body. Further substances once thought to be vitamins have been given numbers (B4, B8, B10, B11 and others) in B-vitamin numbering scheme, however are subsequently discovered to be either not essential for life or manufactured by the body, thus not meeting the two essential qualifiers for a vitamin [6].

Individually, these eight B vitamins play key parts in the body desired to initiate the chemical reactions that upkeep body's lots of tasks. For instance, cells usage B vitamins to create energy from fatty acids, sugar and other nutrients. Thus, deprived of B-complex vitamins, the human body may possibly not function fine of any kind. B vitamins are water-soluble or they can dissolve in water, thus excess B vitamins in body does not use are washed out through urination. The body cannot easily store B vitamins for long periods of time, thus it is especially important to regularly consume B vitamins from diet or supplements to avoid deficiency. Entirely, B vitamins aid to transform the fats, carbohydrates and proteins consumed into energy. B vitamins are likewise required for strong hair, eyes and skin; right working of the liver and nervous system; fit gastrointestinal expanse; creating red blood cells that transport oxygen all over the body; and constructing sex and stress-related hormones in adrenal glands [7].

Each member of the B-complex has a unique structure and performs unique functions in the human body. Their supplement has been used in connection with the subsequent health conditions. Thiamine (vitamin B1), pantothenic acid (vitamin B5) and other B vitamins have all been shown to play a role in wound healing. Supplementing with vitamins B1, B2 and B6 has been reported to provide relief from canker sores. Preliminary reports have claimed that acne rosacea improved; treatment of B-complex vitamin deficiencies may actually reduce alcohol cravings (desires); may reduce feelings of anxiety, perceived stress and tiredness; athletic performance can suffer if these slightly increased requirement needs are not met; helpful to treat people with hives; can prevent the development of tardive dyskinesia; women may benefit from supplementing with B-complex vitamins for premenstrual syndrome symptom relief; postmenopausal women increase their bone density against osteoporosis remarkably; and may depigment skin affected by vitiligo. Human requirements for each B vitamin vary considerably from 3 mcg per day for vitamin B12 to 18 mg per day for vitamin B3 in adult males [8].

The symptoms of a vitamin B deficiency vary depending on which B vitamin a person is deficient resulting in different symptoms. Certain conditions, such as crohn's disease, celiac disease, HIV (human immunodeficiency virus) and alcohol use disorder can prevent the body from absorbing B vitamins effectively, thus increasing the risk for deficiencies. They can range from fatigue and confusion to anemia or a compromised immune system, while skin rashes can also occur. Most people can get all nutrients they need, including B vitamins, by eating a varied diet of lean meats, grains, fruits and vegetables. Some foods are high in several B vitamins and certain foods are particularly high in specific B vitamins. So, eating of a balanced diet is the key to get all nutrients the body needs. Older adults, pregnant women and people with certain health conditions are more likely to have vitamin B deficiencies [9]. To treat a B vitamin deficiency, physician will likely recommend that a person should take supplements or increase intake of certain foods that contain the target vitamin.

2. Indicators of vitamins B scarcities

Neurotropic B vitamins play crucial roles as coenzymes and beyond in the nervous system. Particularly vitamin B1 (thiamine), B6 (pyridoxine), and B12 (cobalamin) contribute essentially to the maintenance of a healthy nervous system. Taking into consideration the present information on the neurotropic vitamins B1, B6 and B12, it is ultimate that a biological interaction come to be obvious in several diverse passageways in nervous system, mainly in the PNS (peripheral nervous system) as illustrated by their collective usage in the management of peripheral neuropathy [10]. This is significant to start bearing in mind about B vitamins as a therapeutic and neuroprotective tactic for peripheral neuropathies and several brain disorders equally, in forthcoming medical studies. **Table 1**, provides an overview on the major implications in overlapping biochemical pathways important for the nervous system, pointing to a synergistic effect as a logical consequence of these overlaps.

The indications of a vitamin B shortage vary depending on which B vitamin a body is deficient. Maximum nutritious syndromes with harmful sound effects on the peripheral and central nervous system are secondary to vitamin insufficiencies, principally those of the B group. Several of these syndromes happen in the site of undernourishment linked with alcoholism. A thorough dialog of whole vitamin B shortages lies inside the space of this chapter. Nevertheless, four of the further common vitamin B deficiencies such as thiamine (B1), cobalamin (B12), niacin (B3) and folate (B9) are specially appraised.

Vitamin	Processes	Coenzyme for	Implication in nervous system
B1 (Thiamine)	Glycolysis Pentose phosphate pathway Krebs cycle (citric acid cycle)	Pyruvate dehydrogenase Transketolase Alpha-ketoglutarate dehydrogenase	Delivers energy to nerve cells that are required for production of nucleic acids, myelin and neurotransmitters
B6 (Pyridoxine)	One-carbon unit metabolism Hcy metabolism Dopamine and serotonin synthesis	Serine-hydroxymethyltransferase Cystathionine-beta-synthase/ lyase Aromatic L-amino acid decarboxylase	Metabolism of DNA/ RNA, amino acids and neurotransmitters
B12 (Cobalamin)	Hcy metabolism Methymalonyl CoA pathway	Methionine synthase Methylmalonyl CoA mutase	Metabolism of amino acids, fatty acids, DNA/ RNA, myelin and neurotransmitters

Table 1.
Outline on key biochemical mechanisms of action for nerve function by vitamins B1, B6 and B12.

2.1 Beriberi

Amongst the B vitamins, one and only of the initial scarcity disorders to be discovered is beriberi, due to an insufficiency of vitamin B1 (thiamine). The signs of beriberi are numbness and weakness in the legs and feet, inflammation, trouble in inhalation, and heart tragedy. Beriberi is ultimately found to a nutrition of refined white rice consumption and as soon as the rice bran is resumed to the food, the indicators are retreated. There are double types of beriberi; wet beriberi disturbs the cardiac-vascular structure and dry beriberi likewise famous as wernicke korsakoff syndrome, upsets the nervous structure. The signs of dry beriberi comprise; difficulty in speaking, confusion, pain, nystagmus (uncontrolled repetitive eye movements), difficulty in walking, tingling or numbness in feet and hands, muscle paralysis or weakness, and nausea. Wet beriberi may result further signs such as increased heart rate, rapidity of breathing and inflammation in legs [11].

2.2 Pellagra

Pellagra is one more illness, which is directed to the finding of a vitamin that is initiated by a shortage of B3 vitamin (niacin). The usual indications of pellagra are; loss of hair, swelling and dermatitis of skin, inflammation of tongue, weakness, insomnia, ataxia, diarrhea, aggression, confusion, dilation of cardiomyopathy and dementia (loss of thinking, remembering or making decisions). A deficiency of niacin in food centrals to reduce of nicotinamide adenine diphosphate production, which is necessary for a several serious metabolic tasks in the body. Uncertainty, if untouched, it may lead to expiry in four to five years. Pellagra is found in populations where the diet is heavily based on corn, and in addition to dietary deficiency, pellagra can be caused by conditions that prevent the absorption of niacin, such as crohn's disease or other inflammatory disorders of the intestine. Alcoholism can also interfere with absorption leading to pellagra [12, 13].

Mammals on their own are not capable to produce B vitamins; for that reason, they might take up these in adequate amounts by way of food. Despite the fact that maximum of these are manufactured by plant life, they may be indirectly consumed by the use of animal derivative diet such as eggs, dairy and meat. Merely, vitamin B12 is not formed by plant life. However, it is produced by means of bacteria, which inhabit colon of humans or foregut of ruminants and as a result only can be set up in animal foodstuffs such as eggs, fish, dairy products or liver. Nevertheless, the vitamin B12 formed in the colon of humans by bacteria is not obtainable for uptake for the reason that adsorption merely further takes place in ileal mucosa through an intrinsic factor-mediated mechanism [14]. The following **Table 2**, provides a synopsis on the particularly common deficiency symptoms of each B vitamin as listed below.

3. Needs, deficiency signs and sources of vitamins B

In the section given underneath, it is looked at each B vitamin in more detail. Vitamin B1 is also called thiamin and vitamin B2 is also called riboflavin, and these vitamins help to convert food into energy. Vitamin B1 has neurological benefits and vitamin B2 helps to maintain proper eyesight. Vitamin B1 deficiency is rare and vitamin B2 deficiency is very rare, which is due to the fact that many foods, such as milk and whole-grain cereals are fortified with these vitamins. Vitamins B1 and B2 deficiencies symptoms include confusion and cracks along the sides of the mouth. It can become an issue with alcoholics people who misuse alcohol,

Vitamin	Name	Deficiency effects
B1	Thiamine	Thiamine deficiency causes beriberi and symptoms of this nervous system disease comprise emotional disturbances, weight loss, weakness and pain in limbs, edema (bodily tissues swelling) periods of irregular heartbeat and wernicke encephalopathy (impaired sensory perception). Heart disaster and passing away could happen in progressive circumstances. Prolonged thiamine shortage may as well cause alcoholic korsakoff syndrome, an irreversible dementia regarded as compensatory confabulation and amnesia.
B2	Riboflavin	Riboflavin insufficiency can cause ariboflavinosis, which might result in great sensitivity to sunlight, cheilosis (cracks in lips), glossitis (swelling of the tongue), angular cheilitis, pharyngitis (sore throat), edema of the pharyngeal and oral mucosa, hyperemia, and pseudo-syphilis or seborrheic dermatitis (predominantly distressing the mouth, and labia majora or scrotum.
B3	Niacin	Niacin shortage, together with a lack of tryptophan, results in pellagra. Signs consist of weakness, dermatitis, aggression, diarrhea, mental confusion and insomnia (sleep disorder). In progressive circumstances, pellagra can lead to dementia and passing away.
B5	Pantothenic acid	Pantothenic acid scarcity may give rise to acne and paresthesia, even though it is rare.
B6	Pyridoxine, pyridoxamine	Vitamin B6 deficiency causes seborrhoeic dermatitis-like eruptions, pink eye and neurological symptoms (epilepsy).
B7	Biotin	Biotin deficit does not normally cause indications in adults other than superficial matters like declined nail and hair development, however can cause reduced development and nervous syndromes in children. Manifold carboxylase scarcity, an innate fault of metabolism, may cause biotin lack even at what time dietetic biotin eating is common.
B9	Folic acid	Folic acid deficit leads to raised stages of homocysteine and macrocytic anemia. Lack in expectant females may cause delivery faults, mainly neural tube defects like anencephaly and spina bifida.
B12	Cobalamins	Vitamin B12 shortage leads to loss of memory and further cognitive insufficiencies, macrocytic anemia, peripheral neuropathy, and elevated methylmalonic acid and homocysteine. It is maximum expected to happen amongst elderly persons, as absorption through gut drops with age and autoimmune disease pernicious anemia is one more common cause. It could likewise result to signs of psychosis and mania. In occasional risky cases, paralysis can take place.

Table 2.
Delivers each B vitamin deficiency that can cause the symptoms in human.

however, presenting symptoms such as confusion and cracks along the sides of the mouth [15].

Most people get their B1 and B2 from fortified breakfast cereals and whole grains. Sources of vitamin B1 include organ meats, lean meats, kidney and liver, fish, eggs, low-fat milk, whole grains; fortified bread, cereal, pasta and rice, nuts and seeds, legumes, including black beans and soybeans, green vegetables including broccoli and spinach, and fortified cereals, grains and bread.

3.1 Vitamin B1 (thiamine)

Thiamine or vitamin B1 is an essential nutrient required by the body. It has many health benefits, such as vital for metabolism, supports brain function, boosts the immune system, protects the heart and helps in digestion [16].

Thiamine essential in the conversion of carbohydrates into glucose, is the preferred source of energy that the body runs to keep metabolism running smoothly. It also helps to break down proteins and fats. The percentage of persons by type 1 or type 2 diabetes ranges from 17 to 79%, who have little thiamine. Readings have established that rising vitamin B1 consumption declines the harshness of signs linked to early stage diabetes [17].

Vitamin B1, similar to other B-complex vitamins, is from time to time termed an 'anti-stress' vitamin for the reason that it can build up the immune system and progress the body's capability to survive tense circumstances [18]. Vitamin B1 looks to aid in the growth of the myelin sheath, which is a coat that wraps round nerves to guard these from harm and passing away. Within the brain, it is necessary together by the nerve cells and other supporting cells in the nervous system [19].

Thiamine deficiency causes beriberi, whose signs comprise swelling, tingling or burning in the feet and hands as well as trouble in breathing because of fluid in the lungs. An inadequate thiamine intake can lead to fatigue, muscle weakness, nerve damage, cognitive complications, interfere with the body's defense against oxidative stress and cardiovascular complications. Sources of thiamine are green peas, beans, lentils, seafood seeds and nuts, soy products, white rice, brown rice, wheat germ, whole-wheat bread, egg, milk, spinach, pecans, cantaloupe and orange.

3.2 Vitamin B2 (riboflavin)

Vitamin B2 or riboflavin is one of eight B vitamins that are essential for human health. It can be found in grains, plants and dairy products. It is crucial for breaking down food components, captivating other nutrients and preserving tissues. Riboflavin aids to alter carbohydrates into adenosine triphosphate (ATP). The human body yields ATP from diet and ATP yields vitality as per the body needs it. The compound ATP is vigorous for storage of energy in muscles [20].

Along with vitamin A, vitamin B is crucial for keeping of the eyes, nerves, muscles and skin healthy; hormone production by the adrenal glands; maintaining a healthy liver; maintaining mucous membranes in digestive system; absorbing and activating iron, folic acid and vitamins B1, B3 and B6; preventing the development of cataracts; converting tryptophan into niacin, an amino acid; and fetal development, especially in areas where vitamin deficiency is common. Migraine headaches typically produce intense pulsing or throbbing pain in one area of the head and mitochondrial dysfunction is thought to play a causal role in some types of migraine. Because riboflavin is required for mitochondrial function, there is potential use of riboflavin to prevent or treat migraine headache [21].

The riboflavin might help to prevent the DNA damage caused by many carcinogens by acting as a coenzyme. The total intakes of riboflavin from both foods and supplements are associated with a lower risk of colorectal cancer, and a significant inverse association between dietary riboflavin intake and lung cancer risk [22].

Pregnant or lactating women who rarely consume meats or dairy products are at risk of riboflavin deficiency, which can have adverse effects on the health of both mothers and their infants, moreover, people who drink excessive amounts of alcohol are at greater risk of vitamin B deficiency. Symptoms and signs of deficiency include dry skin, cracked lips, angular cheilitis or cracks at the corners of the mouth, inflammation of tongue and lining of mouth, red lips, mouth ulcers, sore throat, fluid in mucous membranes, scrotal dermatitis, and Iron-deficiency anemia. Eyes may be sensitive to bright light and they may be watery, itchy or bloodshot [23].

Sources of B2 include fish; chicken; poultry such as turkey, meat such as beef, liver and kidneys; dairy products; eggs; cayenne; asparagus; artichokes; currants; avocados; kelp; fortified cereals; lima beans, peas; navy beans; mushrooms; molasses; parsley; pumpkins; nuts; sweet potatoes; sage; rosehips; cruciferous vegetables such as Brussels sprouts, spinach, broccoli, watercress and dandelion greens; wheat bran; whole-grain breads; enriched breads; and yeast extract.

3.3 Vitamin B3 (nicotinic acid)

Vitamin B3, also called niacin, nicotinamide or nicotinic acid, helps to convert food eaten into energy. It helps the body to use proteins and fats, and keeps the skin, hair and nervous system healthy. It also aids in proper digestion and healthy appetite, and is important for cell development. Other possible benefits of vitamin B3 stem from its potential cholesterol-lowering trusted source, antioxidative and anti-inflammatory properties [24]. A lack of niacin can cause digestive issues, such as nausea and abdominal cramps. Severe deficiency may also cause mental confusion and can result in a condition called pellagra, which causes many symptoms, but the most common are diarrhea, dermatitis and dementia [25].

A person who lacks vitamin B3 may experience symptoms such as headache; depression; memory loss; circulatory problems; a pigmented rash on skin; bright red tongue; rough skin that turns red or brown in the sun; constipation or diarrhea; vomiting; aggressive, paranoid or suicidal behavior; fatigue and hallucinations [26].

A healthful diet can provide all of a person's vitamin B3 needs, and food sources of vitamin B3 include meat including beef, fish and poultry; some legumes, grains and nuts; fortified breads and cereals; sunflower seed; and almond. The foods cooked brown rice, beef liver, grilled chicken breast, turkey breast, dry roasted peanuts, sockeye salmon and enriched breakfast cereal are good sources of vitamin B3. The chicken tacos with peanut sauce are a great way to get niacin in diet.

3.4 Vitamin B5 (pantothenic acid)

Vitamin B5, also called pantothenic acid, is one of the most important vitamins for human life. Vitamin B5 is a medication used in the management and treatment of patients with nutritional deficiencies and related conditions. It is necessary for making blood cells, synthesizing cholesterol, converting the food eaten into energy, and forming sex and stress-related hormones.

The usage of vitamin B5 is prevalent within the field of dermatology to compare efficiency of dexpanthenol (an alcoholic correspondent of D-pantothenic acid) as a substitute usage to atopic dermatitis therapy in contrast to a normal dealing of hydrocortisone. Generally, the reading set up reveals that dexpanthenol is able to possibly deal minor to modest infant atopic dermatitis [27]. Further investigation proposes that dexpanthenol cream may be beneficial in treatment of mucocutaneous sideways special effects, which take place in isotretinoin healing. Isotretinoin treatment is used as per a management for acne, and its mucocutaneous side effects comprise xerosis (abnormally dry skin), cheilitis (inflammation of lips) and dry of mucous membranes. The small clinical drug as a pastille trials is used and or spray to heal wounds in postoperative endotracheal intubation, endoscopic sinus surgery, and tonsillectomy [28].

Vitamin B5 deficiency is associated with the symptoms such as personality changes, fatigue, headache, irritability, nausea, stomach pains, malaise, numbness, muscle cramps, paresthesia, muscle or abdominal cramps, impaired muscle coordination, numbness, sleep disorders, upper respiratory infections, burning feet and an increased sensitivity to insulin [29].

Vitamin B5 is an easy vitamin to incorporate into a good diet and it is found in most vegetables, including members of the cabbage family, white and sweet potatoes, broccoli and whole-grain cereals. Other healthy sources of B5 include organ meats (liver, kidney), eggs, poultry, dairy products, mushrooms, peas, beans, lentils, seeds, nuts, brown rice and oats. It can also be applied to the skin to relieve itchiness and promote healing from skin conditions, such as eczema, diaper rash, poison ivy and insect bites. Dexpanthenol has also been used to prevent and treat skin reactions from radiation therapy [30].

3.5 Vitamin B6 (pyridoxine)

Vitamin B6, also called pyridoxine, helps the body to turn food into energy. It can also help the body to fight infections by supporting the immune system. Pregnant and breastfeeding women need it to help their babies' brains develop normally. Vitamin B6 deficiency is not common, however insufficient amounts of B6 can result in anemia as well as skin disorders, such as an itchy cracks or rash around the mouth. A lack of B6 can also cause confusion, depression, anemia, nausea skin rashes or dermatitis and susceptibility to infections [31].

Symptoms of B6 scarcity comprise receiving sickening from contaminations more frequently for the reason that B6 aids to looking after immune system. Attainment of sores or cracks on skin round the junctions of mouth or a sensitive and swollen teongue, feeling of tingling or numbness on feet and hands termed as paresthesias, fatigue, irritability or anxiety and depression. A red, irritated rash commonly flaky or oily generally looks on face or upper body. Slight parts of skin may swell as well causing in white areas, reduced attention and convulsions [32]. Foods high in vitamin B6 comprise organ meats, fish, poultry, and potatoes and other starchy vegetables and fruits except citrus fruits.

3.6 Vitamin B7 (biotin)

Biotin, also known as vitamin H or B7, is a vitamin that helps the body to metabolize fats, carbohydrates and protein. Vitamin B7 may not be manufactured by means of human cells, however, it is formed by bacteria in the body and exists in various diets. Biotin rehabilitation can help to give round about curative settings and several persons receipt complements to make stronger their hair and nails, however, there is a shortage of indication in supportive to this usage.

Biotin deficiency is rare in humans, because biotin is widely available in foods and the 'good gut bacteria' can normally synthesize more biotin than the body needs. Signs of deficiency include a scaly red rash around the eyes, nose, mouth and genitals; hair loss or alopecia; numbness and tingling in the hands and feet; lethargy; depression; hallucinations; seizures; a loss of bodily movements control known as ataxia; and impaired immune function resulting increased risk of bacterial and fungal infection [33].

Biotin deficiency is most likely to arise in people who smoke, women during pregnancy, infants who consume breast milk with low amounts of biotin, patients receiving prolonged intravenous nutrition and patients with impaired biotin absorption due to an inflammatory bowel disease or other gastrointestinal tract disorder. It may also affect those with some kinds of liver disease and those who use medications for epilepsy, such as phenytoin, phenobarbital or carbamazepine [34].

Foods that are rich in biotin include organ meats, cooked whole eggs, baker's yeast, wheat bran and oysters. Raw eggs contain a protein called avidin that inhibits the absorption of biotin. Eating two or more raw egg whites a day for several months has been linked to biotin deficiency. Many foods, such as fruits and vegetables, contain a small amount of biotin [35].

3.7 Vitamin B9 (folate or folic acid)

Vitamin B9, also called folate or folic acid, is the synthetic form of B9, found in supplements and fortified foods, while folate occurs naturally in foods. Folic acid is crucial for proper brain function and plays an important role in mental and emotional health. It is an important vitamin that works with other B vitamins to metabolize proteins, and aids in the production of DNA and RNA, body's genetic material, and is especially important when cells and tissues are growing rapidly, such as in infancy, adolescence and pregnancy [36].

Meats and organ meats, grains, legumes and green leafy vegetables are elevation in folate. The vitamin is not stockpiled in the body, therefore intensities of folate in the body may become little just afterward a small number of weeks of intake a folate scarce food. As per with other shortage disorders, syndromes of the intestinal structure and alcoholism may add to folate insufficiency. Moreover, kidney dialysis, hemolytic anemia and certain medications may lead to folate scarcity [37].

Indicators of folate insufficiency contain fatigue, poor growth, diarrhea, irritability and a tender or smooth tongue. For women with pregnancy, a lack of folate may likewise rise the danger of neural tube faults in the emerging fetus. Low levels of vitamin B9 can result in the symptoms such as irritability, fatigue, megaloblastic anemia that causes weakness, trouble concentrating, headache, shortness of breath, heart palpitations, swollen tongue, open sores in the mouth and changes in skin, hair or fingernail color.

Pregnant women with a folate deficiency could result in their babies being born with neural tube defects, such as spina bifida including cleft palate, spina bifida and brain damage. Neural tube defects are birth defects caused by abnormal development of the neural tube, a structure that eventually gives rise to the brain and spinal cord. Most people (except pregnant women should be able to get enough folic acid from their diets. Supplementing with high-enough levels of B9 before pregnancy as well as during pregnancy) significantly lowers the risk of giving birth to a baby with neural tube defects [38, 39].

Folate occurs naturally in foods and folic acid is the synthetic form, often found in fortified, processed foods. Food sources of folate contain eggs, beef liver, Brussels sprouts, asparagus dark green leafy vegetables (spinach, turnip greens, lettuce, romaine asparagus, broccoli and Brussels sprouts), oranges and orange juice as well as other fresh fruits and fruit juices, beans including kidney beans, peanuts, and other nuts, sunflower seeds, seafood and peas including black-eyed peas. Foods fortified with folic acid include enriched bread, pasta, flour and rice, corn tortillas and tamales made with fortified flour, fortified breakfast cereals, whole grains, fortified foods and supplements, and spicy roasted beet as a snack or appetizer [40].

3.8 Vitamin B12 (cobalamin)

Vitamin B12 (cobalamin) is only found in animal food sources and helps to regulate the nervous system. It also plays a role in growth and red blood cell formation. Owing to this, persons commonly at danger of B12 deficit comprise lacto-vegetarians. Individuals who have weightiness loss surgical treatment are as well at a great threat for the reason that the operation disturbs absorption of B12 from diet. Further situations, which disturb absorption such as Crohn's disease or celiac disease, may as well result in B12 dearth. Just about 3.2 percent of adults over the age of 50 have a B12 deficiency, and up to 20 percent may have levels of B12 that are at borderline. A vitamin B12 deficiency can lead to disruption in the nervous system and the circulatory system [41].

Vitamin B12 deficiencies can lead to megaloblastic anemia, a condition where the bone marrow produces large abnormally shaped red blood cells that do not function properly. Psychological conditions such as dementia, paranoia, depression and behavioral changes can result from a vitamin B12 deficiency. Neurological damage sometimes cannot be reversed. The primary symptoms of B12 deficiency are mood changes, numbness or tingling in hands, legs and feet, anemia, loss of breath, pale skin, dizziness, blurry vision, difficulty in walking, a swollen tongue, cognitive changes, jaundice, paranoia, hallucinations, soreness of the mouth or tongue, loss of appetite, constipation, weight loss, poor memory, confusion, fatigue or tiredness, and weakness. In addition to animal-based foods, dietary vitamin B12 can be obtained from breads cereals fortified with B12, or through a supplement [42, 43].

Vitamin B12 is found primarily in meat and dairy products, so people on a vegan diet are at risk for deficiency. However, there are foods options to meet their B12 dietary needs. Vegetarian sources include dairy and eggs. Vegan sources of B12 include fortified foods and nutritional yeast. The top sources of vitamin B12 include; beef liver, meat including fish and poultry eggs, milk, cheese, clams, nutritional yeast, fortified breakfast cereals, and other fortified foods including plants, milk and flours [44].

4. Conditions and causes of B vitamins deficiency

Vitamin B denotes to a number of diverse kinds of vitamins, which are collectively recognized as B-complex vitamins. Vitamin B aids to the cells and nerves inside the body and moreover supports by the manufacture of DNA that is biochemical constituent of which genes are prepared. Each of the eight B vitamins plays vital parts within the body and is desired to motivate the biochemical reactions that upkeep body's several roles. For instance, cells usage B vitamins to create vitality from fatty acids, sugar and other nutrients. Therefore, deprived of B-complex vitamins, the human body might not task fine of any kind. B vitamins are soluble in water, can dissolve in water, excess B vitamins body does not use are washed out through urination, body cannot easily store B vitamins for long time periods, so it is particularly important to commonly consume B vitamins either from diet or from supplements to escape scarcity. Deficiencies in these B vitamins can lead to a number of different symptoms over time if the deficiency is not retreated [45, 46].

When it comes to vitamins needed for both a sound body and mind, B vitamins are not something to be ignored. For instance, if body does not get enough of vitamin B12, energy levels throughout the day might fall with mind constantly perhaps turning to or thoughts of sleep. Otherwise consider vitamin B9, a deficiency of which may acquire swollen tongue or sores on mouth amongst other probable indications. There is then vitamin B6 and when intensities of this vital B vitamin are excessively little, at that time there might be observed scaly, oleaginous eruptions on face or upper body. However, these are not the lone signs of B vitamin deficit, but there is a further wide-ranging list of vitamin B scarcity signs [47, 48].

People from all age groups are suffered with depression, severe anxiety and psychiatric disorders. These patients are prescribed costly psychotropic drugs, narcotics or benzodiazepines; however, actually the problem is B12 deficiency. The B12 vitamin is a vital micro-nutrient for healthy brain in children, younger and elders. Its deficiency primarily occurs due to insufficient dietary intake resulting neurological disorders including apathy, anorexia, irritability, growth retardation and developmental regression. Some diseases caused by B12 deficiency are myeloneuropathy, demyelination, alzheimer's disease, atrophy or brain shrinkage, sub-acute combined degeneration, vascular complications, neuropsychiatric abnormalities, infantile seizures and poor fetal brain and cognitive development. A timely and proper supplementation is necessary if it is dietary deficiency [49].

4.1 Conditions of vitamin B deficiency diseases

Vitamin B shortage might upsurge the danger of many syndromes and disorders that may well upset to mental well-being, brain health, heart health and further more. For instance, together B9 and B12 scarcity bases for anemia in many circumstances (a disorder in that body shortages red blood cells, which make it tough for diverse body parts to acquire oxygen they needed). Anemia may too lead to dizziness, fatigue, quickness of inhalation, and even tingling and numbness in several circumstances [50].

Vitamin B shortages are extremely widespread in several emerging nations, particularly wherever foods are little in fruits, vegetables and animal products, and anywhere breakfast cereal are crushed earlier to feeding. Expecting and suckling womenfolk, children and youngsters are greatest at danger of vitamin B deficits [51]. Several B vitamin deficiencies cause homocysteinemia (amino acid produced when proteins are broken down that can contribute to arterial damage and blood clots in blood vessels), notably folic acid, vitamin B12, riboflavin and vitamin B6. Importantly, homocysteinemia is associated with adverse pregnancy outcomes. Severe thiamine (vitamin B1) scarcity may lead to the disorder 'beri-beri', possibly lethal heart miscarriage or peripheral neuropathy. Early symptoms of riboflavin (vitamin B2) scarcity may include burning eyes, mouth hurt, itching, fatigue and weakness. Additional progressive shortage may lead to brain dysfunction [52, 53].

Niacin (vitamin B3) insufficiency may lead to 'pellagra, resulting skin eruptions being an indication, along with depression, diarrhea, vomiting, loss of memory and fatigue. Symptoms of severe pyridoxine (vitamin B6) deficiency consist of neural syndromes (epileptic convulsions), skin modifications and probably anemia. Folate (vitamin B9) shows a vital part in cell duplication and tissue development. Insufficiency results to the hazard of neural tube faults and may similarly impair cognitive function in adult persons. This deficit situation is frequently linked to populations, which ingest in their food lots of cereals that are short in folate, and few fruits and leafy greens, those are striking in it. Deficiency of vitamin B12 causes neurological deterioration, megaloblastic anemia and possible impaired immune function. Deficiency can severely delay the development of infants and young children [54, 55]. Other conditions linked with vitamin B deficiency are included in the ensuing section.

4.1.1 Paresthesias

Paresthesia is an unusual feeling of the skin (pricking, tingling, burning, chilling, numbness) with no physical cause apparently. Paresthesia may be transient (common symptoms of hands, feet, leg and arms) or chronic (problematic with working of neurons or poor circulation) and may have any of dozens of possible underlying causes [56]. Paresthesias are generally painless and can happen anywhere on the body, but most generally occur in the arms and legs. The most familiar kind of paresthesias is the sensation known as 'pins-and-needle' after having a limb 'fall asleep'. A feeling is often experienced around the hands, arms, feet or legs. A lesser familiar and infrequent, but main paresthesias is formication that is like sensation of insects crawling on the skin [57].

4.1.2 Peripheral neuropathy

Peripheral neuropathy may be acute (with sudden onset, rapid progress) or chronic (symptoms begin subtly and progress slowly) and may be reversible or permanent. It is a nervous system condition that is often felt as a stabbing (sharp) or burning pain. Peripheral neuropathy often describes disease affecting the peripheral nerves beyond the brain and spinal cord. Damage to peripheral nerves may impair sensation, movement and gland or organ function depending on which nerves are affected resulting in different symptoms [58].

Neuropathy could basis of painful cramps (shooting pain), fasciculations (fine muscle contracting), bone deterioration, muscle damage, and alterations in skin, hair and nails. Moreover, motor neuropathy can reason of decreased coordination and balance or best generally, muscle fault; sensory neuropathy might affect lack of feeling to vibration and touch, condensed location common sense resulting lesser balance and coordination, decreased feeling to pain and temperature change, unplanned burning or tingling pain, or skin allodynia (intense pain from usually non-painful stimuli, like touch or light); and autonomic neuropathy could yield varied indications, dependent on the affected organs and glands, however general indicators are abnormal heart rate or blood pressure, poor bladder control and decreased capability to perspire routinely [59].

4.1.3 Psychosis

Psychosis is a mental condition in which one's thoughts and perceptions are significantly altered or other symptoms occur. It is a condition of the mind that results in difficulties in determining what is real and what is not real. Psychosis can have serious outcomes, resulting in delusion (unrelenting sense of certainty maintained despite of strong contradictory evidence) and hallucinations (sensory perception in the absence of external stimuli). Other symptoms may include incoherent speech (nonsense speech) and behavior that is inappropriate for the situation. There may also be social withdrawal, sleeping problems, lack of motivation and difficulties in carrying out daily activities [60–62].

4.1.4 Heart attack and stroke

A deficiency in vitamin B12 may heighten the risk of getting a heart attack or stroke. Myocardial infarction commonly called heart attack, takes place while blood movement stops or decreases to a portion of heart resulting harm to heart muscle. The utmost usual indication is chest discomfort or pain that could move into jaw, neck, back, shoulder or arm. Habitually it takes place in the middle or leftward sideway of the chest and continues for more than a little minute [63, 64].

A stroke is a health situation in which a reduced blood movement to the mind leads to cell expiry. There are two foremost kinds of stroke; ischemic, for the reason that of shortage of blood movement and hemorrhagic due to blood loss. Both types results portions of the mind to halt working correctly. Symptoms and signs of a stroke could comprise dizziness, an incapability to move or touch on one sideway of the body, difficulties in speaking or understanding, or damage of visualization to one sideway [65].

4.2 Causes of vitamin B deficiency

Here is a roundup of the four top causes of vitamin B deficiency that are somewhat not to be ignored.

4.2.1 A non-balanced diet

Human body cannot accurately create B vitamins contrasting to proteins that the body constructs by mean of several minor building blocks. However, this is generally not a problematic for the reason that body develops B vitamins from the diet eaten. If any person follows a well-proportioned food that offers to body through the correct level of nutrients, these may aid to escape signs of vitamin B insufficiencies. On the other hand, for a range of causes, from time to time people do not consume the correct equilibrium of diet essential to acquire sufficient of vitamins needed. For instance, if a vegetarian or vegan food is followed, then an individual may not acquire sufficient vitamin B12 as vitamin B12 is set up practically wholly in dairy products and animal-created diets [66].

As such, dietary inadequacies are one of the key causes of vitamin B deficiency. So, whatever diets comprise several B vitamins that rest which B vitamin is in concern (vitamins B6, B9 or B12). A rapid prosperous diets that may be eaten to increase stages of each of these B vitamins is vitamin B6 (fish, legumes, nuts, potatoes, bananas, meat); vitamin B9 (leafy vegetables, citrus fruits, legumes); and vitamin B12 (fish, meat, dairy and animal foodstuffs) [67].

4.2.2 Excessive alcohol consumption

An excessive alcohol consumption whether beer or spirit might have its disadvantages and single of which is vitamin B insufficiency. In a nutshell, alcohol creates kidneys to flush B vitamins out of system greatly further rapidly than normal. This means body does not have whole the period it desires to create usage of these B vitamins, hence they somewhat accurately go to discarded [68].

4.2.3 Various medications

More than a few types of recommended medications are able to knock up the likelihood of a vitamin B scarcity. An improved possibility of vitamin B6 scarcity is marked from penicillamine (Cuprimine), corticosteroids, hydralazine, isoniazid and anticonvulsants. An augmented probability of vitamin B9 shortage comes by sulfasalazine (Azulfidine), methotrexate (Rheumatrex, Trexall), trimethoprim-sulfamethoxazole and phenytoin (Dilantin). An enlarged chance of vitamin B12 shortage rises by long-term antidepressants and antibiotics, antacids, proton pump inhibitors and metformin [69].

4.2.4 Gut malabsorption conditions

In well circumstances, B vitamins are absorbed into bloodstream through the gut. The bloodstream at that moment vehicles these greatly-required vitamins all over body. Hence, if B vitamins do not create way into the bloodstream, straightforwardly they may not be placed into upright usage by the body. Then these correctly can drive erroneous if someone have a gut malabsorption situation like ulcerative colitis, celiac disease or crohn's, disease. These situations check to B vitamins for arriving the bloodstream, considerably reducing blood's vitamin B intensities and possibly damaging safety [70].

For first course, study inspection of vitamin B levels with B vitamins assessment. At that point, if persons are definitely lacking, they may access to healthcare worker on the afterward stages. It is suggested that elder adult persons who have lacking levels of vitamin B should consult to their healthcare supplier as early as probable to get a cure strategy. If there are seen vitamin B scarcity signs, it might be for the reason that peoples do not have sufficient vitamin B in their food. If there are shown symptoms of fatigue, quickness of inhalation, lightheadedness or other shortage signs, then think through scrutiny of B12 level [71].

5. Preventing of B vitamins deficiency

For maximum persons, some fit foods riched by means of a diversity of fruits, meats, vegetables and grains, are sufficient to check a scarcity of B complex vitamins. Ladies who are supposed to get into pregnancy are guided to takings folate complements. Elder persons or those with health situations that rise the danger of B vitamin insufficiency might as well advantage from taking an everyday complement. In the direction of staying healthy, most people do not need to take a supplement in order to get enough B vitamins. There are plenty of delicious foods available to get all the nutrients the body needs naturally. Try to eat a complete diet of meats, grains, fruits and vegetables. If any person does not eat meat, eggs or dairy, he or she can still get vitamin B12 from fortified foods or nutritional yeast to help prevent its deficiency [72, 73].

Supplementation is only a last resort if a person cannot obtain B vitamins through diet nor have certain health conditions that require using of supplements. Over the counter supplements can often treat or prevent deficiency. It is best to check with a physician before taking vitamin supplements. However, supplements may still cause side effects, long-term health effects and interactions with medications a person is taking. If any person suspects to be vitamin B deficient, he or she can contact to physician who might perform a physical examination as well as instruct blood testing. If a woman is pregnant or any person is over 50 years old, they are more likely to need supplements. The risk of overdose is lower than other nutrients because B vitamins are water soluble [74, 75]. For maintaining a good health, healthcare professionals mention that people should acquire a sure quantity of each vitamin per day as given in the ensuing **Table 3**.

It needs to be stressed that vitamin B1, B6 and B12 most likely hold synergistic biochemical roles in the nervous system that is neither of these can replace one of the others. Because of the potential for side effects and interactions with medications, people should take dietary supplements only under the supervision of a knowledgeable health care provider [76].

Vitamin B1 helps body to make main energy-carrying molecule ATP, and prevents complications in the nervous system, brain, muscles, heart, stomach and intestines. It is also involved in the flow of electrolytes into and out of muscle

Vitamins	Intake for adults and children ages 4+	Intake for pregnant or breastfeeding
Thiamin (B1)	1.2 mg	1.4 mg
Riboflavin (B2)	1.3 mg	1.6 mg
Niacin (B3)	16 mg or equivalent	18 mg or equivalent
Pantothenic acid (B5)	5 mg	7 mg
Pyridoxine (B6)	1.7 mg	2 mg
Biotin (B7)	30 mcg	35 mcg
Folate (B9)	400 mcg or equivalent	600 mcg or equivalent
Cobalamin (B12)	2.4 mcg	2.8 mcg

Table 3.
Offering daily intake values of each B vitamin in microgram (mcg) or milligram (mg).

and nerve cells. Vitamin B6 helps the body to build neurotransmitters (like dopamine) that are special chemicals the brain needs to functions; makes red blood cells, aids immune system antibodies to work correctly and lowers the risk of lung cancer. Vitamin B12 helps the body to make and repair genetic material DNA, make red blood cells as well as nerve cells, and supports healthy hair, skin and nails [77–79]. If someone's vitamins B status is not at a normal level, it may be useful to get vitamins level checked and their healthcare provider may recommend a high-dosage supplement or even in some cases injections.

Each B vitamins have their particular distinctive utilities, however they depend upon one another for suitable absorption and the best fitness welfares. Consumption of an advantageous, diverse food can usually offer altogether the B vitamins somebody require. Persons can prevent and treat B vitamin insufficiencies through enhancing their food ingestion of high-vitamin diets or taking vitamin complements. Consult to a physician earlier to take any complements to make certain that they will act together with medicines.

6. Conclusions

B vitamins play vital roles in maintaining of good health and well-being. As the building blocks of a healthy body, vitamin B benefits to the nerves and cells within the body and also helps with the production of DNA (chemical substance that genes are made). B vitamins are important exclusively for womenfolk who are expectant and breastfeeding, and help in fetal's mind growth along with decrease the danger of delivery faults. For pregnant mammies, B vitamins can enhance energy intensities, easiness vomiting and lessen the risk of rising preeclampsia (high blood pressure and possibly protein in urine during pregnancy or after delivery or lower platelets in blood). In men, B vitamins are thought to increase testosterone levels, which decrease with age naturally and may also help men to build muscle and increase strength. For infants of vegan mothers, starting of vitamin supplements immediately after birth helps to prevent vitamin deficiency. For people with nerve damage, vitamin is given by injection into muscle unless the disorder causing the deficiency can be corrected. Blood tests are done periodically to make sure that vitamins level return to and remain normal or sometimes endoscopy diagnosis is done. Treatment of vitamins deficiency consists of high doses of vitamin supplements. If people have the deficiency, but no symptoms, the vitamin may be taken by mouth. Older people with vitamin deficiency can benefit from taking of vitamin supplements because the deficiency usually results from difficulty in absorbing vitamin from meat. They can absorb the vitamin more easily from supplements than from meat. Maximum of multivitamin-mineral foodstuffs comprise the B-complex together with the rest of the vital minerals and vitamins. As these are further wide-ranging than B-complex vitamins only, various vitamin-mineral additions are suggested to mend whole micronutrient eating and avoid insufficiencies. There is a link between fruit and vegetable intake, and stress levels of persons. People who have higher fruit and veggie intakes are less stressed than those with lower intakes, which suggest that diet plays a key role in mental wellbeing (eating at least 400 grams of fruits and vegetables per day). A practitioner knowledgeable in nutrition must be consulted when using of vitamins and always follow label directions before use.

Author details

Muhammad Farhan Sarwar[1], Muhammad Haroon Sarwar[1] and Muhammad Sarwar[2]*

1 Mayo Hospital, King Edward Medical University, Lahore, Pakistan

2 National Institute for Biotechnology and Genetic Engineering (NIBGE), Faisalabad, Pakistan

*Address all correspondence to: drmsarwar64@gmail.com

References

[1] Sarwar MF, Sarwar MH, Sarwar M. Understanding Some of the Best Practices for Discipline of Health Education to the Public on the Sphere. International Journal of Innovation and Research in Educational Sciences. 2015; 2 (1): 1-4.

[2] Khalid MT, Sarwar MF, Sarwar MH, Sarwar M. Current Role of Physiotherapy in Response to Changing Healthcare Needs of the Society. International Journal of Education and Information Technology. 2015; 1 (3): 105-110.

[3] Sarwar MF, Khalid MT, Sarwar MH, Sarwar M. Emerging Responsibility of a Physiotherapist in the Healthcare System for an Individual, Family, Community and Country. International Journal of Education and Information Technology 2015; 1 (3): 99-104.

[4] Lykstad J, Sharma S. Biochemistry, Water Soluble Vitamins. In: StatPearls. Treasure Island (FL): StatPearls Publishing; 2020.

[5] Sarwar MH, Sarwar MF, Sarwar M. Understanding the Significance of Medical Education for Health Care of Community around the Globe. International Journal of Innovation and Research in Educational Sciences. 2014; 1 (2): 149-152.

[6] Sarwar MH, Mirz B, Sarwar MF, Khalid, MT, Sarwar M. Snapshot of Scientific Evidence for Remunerations of Physical Activity on Public Well-Being in Different Settings and Contexts. American Journal of Clinical Neurology and Neurosurgery. 2015; 1 (3): 182-188.

[7] Abu Bakar H, Rauf A, Sarwar MH, Sarwar M. Essential Vitamin and Mineral Nutrients Body Needs and Their Best Food Sources. American Journal of Economics, Finance and Management. 2017; 3 (4): 36-41.

[8] Spinas E, Saggini A, Kritas SK, Cerulli G, Caraffa A, Antinolfi P, Pantalone A, Frydas A, Tei M, Speziali A, Saggini R, Pandolfi F, Conti P. Crosstalk Between Vitamin B and Immunity. J Biol Regul Homeost Agents. 2015; 29(2): 283-288.

[9] Sarwar MH, Sarwar MF, Khalid M T, Sarwar M. The Roles of Pharmacy and Clinical Pharmacy in Providing Healthcare Services to the People. Journal of Pharmacy and Pharmaceutical Sciences, 2015; 3 (1): 1-5.

[10] Calderon-Ospina CA, Nava-Mesa MO. B Vitamins in the nervous system: Current knowledge of the biochemical modes of action and synergies of thiamine, pyridoxine, and cobalamin. CNS Neurosci. Ther. 2020; 26(1): 5-13.

[11] Sarwar MH, Sarwar MF, Khalid MT, Sarwar M. Effects of Eating the Balance Food and Diet to Protect Human Health and Prevent Diseases. American Journal of Circuits, Systems and Signal Processing. 2015;1 (3): 99-104.

[12] Wan P, Moat S, Anstey A. Pellagra: a review with emphasis on photosensitivity. The British Journal of Dermatology. 2011; 164 (6): 1188-1200.

[13] Kennedy D. B vitamins and the brain: mechanisms, dose and efficacy- a review. Nutrients. 2016; 8(2):68.

[14] Zempleni J, Suttie JW, Gregory JF III, Stover PJ. Handbook of Vitamins. CRC Press, Boca Raton, Florida, USA. 2013. p. 605.

[15] Rosalind GS. Riboflavin. In: Principles of Nutritional Assessment (2nd ed). Oxford University Press. 2005. p. 908.

[16] Boehnke C, Reuter U, Flach U, Schuh-Hofer S, Einhäupl KM, Arnold G.

High-dose riboflavin treatment is efficacious in migraine prophylaxis: an open study in a tertiary care centre. European Journal of Neurology. 2004; 11 (7): 475-477.

[17] Lonsdale D. A review of the biochemistry, metabolism and clinical benefits of thiamin(e) and its derivatives. Evid Based Complement Alternat Med. 2006; 3(1): 49-59.

[18] Pletsityi AD. Changes in activity of some mechanisms of specific and nonspecific immunity in vitamin B1 deficiency. Bull Exp Biol Med. 1979; 88: 741-743.

[19] Peter RM, Charles KS, Hiller–Sturmhofel S. The Role of Thiamine Deficiency in Alcoholic Brain Disease. Alcohol Research & Health. 2003; 27(2): 134-142.

[20] Yorns WR, Jr., Hardison HH. Mitochondrial dysfunction in migraine. Semin Pediatr Neurol. 2013; 20: 188-193.

[21] Kabat GC, Miller AB, Jain M, Rohan TE. Dietary intake of selected B vitamins in relation to risk of major cancers in women. Br J Cancer. 2008; 99: 816-821.

[22] Condo M, Posar A, Arbizzani A, Parmeggiani A. Riboflavin prophylaxis in pediatric and adolescent migraine. J Headache Pain. 2009; 10: 361-365.

[23] Jaeger B, Bosch AM. Clinical presentation and outcome of riboflavin transporter deficiency: mini review after five years of experience. J Inherit Metab Dis. 2016; 39(4): 559-564.

[24] Villines TC, Kim AS, Gore RS, Taylor AJ. Niacin: the evidence, clinical use, and future directions. Current Atherosclerosis Reports. 2012; 14 (1): 49-59.

[25] Hegyi J, Schwartz RA, Hegyi V. Pellagra: dermatitis, dementia, and diarrhea. International Journal of Dermatology. 2004; 43 (1): 1-5.

[26] Knip M, Douek IF, Moore WP, Gillmor HA, McLean AE, Bingley PJ, Gale EA. Safety of high-dose nicotinamide: a review. Diabetologia. 2000; 43 (11): 1337-1345.

[27] Udompataikul M, Limpa-o-Vart D. Comparative trial of 5% dexpanthenol in water-in-oil formulation with 1% hydrocortisone ointment in the treatment of childhood atopic dermatitis: a pilot study. J Drugs Dermatol. 2012; 11 (3):366-374.

[28] Romiti R, Romiti N. Dexpanthenol cream significantly improves mucocutaneous side effects associated with isotretinoin therapy. Pediatr Dermatol. 2002; 19 (4): 368.

[29] Gulhas N, Canpolat H, Cicek M, Yologlu S, Togal T, Durmus M, Ozcan Ersoy M. Dexpanthenol pastille and benzydamine hydrochloride spray for the prevention of post-operative sore throat. Acta Anaesthesiol Scand. 2007; 51 (2): 239-243.

[30] Miller JW, Rucker RB. Pantothenic Acid. In: Present Knowledge in Nutrition, BP Marriott; DF Birt; VA Stallings; AA Yates (Eds.), Eleventh Edition. London, United Kingdom: Academic Press (Elsevier). 2020. pp. 273-288.

[31] Rosenberg IH. A history of the isolation and identification of vitamin B (6). Ann Nutr Metab. 2012; 61 (3): 236– 238.

[32] Gibson RS. Assessment of vitamin B_6 status. Principles of Nutritional Assessment (2nd ed.). New York: Oxford University Press. 2005. pp. 575-594.

[33] Penberthy WT, Sadri M, Zempleni J. Biotin. In: Present Knowledge in Nutrition, BP Marriott, DF Birt, VA Stallings, AA Yates (Eds.), Eleventh

Edition. London, United Kingdom: Academic Press (Elsevier). 2020. pp. 289-304.

[34] Canda E, Kalkan US, Çoker M. Biotinidase Deficiency: Prevalence, Impact And Management Strategies. Pediatric Health Med Ther. 2020; 11: 127-133.

[35] Staggs CG, Sealey WM, McCabe BJ, Teague AM, Mock DM. Determination of the biotin content of select foods using accurate and sensitive HPLC/avidin binding. Journal of Food Composition and Analysis. 2004; 17 (6): 767-776.

[36] Paul C. Folic acid in pregnancy. BJOG. 2016; 123 (3): 392.

[37] Smith AD. Folic acid fortification: the good, the bad, and the puzzle of vitamin B-12. The American Journal of Clinical Nutrition. 2007; 85 (1): 3-5.

[38] Ahrens K, Yazdy M, Mitchell A, Werler M. Folic acid intake and spina bifida in the era of dietary folic acid fortification. Epidemiology. 2011; 22 (5): 731-737.

[39] Alpert JE, Mischoulon D, Nierenberg AA, Fava M. Nutrition and depression: focus on folate. Nutrition. 2000; 16: 544-581.

[40] Choi SW, Mason JB. Folate and carcinogenesis: an integrated scheme. J Nutr. 2000; 130: 129-132.

[41] Oakley GP, Adams MJ, Dickinson CM. More folic acid for everyone, now. The Journal of Nutrition. 1996; 126 (3): 751S–755S.

[42] Devalia V, Hamilton MS, Molloy AM. Guidelines for the diagnosis and treatment of cobalamin and folate disorders. British Journal of Haematology. 2014; 166 (4): 496-513.

[43] Visentin M, Diop-Bove N, Zhao R, Goldman ID. The intestinal absorption of folates. Annual Review of Physiology. 2014; 76: 251-274.

[44] McNulty H. Folate requirements for health in different population groups. British Journal of Biomedical Science. 1995; 52 (2): 110-119.

[45] Da Silva VR, Gregory III JF. Vitamin B6. Present Knowledge in Nutrition. In: BP Marriott, DF Birt, VA Stallings, AA Yates (Eds.), (11th ed). London, United Kingdom: Academic Press (Elsevier). 2020. pp. 225-238.

[46] Sarwar MF, Sarwar MH, Sarwar M, Qadri NA, Moghal S. The role of oilseeds nutrition in human health: A critical review. Journal of Cereals and Oilseeds. 2013; 4 (8): 97-100.

[47] Nawaz A, Khattak NN, Khan MS, Hasnain N, Sabri S, Shakir M. Deficiency of vitamin B12 and its relation with neurological disorders: a critical review. The Journal of Basic and Applied Zoology. 2020; 81: 10.

[48] Sarwar M. The theatrical usefulness of olive *Olea europaea* L. (Oleaceae Family) nutrition in human health: A Review. Sky Journal of Medicinal Plant Research. 2013; 2 (1): 1-4.

[49] Selhub J, Jacques PF, Dallal G, Choumenkovitch S, Rogers G. The use of blood concentrations of vitamins and their respective functional indicators to define folate and vitamin B12 status. Food and nutrition bulletin. 2008; 29 (2): S67-S73.

[50] Combs GF. The Vitamins: Fundamental Aspects in Nutrition and Health (3rd ed.). Elsevier, Boston, MA. 2007. pp. 7-33.

[51] Jaweria N, Ayesha N, Sarwar M, Sarwar M. Breastfeeding Benefit from Mom Gives the Gift of a Lifetime to the Baby for Healthy Future. American Journal of Food Science and Health. 2017; 3 (5): 95-101.

[52] Stipanuk, MH. Biochemical, physiological, molecular aspects of human nutrition (2nd ed.). St Louis: Saunders Elsevier. 2006. p. 667.

[53] Bender DA. Nutritional Biochemistry of the Vitamins. Cambridge University Press. 2003. p. 5.

[54] Ashley JM. Manifestations and Measurement of Food Insecurity. In: Food Security in the Developing World, Elsevier Inc. 2016; 19-38.

[55] Belch JJ, McCollum PT, Walker WF, Stonebridge PA. Color atlas of peripheral vascular diseases. Mosby-Wolfe. 1996; 77.

[56] Tihanyi BT, Ferentzi E, Beissner F, Koteles F. The neuropsychophysiology of tingling. Consciousness and Cognition. 2018; 58: 97-110.

[57] Cohen KR, Salbu RL, Frank J, Israel I. Presentation and management of herpes zoster (shingles) in the geriatric population. Pharmacy and Therapeutics. 2013; 38 (4): 217-227.

[58] Hughes RA. Peripheral neuropathy. British Medical Journal. 2002; 324 (7335): 466-469.

[59] Torpy JM, Kincaid JL, Glass RM. JAMA patient page. Peripheral neuropathy. Journal of the American Medical Association. 2010; 303(15): 1556.

[60] Maio DV, Franscell R. Morgue. A Life in Death. St. Martin's Press. 2016. p. 236.

[61] Bogousslavsky J, Boller F. Neurological Disorders in Famous Artists. Karger Medical and Scientific Publishers. 2005. p. 125.

[62] Kennedy DO. B Vitamins and the Brain: Mechanisms, Dose and Efficacy-A Review. Nutrients. 2016; 8(2):68.

[63] Coventry LL, Finn J, Bremner AP. Sex differences in symptom presentation in acute myocardial infarction: a systematic review and meta-analysis. Heart & Lung. 2011; 40 (6): 477-491.

[64] Sarwar MH, Mughal AR, Mughal S, Sarwar M. Concerns of Heart Diseases and Mediations to Encourage Healthful Actions for Their Deterrence. International Journal of Bioinformatics and Biomedical Engineering. 2015; 1 (2): 70-76.

[65] Lonn E. Homocysteine in the prevention of ischemic heart disease, stroke and venous thromboembolism: therapeutic target or just another distraction. Current Opinion in Hematology. 2007; 14 (5): 481-487.

[66] Sarwar MH, Sarwar M, Sarwar M, Qadri, NA, Moghal S. The importance of cereals (Poaceae: Gramineae) nutrition in human health: A review. Journal of Cereals and Oilseeds. 2013; 4 (3): 32-35.

[67] Navarra T. The Encyclopedia of Vitamins, Minerals, and Supplements. Infobase Publishing. 2004. p. 155.

[68] Khan MA, Sarwar MH, Jahanzeb M, Raees, T., Abu Bakar H, Zareef AW, Sarwar M. A Crucial Focus on Special Nutrition Needs During Old Age Adults. American Journal of Food Science and Health. 2017; 3 (3):58-63.

[69] Sarwar MH, Khan MA, Abu Bakar H, Zareef AW, Jahanzeb M, Raees T, Sarwar M. An Essential Motivation on Nutrition Learning for Child Care Health Program. American Journal of Food Science and Health. 2017; 3 (3): 47-52.

[70] Bissett DL, Miyamoto K, Sun P, Li J, Berge CA. Topical niacinamide reduces yellowing, wrinkling, red blotchiness, and hyperpigmented spots in aging facial skin. Int J. Cosmet Sci. 2004; 26 (5): 231-238.

[71] Sarwar MH, Raees AR, Rauf A, Sarwar M. 2019. Great Ways to Encourage Better Health Habits and their Preferment in the Community. Specialty Journal of Medical Research and Health Science. 4 (2): 60-69.

[72] Ayesha N, Jaweria N, Sarwar MH, Sarwar M. Making the Decision to Breastfeed the Baby and Its Advantages for the Women's Health. American Journal of Food Science and Health. 2017; 3 (5): 88-94.

[73] Mirza B, Sarwar MH, Sarwar MF, Khalid MT, Sarwar M. An Overview of Recommendations for Women's Physical Activity Effecting on Health, Knowledge, Attitudes and Behaviour. International Journal of Bioinformatics and Biomedical Engineering. 2015; 1 (3): 366-371.

[74] Bailey RL, West KPJ, Black RE. The epidemiology of global micronutrient deficiencies. Annals of Nutrition & Metabolism. 2015; 66 (2): 22S–33S.

[75] Sechi G, Sechi E, Fois C, Kumar N. Advances in clinical determinants and neurological manifestations of B vitamin deficiency in adults. Nutr Rev. 2016; 74 (5): 281-300.

[76] Spinneker A, Sola R, Lemmen V, Castillo M, Pietrzik K, Gonzalez-Gross M. Vitamin B6 status, deficiency and its consequences-an overview. Nutr Hosp. 2007; 22(1): 7-24.

[77] Craig WJ. Health effects of vegan diets. The American Journal of Clinical Nutrition. 2009; 89 (5): 1627S–1633S.

[78] Wilson SM, Bivins BN, Russell KA, Bailey LB. Oral contraceptive use: impact on folate, vitamin B_6, and vitamin B_{12} status. Nutrition Reviews. 2011: 69 (10): 572-583.

[79] Tayyaba R, Sarwar MH, Sarwar M. Key Characteristics for Good Nutritional Care and Preventing of Its Deficiencies in Women. American Journal of Food Science and Health. 2017; 3 (3): 53-57.

Chapter 6

Retrospective, Perspective and Prospective of B-Complex Vitamins: Encapsulation of Vitamins and Release from Vitamin-Loaded Polymers

Fatma Nur Parin

Abstract

Vitamins are regarded as vital nutrients because, when combined, they performed hundreds of functions in the body. They strengthen bones, heal wounds, and boost your immune system. In addition, they transform food into energy and heal cellular damage. In this regard, B-complex vitamins, such as thiamine, riboflavin, and niacin are soluble vitamins that serve as coenzymes in energy metabolism enzymatic activities which building blocks of a healthy body. However, B-complex vitamins are sensitive to light, pH conditions, and temperature. Consequently, they must be encapsulated before they may be used in pharmaceuticals. Recently, it is mainly focused on reducing drug degradation or loss, increase drug bioavailability, limit adverse effects, and improve drug accumulation in the targeted location. To maintain optimum bioavailability during a defined term of therapy, the fraction of drug dosage released from a controlled release product must be significant enough to adjust for the quantity of active drug metabolized and/or eliminated from the body over the same period. Drug release systems also aim to increase the effectiveness of the drug and treat the damaged area. In this chapter, it is aimed to study the production of the vitamin-loaded polymer systems in various forms, such as micro/nanoparticles, micelle, hydrogel, liposome, and nanofiber, as well as release studies in pharmaceutical and biomedical applications.

Keywords: encapsulatin, vitamin release, vitamin-loaded polymer, B-complex vitamins, nutrients

1. Introduction

Vitamins are a category of chemical substances that are not synthesized naturally by the body and therefore must be supplied in small amounts from food. The B-vitamins are a group of eight water-soluble vitamins that play important, interconnected functions in cellular activity, acting as co-enzymes in a wide range of catabolic and anabolic enzymatic processes [1]. Thiamine (B_1), riboflavin (B_2), niacin (B_3, also known as nicotinamide or nicotinic acid amide), pantothenic acid (B_5), pyridoxine (B_6), biotin (B_7), folic acid or folate (B_9), and cobalamin are the

B vitamins (B_{12}) [2]. Their combined effects are especially significant in many areas of brain function, including energy generation, DNA/RNA synthesis/repair, genomic and non-genomic methylation, and the formation of various neuro-chemicals and signaling molecules [3]. B-complex vitamins are known as energy vitamins and stress fighters. Furthermore, the B-Complex vitamins are usually referred to as "beauty vitamins" since they are necessary for healthy hair, skin, and nails.

Polymeric materials have been regarded as the most effective drug delivery vehicles due to their superior pharmacokinetic characteristics. Several new drug delivery methods have been suggested with the rapid advances of nanotechnology. Polymeric drug delivery is described as a formulation or a technology that allows a therapeutic material to be introduced into the body. The nano-sized drug delivery system, in particular, can be built to overcome the current limitations of some drugs, such as poor bioavailability and significant cytotoxic side effects.

Polymers combined with B-complex vitamins can be bioactive to offer thera-peutic effect on their own, or biodegradable, to enhance release kinetics and reduce carrier aggregation. Thus, B-vitamins encapsulated with polymer are protected

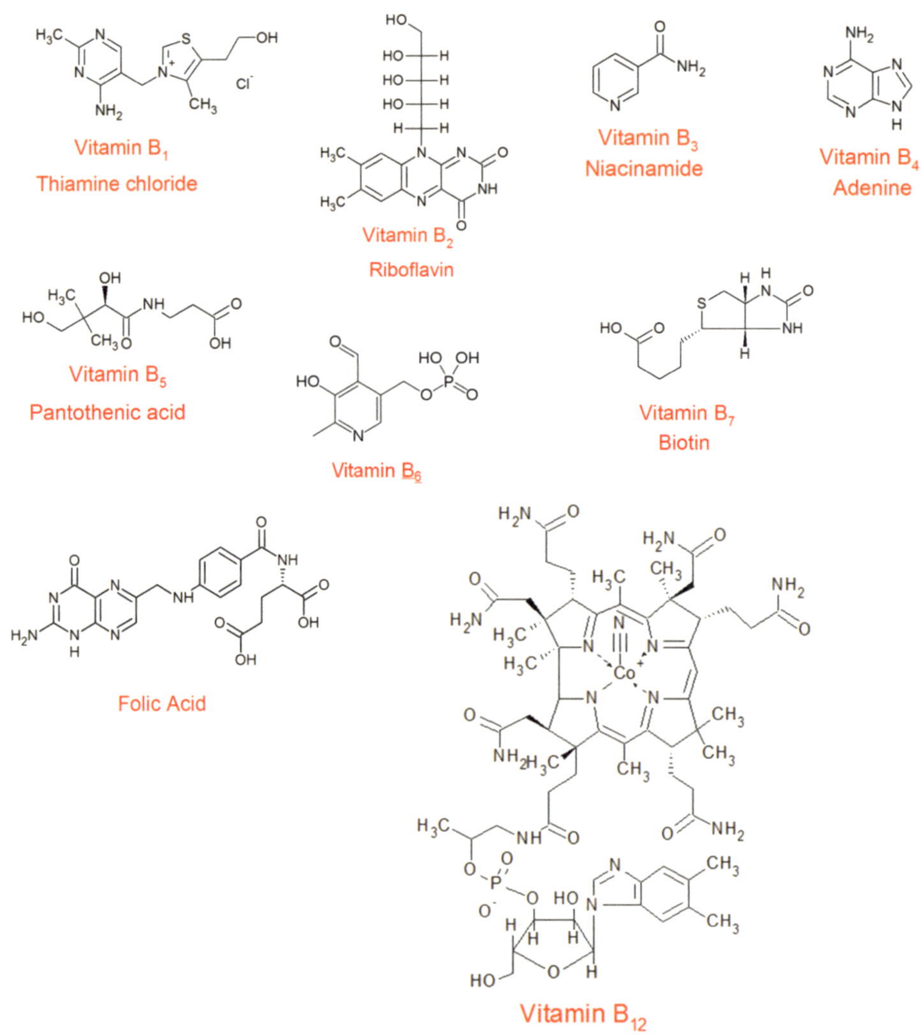

Figure 1.
Chemical structure of the B-complex vitamins.

from degradation caused by environmental effects. Several natural polymers, such as chitosan, starch, dextran, albumin, gelatin, alginate, gums, and also synthetic polymers, such as polylactic acid (PLA), poly-(lactide-co-glycolide) (PLGA), polyanhydrides, and polycaprolactone (PCL), have been utilized in the therapeutic delivery systems [4]. The majority of vitamin B-complex delivery of polymers can be broadly classified as hydrogels, films, nanofibers, beads or micro/nanoparticles, and polymer-drug conjugates, which are discussed in further detail in later sections.

This chapter explains polymer-based materials containing B-complex vitamins briefly and reviews studies in food, biomedical, drug delivery, tissue engineering, wound dressing, cosmetics, and other applications. The importance of vitamin-loaded polymers with unique features in different forms such as outstanding, pH and thermal sensitivity, bioactivity, swelling character, controlled release behavior, biodegradability, good *in vitro* and *in vivo* biocompatibility, and so on is discussed in detail. Finally, future problems and several ideas regarding the production of vitamin-loaded polymers as well as commercial applications have been highlighted (**Figure 1**).

2. Vitamin-loaded polymers

Liposomes, sponges, foams, micro/nanoparticles, fibers, hydrogels, and emulsions have been developed to increase pharmaceutical absorption, stability, penetration, half-life, and bioavailability [5, 6]. From this point, vitamin-loaded polymers have been encapsulated via different techniques such as layer by layer, solvent casting, spray drying, electrospinning/electrospraying, freeze-drying, emulsion polymerization, and complex coacervation. In this regard, vitamin B-complex is encapsulated in different forms with various polymer combinations, enhancing their controlled release and bioavailability. The following sections describe the many applications of vitamin B-complex-releasing polymer forms.

2.1 Hydrogels

Superabsorbent polymers (SAPs) or hydrogels are three-dimensional cross-linked networks of hydrophilic polymers (3D) that can absorb and retain a substantial proportion of liquid (>20%) within their weight [7, 8]. Hydrogels can be used to encapsulate medicines while still preserving bioactivity during gelation and release. Injectable methods can ensure that these medicines are administered locally to the location of the lesion in a continuous and controlled approach.

Natural polymers (e.g., sodium alginate, carboxymethyl cellulose, cellulose, chitosan, gelatin, pectin, starch), synthetic polymers (polyacrylamide, polyacrylic acid, polymethacrylic acid, polyethylene glycol, polyvinyl alcohol), and different acrylates can all be utilized to construct 3D networks [9]. The main factors involved in reversible swelling are based on chemical interactions and physical contact.

There have been several attempts to mix B-complex vitamins with hydrogels for use in medical applications. In this scope, Liu et al. [10] prepared B_{12} vitamin-loaded smart magnetic hydrogels to be utilized in bio-separation, and drug carrier applications. In their study, gelatin has been used as a matrix and Fe_3O_4 nanoparticles are used as magnetic agents. *In vitro* test results showed the crosslinked density of the hybrid gel affected porosity, pore size, and B_{12} release profile, as well. According to UV–VIS analysis, the increased crosslinking of the gelatin matrix via genipin crosslinker led to more vitamin B_{12} absorption at 361 nm [10].

Bajpai and Dubey [11] synthesized VB_{12}-loaded poly(N-vinyl-2-pyrrolidoneco-acrylic acid) hydrogels and evaluated swelling and *in vitro* release behavior. It has been reported pH conditions influence the gel swelling and controlled release profile of vitamins. The gels exhibited nearly 47.8 ± 4.9% swelling in the medium of pH 1.2, while nearly 2164.6 ± 21.8% swelling was observed in the phosphate buffer medium of pH 6.8 due to the functional COO^- groups in acrylic acid [11].

Gong et al. [12] engineered poly(ethylene glycol)-poly(e-caprolactone)-poly(ethylene glycol) (PEG-PCL-PEG) and Pluronic F127 copolymer via sol–gel transition to utilized in biomedical applications as injectable *in situ* gel-forming drug carrier. Because of its high water solubility, VB_{12} was rapidly released from the composite hydrogel. The release results showed that VB_{12} release slowed down due to an increased amount of hydrogel, and there was a 10% decrease in total release with an increased amount of VB_{12} in the hydrogel. The higher Pluronic F127 content caused the higher VB_{12} release amount. Furthermore, cell viability tests revealed that the obtained composite hydrogel copolymers were biocompatible and had low cell cytotoxicity [12].

Ozey [13] synthesized the biocompatible poly(2-hydroxylethyl methacrylate-co-N-allylsuccinamic acid)/B_{12} vitamin hydrogels. *In-vitro* release study indicated around the total release of B_{12} vitamin has 96.9%, 95.8%, and 93.7% release amount in three different media (pH 1.2, PBS and NaCl isotonic serum). As a result, different release media have been found to partially effect the B_{12} release profile. However, it has not seriously affected the total release amount [13].

In a similar study, Maheswari et al. [14] synthesized poly(N-isopropylacrylamide-co-N-vinyl-2-pyrrolidionone) with B_{12} vitamin via radical polymerization. In this study, increased N-vinyl-2-pyrrolidionone (NVP) concentration had led to an increase in pore structure and interconnectivity with pores. The swelling ratio changed with temperature due to using thermo-responsive N-isopropyl-acrylamide (NIPA) monomer. The composite hydrogel had 85% VB_{12} release at 30°C for 10 h. In last, the kinetic results showed the VB_{12}-loaded polymer templates had a non-Fickian diffusion mechanism [14].

Nath et al. [15] developed gelatin-g-poly(acrylic acid-co-acrylamide)/montmorillonite (MMT) clay composite hydrogel containing B_{12} vitamin and they investigated the utility of these materials in pH-responsive drug delivery systems. It was observed that the vitamin B_{12} released from the hybrid hydrogels with 40% (in pH 1.2) and 80% (in pH 7.4) of release amount over 6 h, respectively. Moreover, in comparison to the neat hydrogel, the biodegradability of the vitamin-loaded increased. All samples had no cyctotoxicity effects [15]. Another study of the same research group is about pH- responsive controlled VB_{12} release. It was found almost 50% and 70% VB_{12} were released in two different pH media (1.2 and 7.4) from the hydrogels for 48 h [16]. Fast VB_{12} diffusion was reported in artificial intestinal fluid (AIF, pH 1.2) indicating the pH-dependent cumulative release percentage of VB_{12}. This phenomenon is due to the protonation of a higher number of carboxylic acid (-COOH) groups in VB_{12}at a pH of 7.4. It was revealed to influence the degree of gel swelling, which enhances VB_{12} release, as well as an increase in electrostatic repulsion between the increased amount of ionized -COOH and -OH groups, which leads to increased space expansion inside the network structure.

CMC-xylan/VB_{12}hydrogels had been prepared in different molar ratios by Kundu and Banerjee [17]. According to in-vitro studies, the composite gels indicated a minimum cumulative VB12 release of 28% in pH 1.2. Moreover, 88% in pH 6.8 and 98 in pH 7.4 media, respectively.

Another study contributed to the development of VB_{12} loaded-alginate (Alg) scaffolds by the microfluidic method [18]. In this study, alginate and $CaCl_2$ are used as a template of vitamin, and crosslinker, respectively. The vitamin-loaded hydrogel scaffolds in various alginate concentrations indicated a zero-order kinetic model with more than 80% drug release in 4.5 h.

Gum arabic cross-linked PVA/FA hydrogels were also reported to be a drug delivery carrier with high blood compatibility, good porosity, pH sensitivity and high mechanical properties [19]. The crosslinked hydrogels had a higher swelling ratio at higher pH. Further, *in vitro* release tests indicated the optimized hydrogel had the release FA amount of 78% and 32% at pH 7.4 and 2.1, respectively. It was found the PVA-based hydrogels prevent FA from UV degradation, and therefore, researchers evaluated the hydrogels could be UV-photoprotect material.

In topical/transdermal drug release, active substances have advantages such as localized at the site of infection and reducing their systemic effects [20]. Jung et al. [21] reported that liposomal hydrogels containing VB_{12} could be a candidate for the treatment of atopic dermatitis. VB_{12} derivative adenosyl cobalamin (AdCbl) was loaded into liposomes via a filmhydration method. The results showed the liposomal hydrogels have between 11.5–38.8% loading efficiency of AdCbl. The permeability tests also demonstrated liposomal gel form of AdCblhad 17 times more permeability than in only gel form of AdCbl after 24 h. Consequently, as AdCbl-loaded liposome structures were used, the release amount significantly increased [21].

Folic acid ($C_{19}H_{19}N_7O_6$) is an important member of the B-complex vitamin which also known as vitamin B_9. It consists of the conjugation of one or more L-glutamate units of 4-((pteridine-6-methyl)-amino) benzoic acid [22]. Besides, folic acid is an important factor in the physiological processes of cell metabolism; it is a coenzyme involved in cell growth and development, DNA synthesis and repair, and many metabolic reactions [23].

Folates in their reduced state (without glutamic acid) are chemically unstable and expose oxidatively breakdown at the C-9 and N-10 bonds, resulting in substituted pteridine and p-amino benzoyl glutamate (PABA) moieties. Camacho et al. [24] studied encapsulation of folic acid in copper-alginate hydrogels. This organometalic hydrogels performed as a gastro-resistant material, and slow folic acid release occurred only at pH > 5, particularly under simulated intestinal media (pH 8.2). Besides, the successful materials showed more release in alkaline media (>80 ppm) compared to acidic media (pH 5.4) (~8 ppm) with a zero-order kinetic model [24].

Moreover, in certain commercially available compounds, such as Fruit & Passion Boutiques Inc.'s Hydro Gel Face Masks, the moisturizing effect of these organic polymeric gels is combined with more complex drug-delivery systems designed to release biomolecules such as vitamin C or B_3 [25].

2.2 Films

The film is commonly used in food packaging applications. The edible films and edible coatings are usually used similarly in this field, however, they differ in preparation form [26]. Edible films are formed initially like a solid sheet and then applied to the product surface or between food components, whereas edible coatings are produced directly onto food surfaces once the product makes contact with them. These films and coatings are usually utilized to provide extra barrier protection to increase the shelf life of the food product [27]. Meanwhile, they can also be engineered as vitamin carriers. In this regard to biomaterials, where biopolymers are important due to their biodegradability and biocompatibility, particularly with free-standing devices for wound dressing applications. Coating and film coating are frequently used in applications involving bone implants, cardiovascular devices, wound control and treatment, and oral dissolving strips, as well.

van Dijkhuizen-Radersma et al. [28] have reported polymer composition and crystallinity directly affect the drug release behavior. In the study, VB_{12}-loaded poly(ethylene glycol)/poly(butylene terephthalate) films were synthesized by emulsion polymerization. The release rates and amounts of the resulting films vary between 20% and 100% depending on the copolymer ratios used. Moreover, they determined the total release of $VB_{12}$from the films with 50–100 μm in one day and a sustained release for more than 12 weeks [28].

In recent decades, the importance of thin films in biology and biomaterials research has increased rapidly. Thin films can be coated on a variety of surfaces depends on their physicochemical properties such as wettability, reactivity, conductivity, and corrosion resistance. Thin films are currently being used in the fields of tissue engineering and biomedical industries for osseointegration in dental and orthopedic implants, biodegradable/biobased scaffolds, and biomimetic materials [29]. In this context, Mallakpour and Hatami [30] prepared chitosan nanocomposite films containing folic acid to utilize in tissue engineering as bone and teeth additives. Contact angle measurements showed that the produced films had improved wettability and were more hydrophilic. Furthermore, the bioactivity of these NC films by soaking them in simulated bodily fluid (SBF, pH 7.4), and the pH changes for this solution were observed for 1 month [30].

Banerjee and Ganguly [31] studied on encapsulation of VB_{12} in layered biocomposite films to use in drug delivery, wound healing, and tissue repair applications [31]. In this study, alginate and chitosan was prepared by freeze-dry method as inner and outer layers, respectively. The macrovoids were formed by gas bubbles. Although lyophilized composite films have a high absorption capacity, this form of structure dissolves fast in releasing media. SEM images exhibited the bubbles were homogeneous, having a diameter of almost 0.5 mm, and bubbles were also obtained in multiple layers. Two polymer gel substrates were thought to be different, but the release test showed the diffusivity of biocomposite solute is much lower in crosslinked chitosan than in calcium alginate. In addition, the voids in films did not affect drug release behavior but affect the release amount of drug. Films with void have been found to release more drugs than non-void films.

Orodispersible Film (ODF) is a new progressive dosage form that can provide patients with excellent drug delivery benefits. They are the most revolutionary alternative to conventional dosage forms like pills and capsules [32]. Among all oral drug delivery methods, Oral Film Technology (OFT) has attracted a lot of interest. This dosage form is also known as rapid dissolving film, fast-dissolving film, and orodispersible film [33]. In this scope, Suryawanshi et al. [34] fabricated cyanocobalamin-loaded orodispersible films via a hot-melt extrusion process [34]. The choice of polymer for this process is an important step in the development of ODFs. In this study, Soluplus® (polyvinyl caprolactam-polyvinyl acetate-polyethylene glycol graft copolymer) was utilized as a solvent for drugs that were insoluble in water. Morphological analysis showed the surface of the cyanocobalamin-loaded films was smooth and homogeneous. Besides, 3D micrographs of the film surfaces had a uniform structure. The resulting VB_{12} loaded-films have thickness values varied from 0.21 ± 0.02 to 0.33 ± 0.03 mm. However, the thicknesses of optimized film 0.24 ± 0.02. The average tensile strength of VB_{12} films ranged from 1.53 ± 0.84 to 15.75 ± 0.47 N/cm^2, The optimized film's tensile strength was measured 15.45 ± 0.32 N/cm^2. Contact angle measurement confirmed the optimized films had superior wettability with 24.21°. Moreover, the optimized films have the amount of 99% cyanocobalamin release for 10 hour period.

PVA is a hydrophilic-based synthetic polymer [35]. It includes a carbon chain backbone with hydroxyl groups (-OH); these OH groups can act as a source of hydrogen bonding and therefore assist in the production of polymer composites [36].

Because of its biodegradability, non-toxicity, good film-forming ability, water processability, quick availability, and low cost, PVA is one of the most commonly biodegradable polymers used for different fields of industry [37]. Furthermore, the fabrication of inorganic nanoparticles in polymer matrices has attracted a great deal of interest. Especially, vitamins are used as a biocompatible modifier to form more functional groups on the surfaces of nanoparticles. Many studies on polymer-based metal oxide nanocomposites have been conducted. In this respect, Mallakpour and Shafiee synthesized ZrO_2/PVC nanocomposite films via the solvent casting method [38]. Vitamin B_1 (thiamin) has been used as a ligand for ZrO_2 NPs due to the many functional groups such as amino and hydroxyl groups. The results showed VB_1 is an excellent bioactive agent which provides good dispersion and enhances the interface between NPs and polymer matrix. The modification increased both the surface area of NPs and prevent the formation of zirconium aggregates in films. TEM analysis showed the particle size of ZrO_2-VB_1 NPs has around 36 nm. UV–Vis spectrum revealed that NC films more UV absorption than neat PVC films. The modification did not influence on the crystalline structure of ZrO_2, according to the XRD analysis. Moreover, mechanical testing demonstrated that NC films were more flexible than neat PVC ones. As a result, adding these NPs to hydrophobic materials like PVC increases surface free energy while decreasing contact angle.

In a similar study, Mallakpour and Mansourzadeh fabricated PVA films including CuO nanoparticles modified with vitamin B_1 (thiamine) [39]. In the study, thiamine improved polymer-metal compatibility by creating active regions on the surfaces of metal particles. FTIR analysis showed the new peaks were formed in CuO NPs. The addition of NPs to the PVA matrix further enhanced the optical and mechanical properties.

Buccal delivery has been studied as an alternative to traditional oral delivery for several drugs with limited oral bioavailability. The buccal route also offers the benefit of low enzymatic activity and acceptability by certain sensitizers [40]. The buccal route for drug delivery has received a lot of attention to solve the problem of pre-systemic metabolism caused by gastrointestinal degradation and first pass metabolism [41]. Buccal mucosa has a higher blood supply and is more permeable than oral mucosa. Mohamad et al. [42] designed chitosan/PVA containing VB_{12} buccal hydrogel films. Vitamin-polymer interactions are confirmed by FTIR analysis [42]. According to SEM images, all film samples had homogenous structures without drug agglomeration. In general, the obtained film formulations demonstrated fast drug release due to water absorption by the hydrophilic-based polymers, chitosan, and PVA, which improved wetting, swelling, and penetration of water into the matrix of the film, and therefore increasing drug diffusion phenomena. At the end of 40 minutes, the amount of drug released was 98.59%. Consequently, Higuchi kinetic model has been determined to be the best model for drug release kinetics for films.

Thiamine hydrochloride (THCl) and nicotinic acid (NA)-loaded propylene-glycol (PG) buccal films have been developed using a two-dimensional (2D) inkjet printing method [43]. The rough surface and formation of a pore network in the films were revealed by SEM images. The in vitro release tests revealed that both vitamins were released in a burst in 10 minutes. Additionally, 85%, 98%, and 100% THCl and 78%, 85%, and 100% NA are released from the 1, 5, and 9 print film in 7.5 minutes. The profiles of all formulations for both THCL and NA release were better suited to the first-order kinetic model, with the highest R^2 values.

Acevedo-Fani et al. developed folic acid/polysaccharide-based nanolaminate films [44]. The topology of the nanolaminates improved as folic acid was added, resulting in a uniform and smooth layers. After 7 hours, only 22% of the FA was released from the films at pH 3, however, almost 100% of FA was released at pH 7. This is related to be entrapped folic acid in nanolaminates due to its poor dissolution in an acidic environment. Although researchers have utilized the highly bio-functional folic acid as a great factor for anticancer drug delivery, antibacterial agents, and fluorescence endoscopic detection, there are also studies for usage in electronic applications. Apart from the studies, folic acid has been used in energy storage applications [45]. It was found that FA as the bio-polymeric ligand which was present in different amounts in PVDF, was an effective modifying agent. The results pointed FA particles increased the interaction of β-phase from PVDF's energy storage capability.

2.3 Nanofibers

Nanofibers play a significant role in tissue engineering and drug delivery applications due to their unique properties including low density, high specific surface area, high porosity with<1 μm. Many studies on drug delivery systems indicated nanofibers can be a robust carrier system for therapeutics among other types of nanocarrier systems like hydrogels, beads, films, or others owing to high drug-loading ability, high encapsulation efficiency, target-specific, sustained drug delivery, and ease of processing [46].

Drug release systems are mainly concerned with minimizing drug degradation or loss, enhancing drug bioavailability, avoiding adverse effects, and improving drug accumulation in the targeted site [47]. It is expected to improve the drug's efficacy and treat the damaged site.

A study on increasing the bioavailability of folic acid was conducted by Fonseca et al. [48]. In the study, starch was used as a matrix, and the morphology and release behavior of nanofibers containing varying amounts of FA (5, 10, and 15%) were investigated. Although the diameters of nanofibers formed below 100 nm did not change substantially, it was observed some beads in FA-loaded fibers (5%) higher than in other samples. Moreover, TGA and UV-A radiation results have shown that starch is an effective matrix for FA encapsulation.

Evangelho et al. used an electrospray method to create vitamin B_9-loaded zein nanofibers and vitamin B_9-loaded zein capsules [49]. In the study, thermal and radiation resistance of the VB_9 to determine its availability in food applications. According to SEM micrographs, the addition of VB_9 did not affect on the morphology of the nanofibers and capsules; nevertheless, samples were generated at three different amounts (0.5, 1, and 1.5%, w/w) changed the fiber and capsule diameters. VB_9-loaded zein nanofibers exposed to UV-A radiation for 1 and 24 hours shown significant resistance compared to non-exposed nanofibers. Zein capsules containing 1% VB_9 were also found to be resistant to UV-A. As the mentioned previous section, folic acid is sensitive to UV radiation due to stimulation of the bond between C_9 and N_{10} and this causes the breakdown of the bond and the production of photodegradation products such as P aminobenzoyl-L-glutamic acid, 6-formylpterin, or 6-carboxypterin. It has been presumed that the resistance of VB_9 in both zein fibers and capsule structures may be related to the interaction of vitamins with amino acids in the structure of Zein protein, such as Proline, isoleucine, alanine, phenylalanine, methionine, valine, and leucine, and that this interaction makes it difficult to break down the C_9-N_{10} bond of folic acid. Additionally, TGA analysis revealed that pure folic acid in powdered form degrades considerably faster than folic acid in nanofiber and capsule form.

Amaranth protein/pullulan electrospun/VB_9 fiber structures were fabricated by Aceituno-Medina et al. [50]. The study aimed encapsulate and photoprotection of VB_9, a model bioactive molecule. It has been suggested that the photostability of VB_9 in nanofibers may have better than that produced capsules via conventional techniques such as ionic gelation, coacervation, and spray drying. Optical and electron microscope results showed that VB_9 causes an increase in viscosity, resulting in thicker fibers. VB_9 was extracted from electrospun fibers using PBS for UV–Vis analysis. After UV exposure of non-encapsulated pure VB_9, some changes have occurred in the UV–Vis spectrum, and some characteristic UV absorption peaks of VB_9 are broken down into shorter peaks. It has been associated with p-aminobenzoyl glutamic acid (PGA) and 6-formylpterine (FPT), which are the degradation products of peaks at 275, 278, 310 and 365 nm, relatively.

Some efforts also performed in the incorporation of VB_9 and modified PVA-polyethyleneimine electrospun fibers to use in electronic applications as cancer diagnosis and treatment [51]. The same group prepared polydopamine-VB_9 complexes and the morphological structures of the obtained nanofibers due to the interactions between the polymer and bioactive material. The morphological structures of nanofibers produced by preparing polydopamine-folic acid complexes are predicted to use as graphene-like structures in energy applications such as organic semiconductor material due to ¶-¶ interactions formed between the polymer bioactive material [52]. The same research group also developed multifunctional folic acid-functionalized dendrimers onto electrospun cellulose acetate nanofibers for the specific capture of cancer cells [53].

Other nanofibrous structures based on PVP/dextran octadecyl amine/montmorillonite/VB_9 conjugates have been fabricated by Şimşek et al. [54]. The cytotoxicity results of nanofibers revealed that VB_9 has no negative effect on Vero cells (liver cells), and these fibers could be a pioneer for cancer studies and tissue engineering applications. The electrical properties of the functional nanofibers composed of dextran, PVP, and VB_9 were also investigated by the same study group [55]. As a result, amorphous structures of lower crystallinity and colloidal form of polymer mixtures tend to form the intermolecular hydrogen bond.

Hydrophilic drugs might cause a rapid burst release in the PBS buffer or in vivo conditions due to their high solubility [56]. Madhaiyan et al. prepared cyanocobalamine(vitamin B_{12})-loaded biocompatible PCL nanofibers and carried out plasma treated to nanofibers for periods of 5, 20, 40, and 60 seconds [57]. Contact angle measurements revealed a decrease in surface hydrophilicity from 139° to 108° due to the hydrophilic nature of VB_{12}. Unmodified nanofibers exhibited 18% sudden release in the first 4 hours in the PBS environment and 39% release after 48 hours. Moreover, modified nanofibers with a period of 60 seconds had the highest release value, with a total rate of 95%.

Soy proteins have been used to develop biodegradable and biocompatible nanofibers containing rhodamine B and riboflavin in order to study controlled drug release and desorption processes [58].

Llorens et al. prepared polylactic acid (PLA) nanofibers containing vitamin B_6 and evaluated their release in a hydrophilic Sörensen-ethanol release media and a hydrophobic Tris/Borate/EDTA (TBE) release medium [59]. It was found that within the first 8 hours in the TBE environment, the drug was released quickly from nanofibers; however, in the hydrophilic environment, the drug was released slowly and sustain behavior for several days. From the SEM micrographs, some aggregates formed in PLA nanofibers were related to the presence of vitamin B_6 crystals. As vitamin loading increases, nanofiber diameters increase from 921 nm to 1013 nm, and porosity increases from 69–85%. Cell proliferation results have also shown that these nanofibers have a high antioxidant activity which were produced against free radicals that cause cell damage.

Agarwal et al. produced films with cellulose acetate nanofibers containing nano-ZnO, vitamins B_2 and C for use in oral applications [60]. According to UV–Vis analysis, the release amount of vitamins B_2 and C from nanofibers were measured to be 25% and 95%, respectively. It has been observed that vitamin B_2 is stable in the environment, while vitamin C has a reversible reaction to dehydroascorbic acid and then oxidized to 2,3-diketo-L-gulonic acid via irreversible reaction. Consequently, the study pointed out that cellulose acetate can be used as a carrier in the use of vitamin B_2, which is important for skin disorders, and vitamin C, which supports collagen formation.

Solar UV radiation is the major cause of skin damage owing to the production of reactive oxygen species (ROS), which causes skin collagen imperfection, roughness, allergies, and, therefore, premature aging [61]. Topical formulations have lately attracted great attention as a drug delivery system to the human skin. One approach to combating skin aging is to apply herbal phenolics and antioxidants to the skin. Folic acid (FA) is a significant antioxidant-rich B-complex vitamin that aids in the formation of healthy skin cells [62]. The utilization of folic acid as a beauty patch has been investigated by spraying it onto nanofiber surfaces. However, PVA-based polymers reduced the slow release of folic acid, allowing the process to be completed in a short time. In addition, 3-(4,5-dimethylthiazol-2-yl)-2,5-diphenyltetrazolium bromide (MTT) and neutral red uptake (NRU) assays in L929 have indicated that FA has 120% and 109% cell growth [63]. In a similar study, FA has been directly blended with hydrophilic polymers (PVA, PVP, and gelatin). Micron-seized VB_9 combinations have been created by the electrospraying method for topical applications. The total vitamin release percentages in different polymer matrix including PVA/FA, gelatin/FA, and PVP/FA fibers were 86.88%, 80.20%, and 76.66%, respectively [64].

Vitamin B_2 (riboflavin) is essential for normal reproduction, cell growth and repair, and tissue formation. Furthermore, the vitamin is used by the body to maintain tissues healthy and to repair wounds. The VB_2 (riboflavin)-loaded PCL/gelatin-loaded electrospun fibers have been fabricated to use in biomedical applications [65]. From the same perspective, Heydari et al. [66] developed hydrophobic peracetyl-cyclodextrin (AcCDP) nanofibers containing Vitamin B_2. As expected, the nanofibers exhibited two-phase release profiles involving burst and sustain release in two different release environments (pH 1.2 and pH 7.4). Besides, it was found that they had a 60% and 40% release, respectively, for 170 hours [66].

Vitamins are necessary for cosmetics and other purposes, and transdermal vitamin delivery systems are widely available. Nanoparticulatecombinations offer nearly unlimited delivery possibilities via oral, pulmonary, transdermal, and ocular routes. Vitamins their nano-based combinations would definitely assist in tissue regeneration as well as pharmaceutical delivery [64].

2.4 Beads

Microcapsules or beads produced of biopolymers are of scientific and technical interest, and they have a wide range of potential medical uses, including usage as controlled drug delivery systems. The development of stimuli-responsive microcapsules, especially those composed of biopolymers, is still in its early stages [67].

With relation to hydrogel studies, Bajpai and Tankhiwale [68] prepared vitamin-loaded multilayered polymeric beads using the complex coacervation method. Chitosan and calcium alginate was used as matrix, and VB_2 was used as a model drug, as well. It was found that the beads released around 54% of the drug in the first 3 hours in the artificial gastric fluid (AGF) (pH 1.0), while the remaining drug was released in the mimicking intestinal fluid of pH 7.4 in 6 hours. At last, the concentration of alginate solution used to form the outer layer, as well as the concentration of the ionic crosslinker $CaCl_2$, affects the release profile [68].

Puguan et al. [69] prepared calcium alginate/VB_{12} beads using the dripping technique. In the study, VB_{12} was entrapped in polymers during the gelation process. The in vitro study indicated the total release of VB_{12} complete within 2 hours, and the vitamin loaded-beads had burst release in 30 min. The authors claimed the amount of calcium (crosslinker) ionic strength, and pH factors influence the kinetics of gel formation, as well as the volume and stability of the beads [69].

Vitamins are extremely sensitive, resulting in losing bioactivity during food processing, and storage. Thus, microencapsulation may be used to reduce vitamin loss, allow for a controlled release procedure, and improve their stability. Spray-drying process is versatile and generates high-quality microparticles, and it is superior to other methods in that the product quality in terms of homogenous and cheap [70]. VB_2 has been loaded into 3 different matrix (chitosan, modified chitosan, and sodium alginate) using spray drying method. The total amount of vitamins was released in around 120 min for microparticles produced with chitosan, 15 min for microparticles produced with alginate, and 10 min for microparticles produced with modified chitosan. SEM images and release study showed were associated with a slow release to a rougher surface. For all of the biopolymer-based microparticles with a mean diameter of almost 3 μm were detected [71]. The same group reported to VB_{12}-chitosan microparticles have controlled release behavior in gastric conditions (pH 1.2). The highest release of VB_{12} reached in 10 min. Microstructure of the materials revealed average diameter of particles vary from 3 to 8 μm with a smooth surface and an uniform round shape [72]. Similar to the spray drying of vitamin B_{12}/alginate beads conducted by Abubakr et al. [73]. In the study, as addition of chitosan into the alginate matrix, the release of VB_{12} slowed down. Further chitosan effect the alginate stability and decrease bead permeability. Estevinho and Rocha [74] studied the kinetic models of encapsulated VB_{12} bioactive compounds. Chitosan, modified chitosan, and also sodium alginate were utilized for capsulating matrix [74]. The findings indicated zero-order model was the best fitted for chitosan/VB_{12} particles. Recently, encapsulated the vitamins B_2 and B_3 into six different biopolymers performed using the spray drying method by Carlan et al. [75]. Moreover, encapsulation efficiency reached values higher than 99% for all vitamin-based microcapsules. SEM images demonstrated VB_2 and VB_3 particles had a varying range between 0.10–0.84 mm. Weibull kinetic model was determined suitable model for both vitamin NPs.

The selection of appropriate wall material is critical for achieving the optimum release rate and high encapsulation efficiency of bioactive compounds during the spray drying process. It has been reported that a wall material mixture of polysaccharide-based like gum acacia, cashew nut gum, sodium alginate, sodium carboxymethyl cellulose, and Eudragit RS100 has affected the release kinetics of encapsulated vitamin B_{12} [76]. Bajaj et al. [76] used carbonhydrate-based wall materials to co-encapsulate the vitamin B_{12} and D_3 for the food industry. In the study, the size of smooth microcapsules ranged between 3 and 7 μm without cracking. After the encapsulation process, both vitamins demonstrated higher stability and a change in release rate. Further, the *in vivo* studies showed the maximum concentration of VB_{12} was recorded at 240 min [77].

Madziva et al. [70] produced microcapsules containing folic acid varying in size from 300 to 650 m with a shell material composed of alginate and pectin that may be utilized in food. It was observed that these biopolymers encapsulating the core exhibit significant stability and high encapsulation efficiency in food products for folic acid [70].

Azevedo et al. [78] encapsulated vitamin B_2 into alginate/chitosan matrix by pre-gelation ionotropic method and investigated release behavior in different conditions. The size of nanoparticles has around 120 nm with 56% encapsulation efficiency. The vitamin B_2-loaded nanoparticles are more stable than vitamin B_2-free nanoparticles for 5-week periods [78].

Vitamin compounds, including vitamin B_1 and B_6, have been investigated for controlled drug delivery utilizing various clays to enhance encapsulation efficiency [79–81]. In recent years, gellan gum/laponite clay beads containing $VB_{12}$was fabricated by the ionotropic gelation method [82]. The composite beads had a smoother and regular surface with 2.1 mm. The incorporation of laponite in the bead formulation enhanced drug encapsulation efficiency and delayed the kinetics of the drug in the gastric media. It has been suggested that laponite could be a useful addition in the production of gellan gum beads for long-term drug release.

Several VB_2delivery systems for oral absorption have been developed to solve health problems [83–85]. Riboflavin delivery from ethyl-cellulose-coated barium alginate beads was reported by Bajpai and Sharma [86]. In a pH 1.2 media, the uncoated beads released faster than coated one. Stops et al. observed that calcium alginate beads had fast release VB_2 profile [87]. In another study, Kaygusuz et al. investigated the release of VB_2 from alginate-montmorillonite clay biocomposites [88]. The results showed two different crosslinkers (barium and calcium) utilized had no significant effect on drug release amount, however, barium-alginate beads were more stable than the others. Besides, the *in vitro* study indicated almost 50% VB_2 released from both (barium and calcium) composite beads. In brief, MMT-incorporated calcium and barium alginate beads were appropriate for VB_2 oral applications. The novel biomineralized alginate beads were also designed by Yang et al. [89, 100] to release VB_2. In the study, aliphatic poly(urethane-amine) (PUA) provides thermal sensitivity. Therefore, the VB_2 release was higher at 55°C than at 37°C due to the shrinking of aliphatic PUA at its lower critical solution temperature (LCST) [89].

Carlan et al. [90] researched microencapsulation of VB_1 into different polymers such as carrageenan, chitosan, maltodextrin, modified chitosan, modified starch, pectin, sodium alginate, and xanthan gum. The size of beads ranged from 0.11 to 1.32 μm. The modified starch beads had VB_1 release in 10 min, while VB_1-loaded xanthan gum had in 24 hours. The Weibull kinetic model performed the best on the experimental data [90].

2.5 Others

Carbon nanotubes (CNTs) are made up of micrometer-scale graphene sheets folded into nanoscale cylinders and topped with spherical fullerene [91]. CNTs have found widespread application from optoelectronics and energy storage to drug delivery and biosensor [92, 93]. In recent years, polymer nanocomposites have a worldwide focus to obtain novel polymer materials for practical applications by improving the properties of neat polymers. To improve the reinforcing effect of CNTs in polymer nanocomposites, good dispersion and effective interfacial adhesion between CNTs and the polymer matrix are required (NCs). However, it takes some effort to disperse effectively the CNTs in the composites due to ¶-¶ interactions and strong van der Waals forces between the tubes that cause CNT aggregation in the composites [94]. Therefore, CNT hybrids are used to aid CNTs in disperse inside polymer matrices [95]. A research group decided to improve CNTs functionality and prevent aggregation in composites by treatment with VB_1 [96]. The results showed VB_1 is a biosafe molecule for poly(ester-imide) matrix and CNTs, and it enhanced the morphological properties of polymeric composites. Moreover, the diameter of CNTs increased too, when compared to CNTs-COOH, indicating that a thin coating of VB_1 was bonded to the CNTs via strong interactions.

Cancer is currently one of the most challenging diseases to treat, and its prevalence is on the increase [97, 98]. Targeted anticancer treatments have been widely developed for the past three decades to eliminate cancer cells. Various colloidal carriers, such as lipid nanoparticles, nanofibers, and nanocapsules, nanogels, and polymer-drug conjugates have been investigated for anti-cancer treatments [99]. The application of specific nanomaterials such as CNTs, gold nanoparticles (GNPs), and graphene nanosheets have been considered in the majority of reviews in this field [100]. Due to its low cost, non-toxic, non-immunogenic, low molecular weight, and ease of modification, folic acid (FA) is one of the most widely acknowledged cancer-targeting agents among the different targeting agents. Folate receptors are commonly utilized in epithelial, ovarian, cervical, breast, lung, kidney, colorectal, and brain cancers compared to other agents and B-complex vitamins. Many approaches have been reported for the chemical conjugation of FA to a variety of therapeutic drugs and imaging agents [101]. Yang et al. [89, 100] prepared a folate-conjugated PCL-PEG copolymer to obtain hydrophobic doxorubicin (DOX)-encapsulated active targeting micelles and they combined these micelles with PVA. In conclusion, core-shell nanofibers are produced for the treatment of solid tumors [100]. A study on liver cancer treatment was conducted by Fan and co-workers (2019) [102]. In the study, doxorubicin (DOX)-loaded folic acid-polyethylene glycol-β-cyclodextrin (FA-PEG-β-CD) nanoparticles (NPs) synthesized and this material showed 11.9% drug loading efficiency and 95.2% encapsulation efficiency with 30–60 nm. In another study, folic acid was conjugated with curcumin-encapsulated gum arabic microcapsules in order to drug delivery it to breast cancer cells [103]. Poltavets et al. [101] synthesized docetaxel-loaded PLGA nanoparticles with a folate modification via an emulsion solvent-evaporation. The *in vitro* study revealed these nanoparticles performed non-functionalized nanoparticles and free docetaxel in vitro anticancer activities against HeLa cervical carcinoma cells [104].

3. Future insights

The use of vitamin B-loaded polymers presents a serious progress in the above-mentioned applications and ensures a positive advancement in the upcoming years. Treatments will be more effective and safer owing to the design and functionalization of the various polymeric materials. The potential applications show that the polymeric carriers will progress to a specific active substance to the point where it can be customized to best adapt to a specific component or environment. However, it is important to note that vitamin-loaded polymers have some challenges. First, the number of vitamin-loaded polymers presently accessible for use as the industrial scale is still limited, despite the fact that R&D has advanced from the micro to nano-size scale in the previous decade, exceeding expectations. Secondly, the majority of the tests were improved *in vitro* studies with promising findings, however, the conversion from *in vitro* outcomes to clinical success have been restricted. More clinical trials and data are required to properly understand the mechanism of these polymer carriers. Further, these polymer-carriers must also be biodegradable or have a high capacity to be removed outside the body to minimize accumulation, as well as being non-toxic and non-immunogenic. It is notable to highlight the effect that copolymers or polymer blends might play in adjusting or modifying interactions with the human body in order to control their *in vivo* studies.

In sum, various disadvantages or drawbacks must still be overcome by multiple efforts and focused multidisciplinary scientific collaboration in order to achieve the desired results.

4. Conclusion

Since the beginning of the 2000s, B-complex vitamins have been produced in polymeric materials. Vitamin-loaded polymeric materials have been produced in various structures, such as gel, film, bead, liposome, and fiber forms. Each of these materials containing vitamins has become interesting for many industrial applications. Vitamin-loaded polymer materials have been utilized in medical applications such as implants, additives, tissue engineering, drug carriers, wound dressing, cosmetic applications as a skin-care mask, cancer diagnostic agent, and food applications as food supplements, as well. On the other hand, bioactive polymers are available in smart electronic applications. Some studies also pointed that B vitamins are utilized to obtain nanocomposite materials by modification of certain inorganic nanoparticles. In the studies, it was aimed to increase the bioavailability of B-vitamins by incorporating them in polymers and to develop controlled release behaviors. Furthermore, as the vitamin is introduced to polymer, the morphological, biological and thermal properties of polymeric materials have improved.

Conflict of interest

The authordeclare no conflict of interest.

Author details

Fatma Nur Parin
Polymer Engineering Department, Bursa Technical University, Faculty of Engineering and Natural Sciences, Bursa, Turkey

*Address all correspondence to: nur.parin@btu.edu.tr

References

[1] Kannan R., Vishnupriya V. Relationshşp Between B-vitamins and Bone Health. In Proceedings. 2021;3(1):20.

[2] LeBlanc JG. Introductory chapter: B-group Vitamins. B group Vitamins Current Uses and Perspectives. 2018; 3-5.

[3] Herrmann M, Peter Schmidt J, Umanskaya N, Wagner A, Taban-Shomal O, Widmann T, Colaianni G, Wildemann B, Herrmann W. The role of hyperhomocysteinemia as well as folate, vitamin B(6) and B(12) deficiencies in osteoporosis: A systematic review. Clin. Chem. Lab. Med. 2007;45:1621-1632.

[4] Rani R, Dilbaghi N, Dhingra D, Kumar S. Optimization and evaluation of bioactive drug loaded polymeric nanoparticles for drug delivery. International Journal of Biological Macromolecules. 2015;78:173-179.

[5] Agrahari V, Agrahari V, Mitra AK. Next generation drug delivery: circulatory cells-mediated nanotherapeutic approaches. Expert opinion on drug delivery. 2017;14(3): 285-289.

[6] Singh MK, Varun VK, Behera BK. Cosmetotextiles: State of Art, Fibres& Textiles in Eastern Europe, 2011;19(87): 27-33.

[7] Parın FN, Yıldırım K, Terzioğlu P. Biochar loaded chitosan/gelatin/ poly(ethylene glycol) biocomposite beads: Morphological, thermal and swelling properties. Journal of Innovative Science and Engineering. 2020,4(2):56-68.

[8] Mohammad J, Zohuriaan-Mehr, Kabiri K. Superabsorbent Polymer Materials: A Review. Iranian Polymer Journal. 2008:17(6):451-477.

[9] Behera S, and Mahanwar PA. Superabsorbent polymers in agriculture and other applications: a review. Polymer-Plastics Technology and Materials. 2019:59(4):1-16.

[10] Liu TY, Hu SH, Liu KH, Liu DM, Chen SY. Preparation and characterization of smart magnetic hydrogels and its use for drug release. Journal of Magnetism and Magnetic Materials. 2006:304(1):e397-e399.

[11] Bajpai SK, and Dubey S. In-vitro dissolution studies for release of vitamin B12 from poly (N-vinyl-2-pyrrolidone-co-acrylic acid) hydrogels. Reactive and Functional Polymers. 2005:62(1): 93-104.

[12] Gong CY, Shi S, Dong PW, Zheng XL, Fu SZ, Guo G, Quian ZY. In vitro drug release behavior from a novel thermosensitive composite hydrogel based on Pluronic F127 and poly(ethylene glycol)-poly (ε-caprolactone)-poly (ethylene glycol) copolymer. BMC Biotechnology. 2009;9(1):1-13.

[13] Ozay O. Synthesis and characterization of novel pH-responsive poly (2-hydroxylethyl methacrylate-co-N-allylsuccinamic acid) hydrogels for drug delivery. Journal of Applied Polymer Science. 2014:131(1).

[14] Maheswari B, JagadeeshBabu PE, Agarwal M. Role of N-vinyl-2-pyrrolidinone on the thermoresponsive behavior of PNIPAm hydrogel and its release kinetics using dye and vitamin-B12 as model drug. Journal of Biomaterials Science, Polymer Edition. 2014;*25*(3):269-286.

[15] Nath J, Chowdhury A, Ali I, Dolui SK. Development of a gelatin-g-poly (acrylic acid-co-acrylamide)–montmorillonite superabsorbent hydrogels for in vitro controlled release

of vitamin B12. Journal of Applied Polymer Science, 2019;*136*(22):47596.

[16] Nath J, Saikia PP, Handique J, Gupta K, Dolui SK. Multifunctional mussel-inspired Gelatin and Tannic acid-based hydrogel with pH-controllable release of vitamin B12. Journal of Applied Polymer Science.2020;137(39), 49193.

[17] Kundu D, Banerjee T. Carboxymethyl cellulose–xylan hydrogel: synthesis, characterization, and in vitro release of vitamin B12. ACS omega. 2019:*4*(3), 4793-4803.

[18] Bhasarkar J, Bal D. Kinetic investigation of a controlled drug delivery system based on alginate scaffold with embedded voids. Journal of applied biomaterials & functional materials. 2019:17(2), 22808000 18817462.

[19] Pandit AH, Mazumdar N, Imtiyaz K, Rizvi MMA, Ahmad S. Periodate-modified gum arabic cross-linked PVA hydrogels: a promising approach toward photoprotection and sustained delivery of folic acid. ACS omega. 2019;4(14):16026-16036.

[20] Esentürk İ, Erdal MS, GüngörS. Electrospinning method to produce drug-loaded nanofibers for topical/transdermal drug delivery applications. Journal of Faculty of Pharmacy of Istanbul University. 2016;46(1): 49-69.

[21] Jung SH, Cho YS, Jun SS, Koo JS, Cheon HG, Shin BC. Topical application of liposomal cobalamin hydrogel for atopic dermatitis therapy. Die Pharmazie-An International Journal of Pharmaceutical Sciences. 2011;66(6): 430-435.

[22] Hamed E, Attia MS, BassiounyK. Synthesis, Spectroscopic and Thermal Characterization of Copper(II) and Iron(III) Complexes of Folic Acid and Their Absorption Efficiency in the Blood. 2009;979680:1-6.

[23] Parın F. Yaşlanma Geciktirici B_9Vitamini içeren farklı non-woven kumaşların üretimi ve kontrollü salımının incelenmesi [thesis]. Bursa Teknik Üniversitesi ;2021.

[24] Camacho DH, Uy SJY, Cabrera MJF, Lobregas MOS, Fajardo TJMC. Encapsulation of folic acid in copper-alginate hydrogels and it's slow in vitro release in physiological pH condition. Food Research International. 2019;119: 15-22.

[25] Chirani N, Yahia LH, Gritsch L, Motta FL, Chirani S, FarèS. History and applications of hydrogels. 2015;

[26] Falguera V, Quintero JP, Jiménez A, Muñoz JA, IbarzA. Edible films and coatings: Structures, active functions and trends in their use. Trends in Food Science & Technology. 2011;22(6): 292-303.

[27] De Moraes MA, Da Silva CF, Vieira RS. Biopolymer Membranes and Films: Health, Food, Environment, and Energy Applications. Elsevier. (Eds.). (2020).

[28] Van Dijkhuizen-Radersma R, Péters FLA, Stienstra NA, Grijpma D, Feijen J, De Groot K, Bezemer J. M. Control of vitamin B12 release from poly (ethylene glycol)/poly (butylene terephthalate) multiblock copolymers. Biomaterials. 2002;23(6):1527-1536.

[29] Mozafari M, Ramedani A, Zhang YN, Mills DK. Thin films for tissue engineering applications. In Thin Film Coatings for Biomaterials and Biomedical Applications. 2016;167-195. Woodhead Publishing.

[30] Mallakpour S, HatamiM. Highly capable and cost-effective chitosan nanocomposite films containing folic acid-functionalized layered double

hydroxide and their in vitro bioactivity performance. Materials Chemistry and Physics. 2020;250:123044.

[31] Banerjee A, Ganguly, S. Alginate–chitosan composite hydrogel film with macrovoids in the inner layer for biomedical applications. Journal of Applied Polymer Science. 2019;136(22):47599.

[32] Gupta MS, Kumar TP. Characterization of orodispersible films: an overview of methods and introduction to a new disintegration test apparatus using LDR-LED sensors. Journal of pharmaceutical sciences. 2020;109(10):2925-2942.

[33] Gijare C, Deshpande A. Orodispersible Films: A systematic patent review. Recent patents on drug delivery & formulation. 2018;12(2): 110-120.

[34] Suryawanshi D, Wavhule P, Shinde U, Kamble M, Amin P. Development, optimization and in-vivo evaluation of cyanocobalamin loaded orodispersible films using hot-melt extrusion technology: A quality by design (QbD) approach. Journal of Drug Delivery Science and Technology. 2021;63:102559.

[35] Parin FN, Terzioğlu P, Sicak Y, Yildirim K, ÖztürkM. Pine honey–loaded electrospun poly (vinyl alcohol)/ gelatin nanofibers with antioxidant properties. The Journal of The Textile Institute. 2021;112(4):628-635.

[36] Divya R, Meena, M., Mahadevan, C. K., & Padma, C. M. (2014). Investigation on CuO dispersed PVA polymer films. Journal of Engineering Research and Applications, 4(5), 1-7.

[37] Suganthi S, Vignesh S, Sundar JK, Raj V. Fabrication of PVA polymer films with improved antibacterial activity by fine-tuning via organic acids for food packaging applications. Applied Water Science. 2020;10(4):1-11.

[38] Mallakpour S, ShafieeE. The synthesis of poly (vinyl chloride) nanocomposite films containing ZrO2 nanoparticles modified with vitamin B1 with the aim of improving the mechanical, thermal and optical properties. Designed monomers and polymers. 2017;20(1):378-388.

[39] Mallakpour S, Mansourzadeh S. Application of CuO nanoparticles modified with vitamin B1 for the production of poly (vinyl alcohol)/CuO nanocomposite films with enhanced optical, thermal and mechanical properties. Polymers for Advanced Technologies. 2017:28(12):1823-1830.

[40] Verma S, Kumar N, Sharma PK. Buccal film: an advance technology for oral drug delivery. AdvBiol Res. 2014;8:260-267.

[41] Zhang H, Zhang J, Streisand JB. Oral mucosal drug delivery. Clinical pharmacokinetics. 2002;41(9):661-680.

[42] Mohamad SA, Sarhan HA, Abdelkader H, Mansour HF. Vitamin B12–loaded buccoadhesive films as a noninvasive supplement in vitamin b12 deficiency: in vitro evaluation and in vivo comparative study with intramuscular injection. Journal of pharmaceutical sciences. 2017;106(7):1849-1858.

[43] Eleftheriadis G, MonouPK, Andriotis E, Mitsouli E, Moutafidou N, Markopoulou C, FatourosD. Development and characterization of inkjet printed edible films for buccal delivery of B-complex vitamins. Pharmaceuticals. 2020;13(9):203.

[44] Acevedo-Fani A, Soliva-Fortuny R, Martín-BellosoO. Photo-protection and controlled release of folic acid using edible alginate/chitosan nanolaminates. Journal of Food Engineering. 2018;229:72-82.

[45] Roy S, Thakur P, Hoque NA, Bagchi B, Sepay N, Khatun F, Das S.

Electroactive and high dielectric folic acid/PVDF composite film rooted simplistic organic photovoltaic self-charging energy storage cell with superior energy density and storage capability. ACS applied materials & interfaces. 2017;9(28):24198-24209.

[46] Dhand C, Dwivedi N, Sriram H, Bairagi S, Rana D, Lakshminarayanan R, & Ramakrishna S. Nanofiber composites in drug delivery. In Nanofiber Composites for Biomedical Applications. 2017;199-223. Woodhead Publishing.

[47] Jain KK. Drug Delivery Systems. An Overview of Drug Delivery Systems, USA, Springer.(2008).

[48] Fonsecaa LM, Crizela RL, Silvaa FT, VazFontesa MR, Zavarezea ER, Guerra Dias AR. Starch nanofibers as vehicles for folic acid supplementation: thermal treatment, UVA irradiation, and in vitrosimulation of digestion, Journal of the Science of Food and Agriculture. (2020 https://doi.org/10.1002/jsfa.10809.

[49] Evangelho JA, Crizel RL, Chaves FC, Prietto L, Pinto VZ, Miranda MZ, Dias ARG, Zavareze ER. Thermal and irradiation resistance of folic acid encapsulated in zein ultrafine fibers or nanocapsules produced by electrospinning and electrospraying, Food Research International. 2019;124:137-146.

[50] Aceituno Medina, Marysol; Mendoza, Sandra; Lagarón Cabello, José María ; López-Rubio, Amparo. Development and characterization of food-grade electrospun fibers from amaranth protein and pullulan blends. Food Research International 54: 667-674 (2013), website: https://digital.csic.es/handle/10261/111068.

[51] Fan Z, Zhao Y, Zhu X, Luo Y, Shen M, Shi X. Folic acid modified electrospunpoly(vinyl alcohol)/polyethyleneimine nanofibers for cancer cell capture applications, Chinese Journal of Polymer Science. 2016;34(6):755-765.

[52] Fan H, Yu X, Liu Y, Shi Z, Liu H, Nie Z, Wu D, JinZ. Folic acid–Polydopamine Nanofibers Show Enhanced Ordered-stacking via ¶-¶ Interactions, The Royal Society of Chemistry. 2015;11(23):4621-4629.

[53] Zhao Y, Zhu X, Liu H, Luo Y, Wang S, Shen M, Shi X. Dendrimer-functionalized electrospun cellulose acetate nanofibers for targeted cancer cell capture applications. Journal of Materials Chemistry B. 2014;2(42):7384-7393.

[54] Şimşek M, Rzayev ZMO, Bunyatova U, Khalilova S, TürkM. Multifunctional Electrospun Biocompatible Nanofiber Composites from Water Dispersion Blends of Folic Acid Conjugated PVP/Dextran/ODA-MMT Nanocomposites and Their Responses to Vero cells. Hacettepe Journal of Biology and Chemistry. 2016;44(4):441-450.

[55] Rzayev ZMO, Bunyatova U, Şimnullek M. Multifunctional colloidal nanofiber composites including dextran and folic acid as electro-active platforms, Carbohydrate Polymers. 2017;166:83-92.

[56] Simancas-Herbada R, Fernández-Carballido A, Aparicio-Blanco J, Slowing, K, Rubio-Retama J, López-Cabarcos E, Torres-Suárez AI. Controlled release of highly hydrophilic drugs from novel poly (Magnesium Acrylate) matrix tablets. Pharmaceutics. 2020;12(2):174.

[57] Madhaiyan K, Sridhar R, Sundarrajan S, Venugopal JR, Ramakrishna S. Vitamin B12 loaded polycaprolactone nanofibers: a novel transdermal route for the water soluble energy supplement delivery,

International Journal of Pharmaceutics. 2013;444(1-2):70-76.

[58] Khansari S, Duzyer S, Sinha-Ray S, Hockenberger A, Yarin AL, Pourdeyhimi, B. Two-stage desorption-controlled release of fluorescent dye and vitamin from solution-blown and electrospun nanofiber mats containing porogens. Molecular pharmaceutics. (2013); 10(12):4509-4526.

[59] Llorens E, J.del Valle L, Díaz A, Casas MT, Puiggalí J. Polylactide Nanofibers Loaded with Vitamin B6 and Polyphenols as Bioactive Platform for Tissue Engineering, Macromolecular Research. 2013;21(7):775-787.

[60] Agarwal A, Anil J, Arputharaj A, Manik B, Kartick S, Prasad S, Charlene D, NadanathangamV. Performance Characteristics of Electrospun Cellulose Acetate Nanofiber Mat Embedded with NanoZno/ Vitamins, International Journal of Nanotechnology and Application. 2016;6(3);1-12.

[61] Gisondi P, Fantuzzi F, Malerba M, GirolomoniG. Folic acid in general medicine and dermatology, Journal of Dermatological Treatment. 2007;18(3):138-146.

[62] Parın FN, Aydemir Çİ, Taner G, YıldırımK. Co-electrospun-electrosprayed PVA/folic acid nanofibers for transdermal drug delivery: Preparation, characterization, and in vitro cytocompatibility. Journal of Industrial Textiles. 2021: 1528083721997185.

[63] Parın FN, Yıldırım K. Preparation and characterisation of vitamin-loaded electrospunnanofibres as promising transdermal patches. Fibres& Textiles in Eastern Europe. (2021).

[64] Sridhar R, Madhaiyan K, Sundarrajan S, Góra A, Venugopal JR, Ramakrishna S. Cross-linking of protein scaffolds for therapeutic applications: PCL nanofibers delivering riboflavin for protein cross-linking. Journal of Materials Chemistry B. 2014;2(12):1626-1633.

[65] Zhao Q, Li B. pH-controlled drug loading and release from biodegradable microcapsules. Nanomedicine: Nanotechnology, Biology and Medicine. 2008;4(4):302-310.

[66] Heydari, A., Mehrabi, F., Shamspur, T., Sheibani, H., Mostafavi, A. (2018). Encapsulation and Controlled Release of Vitamin B2 Using Paracetyl-beta-cyclodextrin Polymer-Based Electrospun Nanofiber Scaffold, Pharmaceutical Chemistry Journal, 52(1), 19-25.

[67] Klinger D, Landfester K. Stimuli-responsive microgels for the loading and release of functional compounds: Fundamental concepts and applications. Polymer. 2012;53(23):5209-5231.

[68] BajpaiSK, TankhiwaleR. Investigation of dynamic release of vitamin B2 from calcium alginate/ chitosan multilayered beads: Part II. Reactive and Functional Polymers. 2006;66(12):1565-1574.

[69] Puguan JMC, Yu X, Kim H. Characterization of structure, physico-chemical properties and diffusion behavior of Ca-Alginate gel beads prepared by different gelation methods. Journal of colloid and interface science. 2014;432:109-116.

[70] Madziva H, Kailasapathy K, Phillips M. Alginate–pectin microcapsules as a potential for folic acid delivery in foods. Journal of Microencapsulation. (2005);22(4): 343-351.

[71] Estevinho BN, Carlan I, Blaga A, Rocha F. Soluble vitamins (vitamin B12 and vitamin C) microencapsulated with different biopolymers by a spray drying

process. Powder Technology. 2016;289:71-78.

[72] Carlan IC, Estevinho BN, Rocha F. Study of microencapsulation and controlled release of modified chitosan microparticles containing vitamin B12. Powder Technology. 2017;318:162-169.

[73] Abubakr N, Jayemanne A, Audrey N, Lin SX, Chen XD Effects of encapsulation process parameters of calcium alginate beads on Vitamin B12 drug release kinetics. Asia-Pacific Journal of Chemical Engineering. 2010;5(5):804-810.

[74] Estevinho BN, Rocha F. Kinetic models applied to soluble vitamins delivery systems prepared by spray drying. Drying Technology. 2017;35(10):1249-1257.

[75] Carlan IC, Estevinho BN, Rocha F. Innovation and improvement in food fortification: Microencapsulation of vitamin B2 and B3 by a spray-drying method and evaluation of the simulated release profiles. Journal of Dispersion Science and Technology. 2021;1-13.

[76] Bajaj SR, Marathe SJ, Singhal RS. Co-encapsulation of vitamins B12 and D3 using spray drying: Wall material optimization, product characterization, and release kinetics. Food Chemistry. 2021;335:127642.

[77] Oliveira AM, Guimarães KL, Cerize N, Tunussi AS, Poço JG. Nano spray drying as an innovative technology for encapsulating hydrophilic active pharmaceutical ingredients (API). J. Nanomed. Nanotechnol. 2013;4(6).

[78] Azevedo MA, Bourbon AI, Vicente AA, Cerqueira MA. Alginate/chitosan nanoparticles for encapsulation and controlled release of vitamin B2. International Journal of Biological Macromolecules. 2014;71:141-146.

[79] Joshi GV, Patel HA, Kevadiya BD, Bajaj HC. Montmorillonite intercalated with vitamin B1 as drug carrier. Applied Clay Science. 2009;45(4):248-253.

[80] Joshi GV, Patel HA, Bajaj HC, Jasra RV. Intercalation and controlled release of vitamin B 6 from montmorillonite–vitamin B 6 hybrid. Colloid and Polymer Science. 2009;287(9):1071-1076.

[81] Kevadiya BD, Joshi GV, Patel HA, Ingole PG, Mody HM, Bajaj HC. Montmorillonite-alginate nanocomposites as a drug delivery system: intercalation and in vitro release of vitamin B1 and vitamin B6. Journal of biomaterials applications. 2010;25(2): 161-177.

[82] Adrover, A, Paolicelli P, Petralito S, Di Muzio L, Trilli, J, Cesa S, Casadei MA. Gellan gum/laponite beads for the modified release of drugs: Experimental and modeling study of gastrointestinal release. Pharmaceutics. 2019;11(4):187.

[83] Bajpai SK, SaxenaS. Dynamic release of riboflavin from a starch-based semi IPN via partial enzymatic degradation: part II. Reactive and Functional Polymers. 2004;61(1): 115-129.

[84] Yao H, Xu L, Han F, Che X, Dong Y, Wei M, Li S. A novel riboflavin gastro-mucoadhesive delivery system based on ion-exchange fiber. International journal of pharmaceutics. 2008;364(1): 21-26.

[85] Seidenberger T, Siepmann J, Bley H, Maeder K, Siepmann F. Simultaneous controlled vitamin release from multiparticulates: Theory and experiment. International journal of pharmaceutics. 2011;412(1-2):68-76.

[86] Bajpai SK, Sharma S. Dynamic release of riboflavin from ethyl cellulose coated barium alginate beads for

gastrointestinal drug delivery: an in vitro study. Journal of Macromolecular Science, Part A: Pure and Applied Chemistry. 2005;42(5): 649-661.7

[87] Stops F, Fell JT, Collett JH, Martini LG. Floating dosage forms to prolong gastro-retention—The characterisation of calcium alginate beads. International journal of pharmaceutics, 2008;350(1-2): 301-311.

[88] Kaygusuz H, Uysal M, Adımcılar V, Erim FB. Natural alginate biopolymer montmorillonite clay composites for vitamin B2 delivery. Journal of Bioactive and Compatible Polymers. 2015;30(1):48-56.

[89] Yang L, Shi J, Zhou X, Cao S. Hierarchically organization of biomineralized alginate beads for dual stimuli-responsive drug delivery. International journal of biological macromolecules. 2015;73:1-8.

[90] Carlan IC, Estevinho BN, Rocha F. Production of vitamin B1 microparticles by a spray drying process using different biopolymers as wall materials. The Canadian Journal of Chemical Engineering. 2020;98(8):1682-1695.

[91] Peretz S, Regev O. Carbon nanotubes as nanocarriers in medicine. Current Opinion in colloid & interface science. 2012;17(6):360-368.

[92] Hua J, Wang Z, Xu L, Wang X, Zhao J, Li F. Preparation polystyrene/ multiwalled carbon nanotubes nanocomposites by copolymerization of styrene and styryl-functionalized multiwalled carbon nanotubes. Materials Chemistry and Physics. 2013;137(3):694-698.

[93] Othman RN, Wilkinson AN. Carbon nanotube hybrids and their polymer nanocomposites. In Synthesis, Technology and Applications of Carbon Nanomaterials. 2019;29-60. Elsevier.

[94] Mallakpour S, Soltanian S. Morphology and thermal properties of nanocomposites based on chiral poly (ester-imide) matrix reinforced by vitamin B1 functionalized multiwalled carbon nanotubes. Journal of Composite Materials. 2017;51(16):2291-2300.

[95] Asghari, S, Rezaei Z, Mahmoudifard M. Electrospun nanofibers: A promising horizon toward the detection and treatment of cancer. Analyst. 2020; 145(8):2854-2872.

[96] Pal K, Roy S, Parida PK, Dutta A, Bardhan S, Das S, Karmakar P. Folic acid conjugated curcumin loaded biopolymeric gum acacia microsphere for triple negative breast cancer therapy in in vitro and in vivo model. Materials Science and Engineering: C. 2019;*95*: 204-216.

[97] Peer D, Karp JM, Hong S, Farokhzad OC, Margalit R, Langer R. Nanocarriers as an emerging platform for cancer therapy. Nature nanotechnology. 2007;*2*(12):751-760.

[98] Alwarappan S, Erdem A, Liu C, Li CZ. Probing the electrochemical properties of graphene nanosheets for biosensing applications. The Journal of Physical Chemistry C. 2009;*113*(20): 8853-8857.

[99] Kumar A, Lale SV, Alex MA, Choudhary V, Koul V. Folic acid and trastuzumab conjugated redox responsive random multiblock copolymeric nanocarriers for breast cancer therapy: In-vitro and in-vivo studies. Colloids and Surfaces B: Biointerfaces. 2017;*149*:369-378.

[100] Yang, G., Wang, J., Wang, Y., Li, L., Guo, X., & Zhou, S. (2015). An implantable active-targeting micelle-in-nanofiber device for efficient and safe cancer therapy. ACS nano. 9(2), 1161-1174.

[101] Poltavets YI, Zhirnik AS, Zavarzina VV, Semochkina YP,

Shuvatova VG, Krasheninnikova AA, Posypanova GA. In vitro anticancer activity of folate-modified docetaxel-loaded PLGA nanoparticles against drug-sensitive and multidrug-resistant cancer cells. Cancer Nanotechnology. 2019;10(1):1-17.

[102] Ngo CL, Le QT, Ngo TT, Nguyen DN, Vu MT. Surface modification and functionalization of carbon nanotube with some organic compounds. Advances in Natural Sciences: Nanoscience and Nanotechnology. 2013;4(3):035017.

[103] Hua J, Wang Z, Xu L, Wang X, Zhao J, Li F. Preparation polystyrene/multiwalled carbon nanotubes nanocomposites by copolymerization of styrene and styryl-functionalized multiwalled carbon nanotubes. Materials Chemistry and Physics. 2013;137(3):694-698.

[104] Mallakpour S, Soltanian S. Morphology and thermal properties of nanocomposites based on chiral poly (ester-imide) matrix reinforced by vitamin B1 functionalized multiwalled carbon nanotubes. Journal of Composite Materials. 2017;51(16):2291-2300.

Chapter 7

Vitamin B6 and Related Inborn Errors of Metabolism

Hilal H. Al-Shekaili, Clara van Karnebeek and Blair R. Leavitt

Abstract

Vitamin B6 (vitB6) is a generic term that comprises six interconvertible pyridine compounds. These vitB6 compounds (also called vitamers) are pyridoxine (PN), pyridoxamine (PM), pyridoxal (PL) and their 5′-phosphorylated forms pyridoxine 5′-phosphate (PNP), pyridoxamine 5′-phosphate (PMP) and pyridoxal 5′-phosphate (PLP). VitB6 is an essential nutrient for all living organisms, but only microorganisms and plants can carry out *de novo* synthesis of this vitamin. Other organisms obtain vitB6 from dietary sources and interconvert its different forms according to their needs via a biochemical pathway known as the salvage pathway. PLP is the biologically active form of vitB6 which is important for maintaining the biochemical homeostasis of the body. In the human body, PLP serves as a cofactor for more than 140 enzymatic reactions, mainly associated with synthesis, degradation and interconversion of amino acids and neurotransmitter metabolism. PLP-dependent enzymes are also involved in various physiological processes, including biologically active amine biosynthesis, lipid metabolism, heme synthesis, nucleic acid synthesis, protein and polyamine synthesis and several other metabolic pathways. PLP is an important vitamer for normal brain function since it is required as a coenzyme for the synthesis of several neurotransmitters including D-serine, D-aspartate, L-glutamate, glycine, γ-aminobutyric acid (GABA), serotonin, epinephrine, norepinephrine, histamine and dopamine. Intracellular levels of PLP are tightly regulated and conditions that disrupt this homeostatic regulation can cause disease. In humans, genetic and dietary (intake of high doses of vitB6) conditions leading to increase in PLP levels is known to cause motor and sensory neuropathies. Deficiency of PLP in the cell is also implicated in several diseases, the most notable example of which are the vitB6-dependent epileptic encephalopathies. VitB6-dependent epileptic encephalopathies (B6EEs) are a clinically and genetically heterogeneous group of rare inherited metabolic disorders. These debilitating conditions are characterized by recurrent seizures in the prenatal, neonatal, or postnatal period, which are typically resistant to conventional anticonvulsant treatment but are well-controlled by the administration of PN or PLP. In addition to seizures, children affected with B6EEs may also suffer from developmental and/or intellectual disabilities, along with structural brain abnormalities. Five main types of B6EEs are known to date, these are: PN-dependent epilepsy due to ALDH7A1 (antiquitin) deficiency (PDE-ALDH7A1) (MIM: 266100), hyperprolinemia type 2 (MIM: 239500), PLP-dependent epilepsy due to PNPO deficiency (MIM: 610090), hypophosphatasia (MIM: 241500) and PLPBP deficiency (MIM: 617290). This chapter provides a review of vitB6 and its different vitamers, their absorption and

IntechOpen

metabolic pathways in the human body, the diverse physiological roles of vitB6, PLP homeostasis and its importance for human health. Finally, the chapter reviews the inherited neurological disorders affecting PLP homeostasis with a special focus on vitB6-dependent epileptic encephalopathies (B6EEs), their different subtypes, the pathophysiological mechanism underlying each type, clinical and biochemical features and current treatment strategies.

Keywords: vitamin B6 (vitB6), Salvage pathway, PLP-dependent enzymes, inherited vitB6-dependent epilepsies

1. Introduction

Vitamin B6 (vitB6) is a generic term that refers to a group of six interconvertible chemical compounds that share a pyridine ring in their centre. These vitB6 compounds (also called vitamers) are pyridoxine (PN), pyridoxamine (PM), pyridoxal (PL) and their 5′-phosphorylated forms pyridoxine 5′-phosphate (PNP), pyridoxamine 5′-phosphate (PMP) and pyridoxal 5′-phosphate PLP) [1] (**Figure 1**). VitB6 is required by all living organisms for their survival, but only microorganisms and plants can carry out *de novo* synthesis of this vitamin. Other organisms including humans acquire vitB6 from exogenous sources and interconvert its different forms according to their needs using a biochemical pathway known as the salvage pathway [1, 3].

1.1 Metabolism of vitB6

Among the six vitB6 compounds, PLP is the biologically active and most important vitamer since it is required as a cofactor for a multitude of enzymes in the body. Humans and other mammals obtain PLP directly from diet or through synthesis from other vitameric forms ingested with food or recycled from degraded PLP-dependent enzymes via the salvage pathway [1, 4] (**Figure 2**). The central enzyme in this pathway is PNP oxidase (PNPO), a flavin mononucleotide (FMN)-dependent enzyme that is capable of converting PNP or PMP to the active cofactor

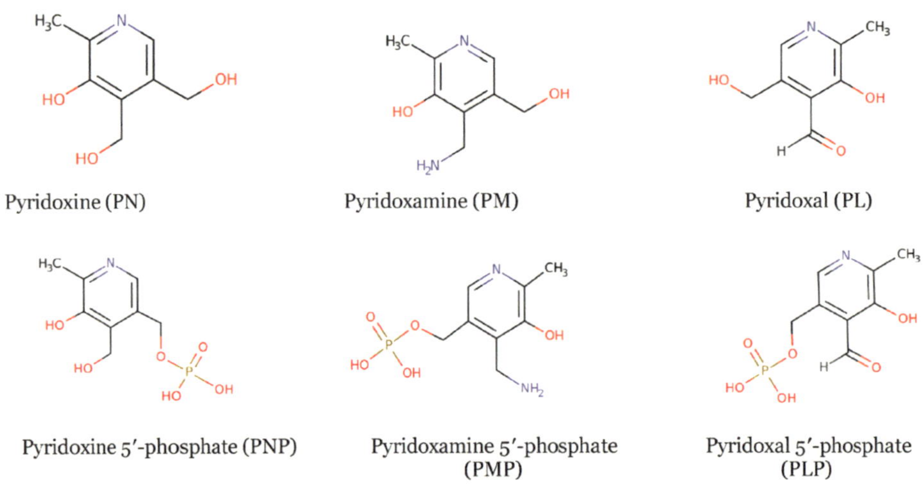

Figure 1.
Chemical structures of the six vitamin B6 vitamers. Colored atoms designate oxygen or hydroxyl group (red), nitrogen or amine group (blue) and phosphorus (brown). (Retrieved from [2]).

DOI: http://dx.doi.org/10.5772/intechopen.99751

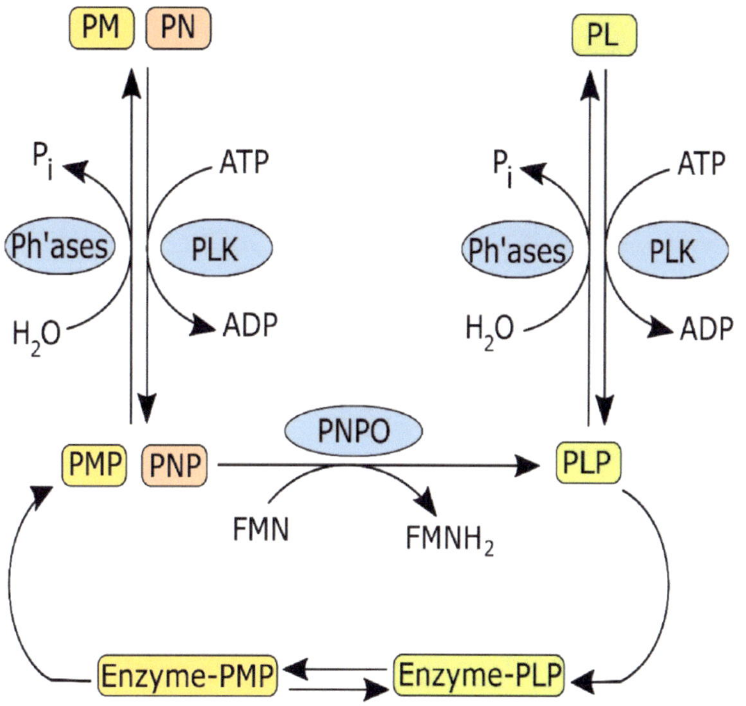

Figure 2.
The PLP salvage pathway. Phosphorylated vitamers are converted to PLP by the enzyme PNPO. PLP is also recycled from degraded holo-B6 enzymes through PMP as an intermediate step. ADP: adenosine diphosphate, ATP: adenosine triphosphate, $FMNH_2$: flavin mononucleotide reduced form, Ph'ases: phosphatases, Pi: inorganic phosphate. (Based on [5–7]).

PLP [1]. Other important enzymes in the salvage pathway are PL kinase (PLK) and a number of different phosphatases [5].

VitB6 vitamers are widely available in animal and plant food sources. PLP and in a lesser amount, PMP are present as such in animal-derived foods, mainly associated with muscle glycogen phosphorylase, while plant foods are more enriched in PN, PNP and PN glucosides [1, 4, 8].

After being ingested, phosphorylated vitamers (PLP, PNP and PMP) undergo dephosphorylation by the ecto-enzyme tissue-specific intestinal phosphatase (IP) [5], whereas PN glucoside (PNG) vitamers from plants are hydrolyzed by a glucosidase before absorption [1, 10]. Absorbed vitamers are carried by the portal circulation to the liver where they are phosphorylated by PLK [5]. Inside liver cells, PNP and PMP are oxidized by PNPO to form PLP, which is then released to the circulation bound to lysine-190 residue of albumin (**Figure 3**) [9–11]. Binding of PLP to albumin is thought to protect the cofactor from hydrolysis and other reactions [11]. About 60% of circulating vitB6 is in the form of albumin-bound PLP, while PN, PM and PL constitutes the remaining proportion [5].

Prior to delivering the circulating PLP to different tissues, it is dephosphorylated to PL by the ecto-enzyme tissue nonspecific alkaline phosphatase (TNSALP) to enable entry into the cells and through the blood–brain barrier. Inside the cell, PL is re-converted by PLK to PLP, which now can be used as a cofactor in many biochemical reactions (**Figure 3**) [1, 5, 9]. Degradation of PLP-bound enzymes (holo-B6 enzymes) can generate PMP, which is then oxidized back to PLP by the action of PNPO [6] (**Figures 2** and **3**).

Besides the liver, it has been shown that the intestine also contributes an important role in vitB6 metabolism. *In vitro* studies utilizing human intestinal

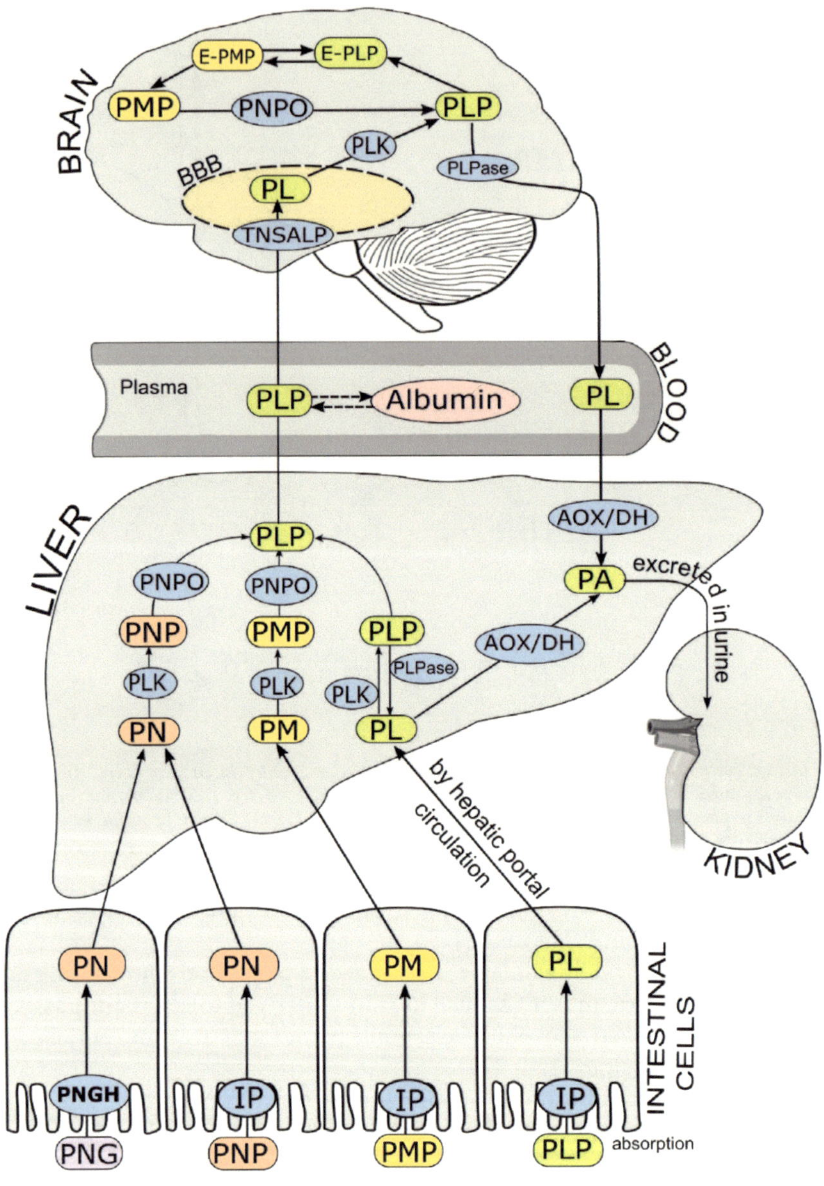

Figure 3.
Metabolism of vitB6 vitamers in different tissues of the body. PNGH: PNG hydrolase; PLPase: PLP phosphatase; AOX/DH: aldehyde oxidase/dehydrogenase; BBB: blood–brain barrier; PA: pyridoxic acid; E-PLP: enzyme-bound PLP; E-PMP: enzyme-bound PMP. (Based on [6, 7]).

epithelial Caco-2 cells [12] demonstrated that, after incubation of these cells with multiple vitB6 vitamers, PL was the only vitamer detected at the basolateral side which indicated that all other vitamers were converted to PL inside the intestinal cells. Excretion of PN and PM at the basolateral side was only detected when the enterocytes were incubated with high concentrations of these vitamers. The authors suggested that under normal dietary intakes, PN and PM are converted to PL by the enterocytes and PL becomes the principal vitamer that reaches the portal circulation. All other organs including the liver can then obtain PL from the circulation and only require PLK to produce PLP. Under high vitB6 intakes, however, the ingested amounts of PN or PM may surpass the intestine's capacity to fully metabolize these vitamers. In this case, PN and PM will be released to the portal circulation and will subsequently be converted to PLP in the liver. The study also showed the expression of a full battery of the salvage enzymes in Caco-2 cells as well as in lysates of human intestine, adding further evidence for a major role of the intestine in vitB6 metabolism [12]. Earlier works in mice [13, 14] have also pointed to a similar role of the intestine. In these studies, following oral administration of radiolabeled PN, labeled PL and PLP were detected in the mouse intestine and portal circulation indicating involvement of the intestine in converting dietary vitamers to circulating PL [13, 14].

1.2 Catabolism of vitamin B6

At the other end of vitB6 metabolism, little is known about the catabolic pathways in humans or other mammals. In contrast, these mechanisms are well established in microorganisms [3, 11, 15]. In humans and other mammals, the primary product of the degradation of PLP (and all other vitB6 vitamers) is 4-pyridoxic acid (4-PA). This compound, which is excreted in urine, is generated in two steps. In the first one, PLP is hydrolyzed to PL by the action of an intracellular enzyme known as PLP phosphatase (PLPase). In the following step, PL is oxidized to 4-PA by a non-specific aldehyde oxidase (AOX) or aldehyde dehydrogenase (**Figure 3**) [3, 6, 12, 15, 16]. In microorganisms, 4-PA is further degraded to other metabolites that can be utilized by the cell in various biochemical processes [15]. Some microbial vitB6 catabolic products such as 5-pyridoxic acid (5-PA), 5-pyridoxolactone [17] and 4-pyridoxolactone [17, 18] have been also discovered in human individuals under consumption of high amounts of vitB6. Several other PN derivatives have been identified in humans and/or other mammalian species, but their biochemical pathways and precise functions have not yet been unraveled.

For example, Coburn and Mahuren [19] detected pyridoxine 3-sulfate, pyridoxal 3-sulfate and *N*-methylpyridoxine in the urine of domestic cats, and, interestingly, these chemicals were excreted at concentrations higher than 4-PA, even with moderate intake of PN. Other studies reported the discovery of multiple PM derivatives in urine samples from PM-administered diabetic and obese rats [20, 21]. Moreover, at least nine unidentified vitB6 metabolites were detected in human urine after oral administration of radiolabeled PN [17, 19].

Oxidation of PN at the 5′ position, followed by sequential dehydrogenation to form 5-PA, is known to exist only in the PN catabolic pathway of some bacterial species like *Pseudomonas* IA and *Arthrobacter* Cr-7, where the enzymes catalyzing these reactions have been characterized [15]. Similar reactions have been proposed to occur in mammals based on experimental clues. The first one was provided by the study of Coburn and colleagues [22] who showed that healthy men who ingested a structural analog of PN, 4′-deoxypyridoxine, excreted 4'-Deoxy-5-pyridoxic acid in their urine. A similar experiment was carried out in guinea pigs [23], and the results indicated that these animals were also able to convert 4′-deoxypyridoxine to 4′-deoxy-5-pyridoxic acid. All together, these studies provided evidence for the possible existence of alternative but currently undiscovered catabolic routes of PN in humans and other mammals.

1.3 Vitamin B6 transportation across cellular membrane

Multiple experimental evidence suggests that, as with most water-soluble vitamins [24], the transportation of vitB6 across mammalian cell membrane is carrier-mediated. Studies in cultured human intestinal [12, 25], colonic [26], and renal cells [27] and animal-derived renal proximal tubular cells [28] demonstrated the presence of an efficient and specific carrier-facilitated mechanism for cellular uptake of vitB6. Such a specific transporting membrane carrier was employed to produce a high affinity gene delivery system into cancer cells using a vitB6-coupled vector [29]. However, the molecular identity of vitB6 transporter protein in mammals has remained elusive [12, 30]. Among eukaryotes, the only vitB6 transporters identified so far are the yeast transporters, Tpn1p [31] and Bsu1 [32], and, recently, PUP1 in plant species *Arabidopsis* (first to be identified in plants) [33].

1.4 Physiological roles of vitamin B6

PLP, the coenzymatically active form of vitamin B6, plays an important role in maintaining the biochemical homeostasis of the body [34]. In the human body, PLP is an essential cofactor for more than 140 distinct enzymatic activities, mainly associated with synthesis, degradation and interconversion of amino acids as well as with neurotransmitter metabolism [35–38]. PLP-dependent enzymes are also involved in a multitude of other cellular processes, including biologically active amine biosynthesis, lipid metabolism, heme synthesis, nucleic acid synthesis, protein and polyamine synthesis and several other metabolic pathways (**Figure 4**) [5, 6]. Furthermore, PLP is important in energy homeostasis through glycogen degradation and gluconeogenesis, since PLP is a cofactor for glycogen phosphorylase and gluconeogenic transaminases [36, 41]. In folate-mediated one-carbon metabolism (FOCM), PLP is required as a cofactor for the enzyme serine hydroxymethyltransferase, both its cytoplasmic (SHMT1) and mitochondrial (SHMT2) isoforms. FOCM is an important pathway that is involved in a number of physiological processes such as DNA methylation, redox homeostasis and purines and thymidine biosynthesis [36, 42].

As a coenzyme for the synthesis of several neurotransmitters including D-serine, D-aspartate, L-glutamate, glycine, γ-aminobutyric acid (GABA),

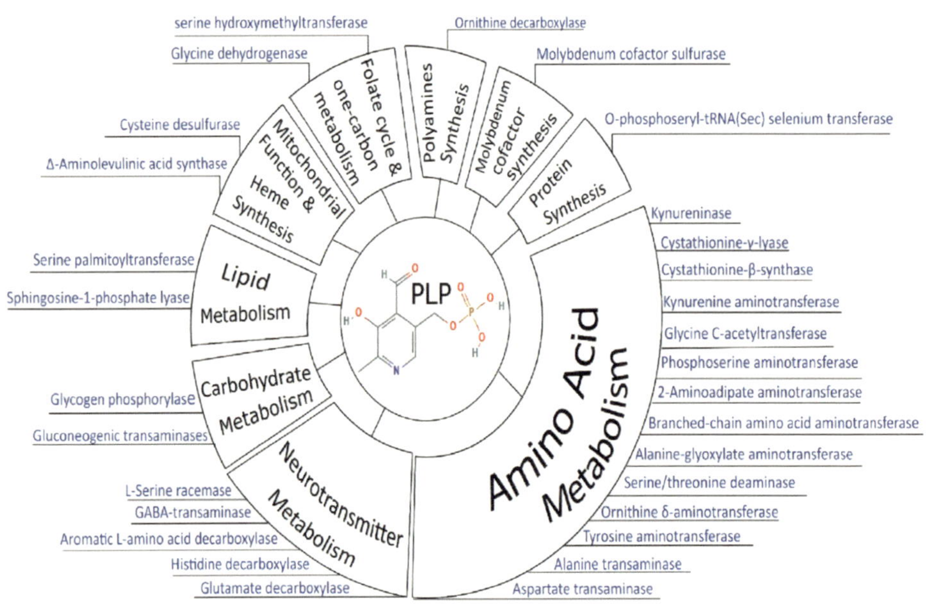

Figure 4.
The diverse cellular functions of PLP. Names in blue are the PLP-dependent enzymes involved in each metabolic process. Some enzymes can be implicated in mutiple processes. An example is branched-chain amino acid aminotransferase which can fall under amino acid and neurotransmitter metabolism. Glycine dehydrogenase can be classified under folate cycle and amino acid and neurotransmitter metabolism. (Based on [6, 39]; PLP chemical structure was retrieved from [40]).

serotonin, epinephrine, norepinephrine, histamine and dopamine, PLP is an important vitamer for normal brain function [5, 43]. For example, GABA, the major inhibitory neurotransmitter in the central nervous system (CNS), is synthesized from L-glutamate by the PLP-dependent enzyme glutamate decarboxylase (GAD). Moreover, PLP is a cofactor for branched-chain amino acid aminotransferase (BCAT) which catalyzes the synthesis of L-glutamate, the major excitatory neurotransmitter, from branched-chain amino acids like leucine and valine [5].

Another important PLP-dependent enzyme in the brain is aromatic L-amino acid decarboxylase (AADC), which catalyzes the final steps in the biosynthetic pathways of serotonin and dopamine (**Figure 5**) [5, 36]. These neurotransmitters also serve as precursors for other important compounds in the brain, specifically melatonin, norepinephrine and epinephrine (**Figure 5**) [5, 46].

In addition to its role as an enzymatic cofactor, PLP has been shown to play a role in preventing DNA damage [47] and in modulating the activity and expression of steroid hormone receptors [6, 48]. vitB6 has also been described as an efficient antioxidant in plants and fungi, with the ability of its different vitamers to quench reactive oxygen species [1, 49, 50].

1.5 PLP homeostasis and its importance for human health

PLP is a highly reactive compound because of its aldehyde group at the 4′ position which can undergo spontaneous complexation with other molecules within the cell [1, 9]. It may bind with amino groups in proteins and disrupt their structure [6]. For example, it has been shown that PLP can react with the lysine residue in the

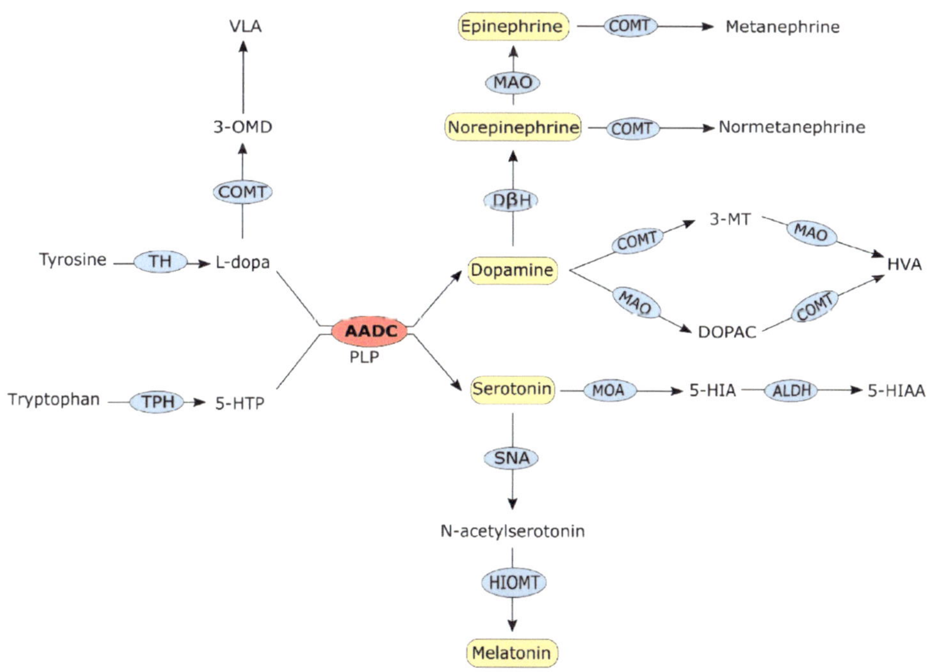

Figure 5.
Biosynthetic pathway for biogenic amine neurotransmitters and melatonin. The PLP-dependent enzyme, AADC, catalyzes a central step in this pathway. TH: tyrosine hydroxylase; TPH: tryptophan hydroxylase; L-dopa: levodopa; 5-HTP: 5-hydroxytryptophan; DβH: dopamine β-hydroxylase; MAO: monoamine oxidase; SNA: serotonin N-acetylase; HIOMT: hydroxyindole O-methyltransferase; COMT: catechol-O-methyltransferase; ALDH: Aldehyde dehydrogenase; 3-OMD: 3-O-methyldopa; VLA: vanillactic acid; 5-HIA: 5-hydroxyindole acetaldehyde; 5-HIAA: 5-hydroxyindoleacetic acid; 3-MT: 3-methoxytyramine; HVA: homovanillic acid; DOPAC: 3,4-dihydroxyphenylacetic acid. (Based on [44, 45]).

active site of human DNA topoisomerase I, causing its inhibition [51, 52]. Through a chemical reaction known as Knoevenagel condensation, PLP can also react with intermediate metabolites like Δ^1-pyrroline 5-carboxylate and Δ^1-piperideine 6-carboxylate, which form the molecular basis of PLP depletion in the neurometabolic diseases ALDH7A1 deficiency and hyperprolinaemia type II, respectively [6]. Because of its high reactivity and to prevent toxic accumulation of this cofactor, the intracellular pool of free PLP is maintained at very low concentration (about 1 μM in eukaryotic cells) [1, 5, 6]. It is therefore likely that PLP production in the cell is tightly regulated [5], and experimental work indicates the presence of an efficient mechanism that maintains intracellular PLP levels within optimum levels [12]. However, how the concentration of PLP is controlled in mammalian tissues is not entirely understood [3, 34].

A number of mechanisms have been proposed that help in PLP homeostasis. First, both enzymes that produce PLP, PLK and PNPO, are inhibited by their product PLP and its rate of synthesis can, therefore, be controlled by this feedback inhibition [1, 5, 6]. Enzymes that degrade PLP and PL, like PLPase and AOX, respectively, have also been proposed as a mechanism that keeps free PLP at low level within the cell [1, 5, 6]. Proteins that are known to naturally bind PLP, like muscle glycogen phosphorylase, plasma albumin and hemoglobin in red blood cells, contribute to reducing the amount of free reactive PLP [6]. In addition to its catalytic role in PLP synthesis, a recent study [53] demonstrated that PNPO forms a tight a binding with PLP at a noncatalytic site *in vitro*. The study further showed that PLP-bound PNPO interacts with several PLP-dependent enzymes and hypothesized that it may serve as a safe carrier of the reactive cofactor to its dependent enzymes [53]. Another more recently proposed PLP carrier protein is known as PLPHP or PLP Homeostasis Protein (*described in detail in Section 2.5*).

Conditions that disrupt cellular PLP homeostasis can cause disease. For example, inactivation of PLPP in mice led to increase in PLP levels, anxiety and motor deficits [54]. In humans, intake of high doses of vitB6 is known to cause motor and sensory neuropathies [1, 5]. Deficiency of PLP in the cell is also implicated in several pathologies, most notably the so-called vitB6-dependent epileptic encephalopathies [1, 5, 9, 37].

2. VitB6-dependent epileptic encephalopathies

VitB6-dependent epileptic encephalopathies (B6EEs) represent a clinically and genetically heterogeneous group of rare inherited metabolic diseases [55, 56]. These debilitating conditions are characterized by recurrent seizures in the prenatal, neonatal, or postnatal period, which are typically resistant to conventional anticonvulsant treatment but well-controlled by the administration of PN or PLP [56–59]. In addition to seizures, children affected with B6EEs may also suffer from developmental and/or intellectual disabilities, along with structural brain abnormalities [60]. The 5 principal types of B6EEs: PN-dependent epilepsy due to ALDH7A1 (antiquitin) deficiency (PDE-ALDH7A1) (MIM: 266100), hyperprolinemia type 2 (MIM: 239500), PLP-dependent epilepsy due to PNPO deficiency (MIM: 610090), hypophosphatasia (MIM: 241500) and PLPBP deficiency (MIM: 617290) [6, 9, 60, 61] (**Table 1**). According to the underlying pathobiochemical mechanism, these forms of B6EEs can be categorized into: 1) defects in amino acid catabolic pathways causing buildup of byproducts that react with PLP (PDE-ALDH7A1 and hyperprolinemia type 2), 2) defects in the vitB6 salvage pathway (PNPO deficiency), and 3) defects in cellular uptake of PLP (hypophosphatasia) [6, 9] (**Table 1**). In the most

Disease name	PN-dependent epilepsy (PDE-ALDH7A1)	PLP-dependent epilepsy	Hyperprolinemia type 2	Hypophosphatasia	PLPBP deficiency
Affected gene	*ALDH7A1*	*PNPO*	*ALDH4A1*	*ALPL*	*PLPBP*
Affected enzyme or protein/pathway(s)	α-AASA dehydrogenase/ lysine catabolism pathway	PNP oxidase/vitB6 salvage pathway	P5C dehydrogenase/Proline catabolism pathway	TNSALP/Extracellular dephosphorylation of PLP, Bone mineralization	PLPHP/PLP homeostasis
Pathophysiological mechanism of PLP deficiency	Accumulating lysine metabolite, P6C, reacts with and inactivates PLP	PNPO is required for intracellular production of PLP from PNP/PMP	Accumulating proline metabolite, P5C, reacts with and inactivates PLP	TNSALP is required for extracellular conversion of PLP to PL to enable its cellular uptake	PLPHP is required for maintaining cellular PLP homeostasis
Main clinical features	Neonatal seizures, DD/ID	Neonatal seizures, DD/ ID	Infantile seizures, DD/ID, ataxia	Rickets, Osteomalacia, Neonatal seizures	Neonatal seizures, DD/ID
Biomarkers (biofluid)	High α-AASA (U/P), P6C (P), PIP (P)	High PM, PM/PA ratio (P)	High proline (P), P5C (U)	Low ALP (P), high PLP (P), high PEA (U)	No specific biomarker
Commonly used vitB6 treatment	PN	PLP	PN	PN	PN
References	[9, 60, 62]	[9, 57, 63]	[6, 60, 63, 64]	[9, 60, 65, 66]	[7, 67, 68]

Abbreviations: α-AASA: α-aminoadipic semialdehyde; P6C: Δ^1-piperideine-6-carboxylic acid; DD: developmental delay; ID: intellectual disability; U: urine; P: plasma; PIP: pipecolic acid; P5C: pyrroline 5-carboxylic acid; ALP: alkaline phosphatase; PEA: phosphatidylethanolamine GPI: glycosyl phosphatidylinositol.

Table 1.
Summary of the genetic, biochemical and clinical features of B6EEs.

recently discovered type, PLPBP deficiency, the exact mechanism that disrupts PLP homeostasis is not fully understood [7].

2.1 PN-dependent epilepsy (ALDH7A1 deficiency)

2.1.1 Disease mechanism

PN-dependent epilepsy (PDE-ALDH7A1) is caused by homozygous or compound heterozygous mutations in the *ALDH7A1* gene (also known as antiquitin, ATQ). *ALDH7A1* codes for α-aminoadipic semialdehyde dehydrogenase, an enzyme that functions within the lysine catabolism pathway in the brain and peripheral tissues [62]. In PDE-ALDH7A1, loss of the enzyme's function leads to the accumulation of three upstream lysine catabolites: Δ^1-piperideine-6-carboxylic acid (P6C), α-aminoadipic semialdehyde (α-AASA) and pipecolic acid (PIP) [60] (**Figure 6**). Through a chemical reaction known as Knoevenagel condensation, accumulating P6C spontaneously conjugates with PLP, forming inactive complex products and causing cellular deficiency of this important cofactor [62] (**Figure 6**). Seizures are thought to occur because PLP is required for neurotransmitter metabolism, particularly for the synthesis of GABA from glutamate [71].

2.1.2 Clinical features

The main clinical manifestation of PDE-ALDH7A1 is recurrent perinatal-onset seizures that are resistant to conventional anticonvulsant treatment, but which show remarkable response to the administration of high doses of PN [60, 72]. Seizures usually relapse when PN treatment is discontinued, either incidentally or for diagnostic purposes [60]. In some cases, the mother of an affected child has described abnormal fetal movements during pregnancy, suggestive of pre-natal onset of seizures [55, 73–75]. In atypical cases, seizure onset can be delayed to up to 3 years of age [60], and in one exceptional case, Srinivasaraghavan et al. [76] reported an Indian female with genetically proven PDE-ALDH7A1 in whom seizures did not start until the age of 17 years (juvenile onset).

In addition to seizures, most PDE-ALDH7A1 patients (about 75%) also suffer from developmental delay and moderate to severe intellectual disability [60, 72, 77]. In addition, as revealed by neuroimaging analysis, a spectrum of structural brain defects have been described in affected children with anomalies of corpus callosum (agenesis/hypoplasia/dysplasia) and white matter being common features [75, 77–79]. Motor deficits (hypotonia/hypertonia/dystonia), irritability, autism spectrum disorder (ASD), attention-deficit hyperactivity disorder (ADHD) and anxiety are additional features reported in patients [75, 77, 80].

The phenotypic spectrum of PDE-ALDH7A1 may also include non-neuronal features, but these are less frequently observed in patients. Reported examples are ocular problems, hypoglycemia, hypothyroidism, lactic acidosis, profound electrolyte disturbances, diabetes insipidus, coagulopathy, anemia, respiratory distress and hypotension [60, 77, 79, 81, 82].

2.1.3 Biochemical features and diagnostic biomarkers

In PDE-ALDH7A1, blockade of the ATQ-catalyzed step in the lysine catabolism pathway leads to accumulation of 3 upstream metabolites, P6C, α-AASA and PIP, as discovered by screening of patients' body fluids. Presence of these metabolites in supraphysiological levels is considered the hallmark biochemical feature of ATQ

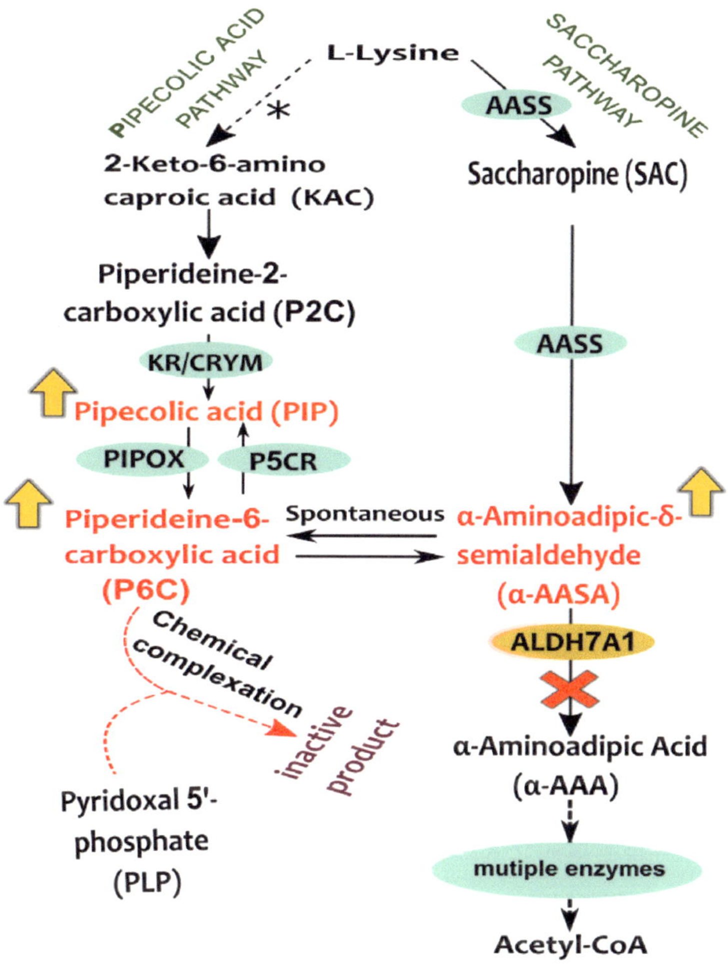

Figure 6.
*Pipecolic acid (left) and saccharopine (right) pathways for L-lysine catabolism in mammals. The two pathways converge at the step of α-AASA/P6C synthesis. ALDH7A1 catalyzes the step indicated by the red "X". Inactivation of the enzyme in PDE-ALDH7A1 causes buildup of its two substrates, P6C and α-AASA, as well as of PIP (the 3 biomarkers in patients). Accumulating P6C condenses with PLP, forming an inactive product and leading to depletion of the cofactor. *The nature of the first step of pipecolic acid pathway is undetermined. AASS: aminoadipic semialdehyde synthase; LKR: lysine-ketoglutarate reductase; SDH: saccharopine dehydrogenase; AADAT: 2-aminoadipate aminotransferase; KR: ketimine reductase; CRYM: Mu-crystallin homolog; PIPOX: pipecolic acid oxidase; P5CR: piperideine-5-carboxilic reductase. (Based on [69, 70]).*

deficiency and have been utilized as diagnostic biomarkers [72]. Recently, two additional lysine metabolites discovered to accumulate in patients have been suggested as novel biomarkers. The first one is 6-oxopipecolate (6-oxo-PIP), which was found to be present in large concentrations in plasma, urine, and CSF of ATQ deficiency patients [83, 84]. By means of an untargeted metabolomics approach, Engelke et al. [83] identified another novel metabolite, 6-(2-oxopropyl)piperidine-2-carboxylic acid (2-OPP), that accumulated in biofluids of affected individuals.

Because P6C inactivates PLP and causes cellular depletion of this enzymatic cofactor, a number of biochemical abnormalities occur that are associated with secondary deficiencies of PLP-dependent enzymes, mainly affecting amino acid metabolism. **Table 2** lists some amino acid changes reported in PDE-ALDH7A1 patients and possible links to PLP-dependent enzymes in their metabolic pathways.

Amino acid (tissue/fluid, change)*	Implicated PLP-dependent enzyme(s)**	Enzyme's function**
Glycine (CSF & plasma, ↑)	Glycine dehydrogenase (decarboxylating)	Important component of the glycine cleavage system
Threonine (CSF, ↑)	Glycine C-acetyltransferase	Catalyzes the second step in the pathway that converts threonine to glycine
	Threonine deaminase	Catalyzes the first step in the catabolic pathway of threonine [85]
Serine (plasma, ↑)	• Serine dehydratase • Serine hydroxymethyltransferase [86]	Involved in breakdown/ conversion of serine to other metabolites
Alanine (CSF & plasma, ↑)	• Alanine-glyoxylate aminotransferase • Alanine transaminase	Involved in breakdown/ conversion of alanine to other metabolites
Phenylalanine (CSF, ↑)	Aromatic L-amino acid decarboxylase	Converts phenylalanine to phenethylamine
Arginine (CSF, ↓)	Ornithine δ-aminotransferase	Catalyzes the formation of ornithine, an indirect precursor for arginine synthesis [57, 87]
Histidine (CSF, ↑)	Histidine decarboxylase	Converts histidine to histamine

**Amino acid changes were retrieved from the case series of Mills et al. [75] and Yuzyuk et al. [85].*
↑: elevated, ↓: lowered.
***Unless another source is specified, information on PLP-dependent enzymes and their catalytic activities were collectively retrieved from the review of Wilson et al. [6] and KEGG pathway database [88].*

Table 2.
Amino acid changes in PDE-ALDH7A1 and related PLP-dependent enzymes.

2.1.4 Treatment and its outcome

In patients with PDE-ALDH7A1, seizures are effectively controlled by PN treatment in about 90% of cases [6]. Patients require life-long intake of pharmacological doses of PN for seizure control as PN withdrawal leads to seizure recurrence [60]. In a subset of patients with ATQ deficiency, better seizure control is achieved when folinic acid is added to the PN regimen (known as folinic acid-responsive seizures or FARS) [60]. The subset of FARS patients can be distinguished by the appearance of a characteristic peak (Peak X) on CSF biogenic amine neurotransmitter analysis [60, 89].

Despite effective control of seizures with PN, treatment outcome is usually still poor, and a large proportion of children with PDE-ALDH7A1 have neurodevelopmental impairments [77]. It has been suggested that PN treatment alone cannot prevent the accumulation of high levels of lysine metabolites (P6C, α-AASA and PIP) in the brain which may have neurotoxic effects [90].

To limit the accumulation of these metabolites, substrate (lysine) reduction therapies have been implemented. These consisted of lysine-restricted diet [91], arginine supplementation [92] and triple therapy [93]. Arginine is a natural antagonist of lysine because the two amino acids use the same transporter (known as the y + system) for their transportation across the BBB. Therefore, it was suggested that arginine could compete with lysine and limit its entry to the brain [72, 92]. Triple therapy refers to a combination therapy of lysine-restriction and arginine supplementation (in addition to PN treatment, therefore it was termed "triple therapy") [93]. Clinical trials using these dietary therapies reported reduction in lysine metabolite levels and improvements in the neurodevelopmental outcome in most treated patients [79, 85, 91, 93–96].

2.2 PLP-dependent epilepsy (PNPO deficiency)

2.2.1 Disease mechanism

PNPO catalyzes the rate-limiting step in the biosynthetic pathway of PLP from other vitB6 vitamers (salvage pathway, **Figure 2**). Patients affected with pathogenic variants in its encoding gene, *PNPO*, have reduced activity of PNPO which leads to dysfunction of the salvage pathway and inability of the patients to produce adequate amounts of PLP [86].

2.2.2 Clinical features

Similar to PDE-ALDH7A1, PNPO deficiency is characterized by early onset, drug-resistant epileptic encephalopathy [87]. Since the disease gene discovery in 2005 [57], about 90 cases of PNPO deficiency have been reported in the medical literature with a phenotypic spectrum that extends from early postnatal lethality to milder forms with well-controlled seizures and normal neurodevelopmental outcome [88, 97–99]. Prematurity is observed in about 50% of the PNPO deficiency cases [88]. Seizures usually start very early after birth (within the first day of life in about 60% of the cases), but can also have a later onset within the first 6 months of life [86, 88]. *In utero* onset of seizures have been suspected in some of the documented cases [87]. PNPO-deficient patients may also suffer from variable degrees of morphological brain defects, most commonly diffuse brain atrophy, and neurodevelopmental deficits [88] as well as systemic co-morbidities such as lactic acidosis, hypoglycaemia, coagulopathy, anemia and ocular and cardiac problems [6, 37, 86, 98].

2.2.3 Biochemical features and diagnostic biomarkers

PNPO deficiency is associated with a number of biochemical alterations most commonly affecting biogenic amine neurotransmitters. The PLP-dependent enzyme, AADC, plays a central role in the biosynthetic pathway of these neurotransmitters (**Figure 5**). A number of amine neurotransmitter metabolites in this pathway were found to be present at abnormal levels in PNPO-deficient patients, suggesting an impaired flux through the AADC catalyzed step. For example, elevated levels of 3-*O*-methyldopa (3-OMD) and vanillactic acid (VLA) have been frequently detected in patients' CSF and urine samples, respectively [6, 88]. Both compounds are metabolites of L-dopa, the direct precursor of dopamine, which are generated upstream of AADC [44] (**Figure 5**). On the other hand, low CSF concentrations have been detected for metabolites downstream to AADC, namely, homovanillic acid (HVA) and 5-hydroxyindoleacetic acid (5-HIAA) [88, 100], the catabolic products of dopamine and serotonin, respectively (**Figure 5**) [44].

The biochemical spectrum of PNPO deficiency also includes amino acid and vitB6 vitamer perturbations. Elevated concentrations of threonine, glycine, histidine and taurine and low concentrations of arginine in CSF and/or plasma have been all reported in patients [6, 10, 88]. Unlike PDE-ALDH7A1, systemic PLP deficiency is a typical finding in PNPO deficiency as evidenced by the detection of low PLP levels in pre-treatment patient samples (CSF and/or plasma) [37, 88, 101]. Another common vitamer finding is the accumulation of PM, the precursor of PNPO substrate, detected in both pre- and post-treatment plasma samples [101, 102].

Currently there is no specific diagnostic biomarker for PNPO deficiency and genetic testing of the *PNPO* gene is required to establish diagnosis [86]. Altered biogenic amine profile along with low PLP and/or elevated PM in patient biofluids have been proposed as indicators of PNPO deficiency [6, 10, 88]. In a small cohort of patients, Mathis et al. [101] noted a consistently elevated plasma PM/PA ratio irrespective of vitB6 treatment status. This distinct vitamer profile was only observed in PNPO-deficient patients but not in other vitB6EE forms and was therefore suggested to be a candidate biomarker for PNPO deficiency, but this is yet to be validated in a larger cohort of patients [101]. Recently, a new and rapid mass spectrometry-based method has been developed for diagnosis of PNPO deficiency in dried blood spots which relies on measurement of enzyme activity [103].

2.2.4 Treatment and its outcome

Seizures are usually controlled by supplementation of pharmacological doses of PLP or PN. Based on early reports [57, 104, 105], PNPO deficiency has for some time been viewed as a disease that is only treatable by PLP but not PN (and hence was given the name "PLP-dependent epilepsy"). This was also consistent with the notion that the defective enzyme, PNPO, in these patients is unable to convert supplemented PN to PLP which explains the lack of response to PN treatment. However, it was later found that a subset of affected children (about 40% of cases [6]) show better clinical response to PN while PLP may in fact exacerbate their seizures [87, 106]. Mills et al. [87] suggested that certain genotypes (namely R225H/C and D33V) seem to be more likely to benefit from PN treatment. This was attributed to possible residual enzyme activity that is associated with these *PNPO* mutations and that PN may also have a chaperone-like stabilizing effect on the mutant protein. PLP, on the other hand, may exert an inhibitory effect on the protein and abolish its presumed residual activity leading to more deleterious consequences [106, 107]. Based on treatment response, PNPO-deficient patients appear to fall into at least 3 groups; patients who respond to PLP but are refractory to PN, patients who respond to both vitamers (PLP and PN) and patients who respond to PN but decline upon switching to PLP [86].

In some patients, better seizure control was achieved by adjunct treatments like anti-seizure drugs [106] and/or riboflavin [108] in combination with vitB6 therapy. Riboflavin is a precursor of flavin mononucleotide (FMN), the cofactor of PNPO, and therefore may enhance residual enzyme activity [87]. There were multiple reports of liver problems in patients receiving PLP treatment, and these were linked to possible toxic effects of chronic PLP administration, an observation that warrants careful mentoring of PLP-treated patients [109–111].

Neurodevelopmental outcome is still poor in a large proportion of affected children. A recent literature survey of 87 cases of PNPO deficiency [88] found that 56% of patients suffered developmental and/or intellectual deficits in spite of adequate seizure control with vitB6 therapy. Other reports suggested that early diagnosis and initiation of treatment could lead to normal developmental outcome [4, 109, 112].

2.3 Hyperprolinemia type 2

2.3.1 Disease mechanism

The genetic cause of hyperprolinemia type 2 (HP2), first identified in an Irish traveler family [64], was found to be due to recessive mutations in *ALDH4A1*. The gene codes for pyrroline 5-carboxylate dehydrogenase (P5CD), an enzyme that catalyzes an intermediate step in the proline degradation pathway [6] (**Figure 7**). In a pathobiochemical mechanism similar to PDE-ALDH7A1, deficiency of P5CD leads to accumulation of pyrroline 5-carboxylate (P5C), an intermediate metabolite that undergo a spontaneous Knoevenagel type of reaction with PLP leading to reduced bioavailability of the cofactor (**Figure 7**) [63].

2.3.2 Clinical features

The clinical manifestation of HP2 is variable [114] and asymptomatic cases have been described [115]. Seizures are the most common clinical fining in HP2 which occur in about 50% of the cases [6, 114]. They are often triggered by febrile illness and have variable age of onset; commonly occurring during infancy or childhood but can also be up to late adulthood (63 years in one HP2 case [116]) [6, 114, 117, 118]. Intellectual and neuropsychiatric abnormalities have also been described in some HP2 patients. In the original HP2 Irish traveler family, 9 out of the 13 affected individuals developed seizures and two of them had intellectual disability [118]. Van de Ven [119] reported 5 HP2 patients; all presented with seizures, 3 had intellectual disability and 4 suffered behavioral problems.

2.3.3 Biochemical features and diagnostic biomarkers

The key biochemical features of HP2 are elevated plasma and urinary levels of proline (about 10–15 folds higher in plasma) and P5C. A combination of both biomarkers is diagnostic of HP2 and distinguishes it from hyperprolinemia type 1 [119, 120]. Walker and Mills [121] identified a new metabolite,

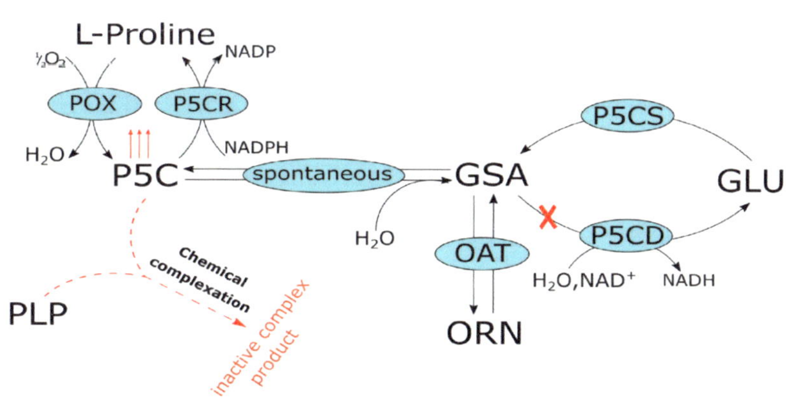

Figure 7.
L-Proline metabolic pathway. In HP2, inactivation of P5CD causes accumulation of the upstream metabolite P5C (red arrows). P5C spontaneously condenses with the enzymatic cofactor PLP leading to the formation of inactive adducts and depletion of the cofactor. GSA: glutamic-gamma-semialdehyde, ORN: ornithine, NAD(P): nicotinamide adenine dinucleotide (phosphate), POX: proline oxidase, P5CR: P5C reductase, P5C: pyrroline 5-carboxylate, OAT: ornithine aminotransferase, P5CD: P5C dehydrogenase, P5CS: P5C synthase. (Based on [63, 113]).

N-(pyrrole-2-carboxyl) glycine, that accumulated in urine of HP2 subject. They subsequently confirmed the presence of this compound in another 4 patients and suggested its use as a diagnostic biomarker for HP2. Other metabolic alterations reported in HP2 patients include increased plasma concentrations of lactate [116, 119], glycine [115, 120], ornithine [120], and alanine [119] and urinary xanthurenic acid; probably secondary to PLP deficiency [59]. VitB6 was previously analyzed in 5 HP2 patients [59, 116, 119] and found to be decreased in 3 patients and at low normal levels in the other two.

2.3.4 Treatment and its outcome

VitB6 supplementation has been used to treat HP2 associated seizures with variable response. Most of the case studies reported effective control of seizures with vitB6, either alone or in conjugation with anti-seizure medications [59, 114, 116], while few described irresponsiveness to vitB6 therapy [119]. Van de Ven et al. [119] assessed the long-term clinical outcome in 4 HP2 patients treated with vitB6 and/or anti-seizure medications. Seizures resolved spontaneously in 3 patients by the age of 12–18 years, however, neurobehavioral problems were persistent in most patients despite therapy. The clinical course was non-progressive and did not correlate with the vitB6 dose and vitB6 therapy [119].

2.4 Hypophosphatasia

2.4.1 Disease mechanism

Hypophosphatasia (HPP) results from autosomal recessive or dominant mutations affecting *ALPL*, the gene encoding tissue nonspecific alkaline phosphatase (TNSALP). TNSALP is an ecto-enzyme that is highly expressed in bone, liver, kidney and developing teeth [122, 123]. The enzyme catalyzes extracellular dephosphorylation of multiple substrates including inorganic pyrophosphate (PPi), phosphoethanolamine (PEA), and PLP (**Figure 3**) [122, 124]. On the osteoblast membrane, TNSALP hydrolyzes PPi into inorganic phosphate (Pi). Together with calcium ions (Ca^{2+}), Pi is required for the synthesis of hydroxyapatite (HA) which is the major inorganic constituent of bones and teeth. In HPP, TNSALP deficiency leads to extracellular accumulation of PPi which impairs the formation of HA and proper bone mineralization leading to an array of skeletal abnormalities [122, 123]. TNSALP is also required for the extracellular hydrolysis of PLP to PL to facilitate its entry into the cell which explains the occurrence of intracellular PLP deficiency and vitB6-dependent seizures in some forms of HPP [123, 124].

2.4.2 Clinical features

There is a remarkable heterogeneity in the clinical presentation of HPP and 5 principal clinical types have been recognized based on skeletal disease features and age of onset. In order of escalating severity, these types are "odonto", "adult", "childhood", "infantile", and "perinatal" HPP [124, 125]. The severe forms (infantile and perinatal) show autosomal recessive inheritance, while in the milder forms both autosomal dominant or recessive inheritance has been described [124, 126]. Defective mentalization of bone and/or teeth is the clinical hallmark feature of HPP in all of these types [127]. Seizures are the most well described extra-skeletal feature of HPP and are exclusively observed in the infantile and perinatal types [128]. According to a recent metanalysis [128], seizures occurred in about 20% of patients with pediatric-onset HPP.

Odonto-HPP is the mildest form and can manifest at any age. It involves minor dental problems like premature shedding of deciduous teeth without any other symptoms [123, 124]. Adult HPP typically manifest during middle age or later and can cause debilitating symptoms like osteomalacia leading to bone fractures, chondrocalcinosis, musculoskeletal pain and loss of dentition. Some patients also suffer from pseudogout due to increased extracellular concentrations of PPi [123, 126]. Childhood HPP presents after the age of 6 months and common features include rickets and premature loss of deciduous teeth. Severe forms are also associated with muscle weakness causing delay in walking and abnormal gait [125]. Infantile HPP is a severe type and can lead to death in about 50% of affected infants [123]. It is diagnosed before 6 months of age and features delayed postnatal development, failure to thrive, hypotonia along with rachitic deformities [125]. Hypercalcemia and hypercalciuria are frequently seen and may lead to renal failure [126]. In rapidly progressive cases, rickets causes thoracic deformity and death may ensue due to respiratory insufficiency [125, 129]. VitB6-dependent seizures may develop, sometimes preceding the skeletal features, and usually predict a fatal outcome [123, 125]. Perinatal HPP is the most severe type in which the symptoms start *in utero* or at birth and almost always lead to lethal outcome. Skeletal hypomineralization is profound and causes deformities such as caput membranaceum, wide fontanels and short limb dwarfism [123, 125, 130]. Chest malformation followed by pulmonary compromise is also a common fatal consequence of the rachitic disease [126]. Additional features described in this extreme form of HPP comprise vitB6-dependent seizures, apnea, irritability, myelophthisic anemia and intracranial hemorrhage [123].

2.4.3 Biochemical features and diagnostic biomarkers

HPP can be diagnosed by the presence of pathognomonic skeletal radiographic changes along with characteristic biochemical features. The most commonly used biochemical marker for HPP is low serum alkaline phosphatase activity which consistently observed in all forms of HPP [126]. Other reported biochemical findings in HPP include increased levels of TNSALP substrates PPi and PEA in urine, elevated Pi in plasma, hypercalciuria and/or hypercalcemia and high urinary levels of phosphoserine [6, 123, 124, 126]. These features can only be used to support the diagnosis of HPP because they may not be present in all HPP forms and are sometimes observed in other skeletal diseases. A more sensitive and specific biomarker for HPP is elevated serum levels of PLP, which has been detected even in the mildest form of HPP (odonto-HPP) and the degree of PLP elevation seems to correlate with disease severity [123, 131].

2.4.4 Treatment and its outcome

HPP-related seizures are usually responsive to PN supplementation [56, 129]. Effective treatment against the skeletal manifestations was lacking until the advent of Asfotase alfa, an enzyme-replacement therapy that was approved in 2015 [124, 126]. Asfotase alfa is recombinant, fusion protein consisting of the catalytic ectodomain of human TNSALP, the Fc fragment of human immunoglobulin G1 (IgG1) and a deca-aspartate motif for bone targeting [123, 131, 132]. Clinical trials have demonstrated the long-term safety and efficacy of Asfotase alfa in preventing life-threatening complications of HPP [123, 132, 133]. HPP patients, including those with severe forms, treated with Asfotase alfa showed marked improvements in all clinical aspects (radiography, pulmonary, neurodevelopmental and motor functions) along with resolution of pain and disability

[123, 126, 133]. At the biochemical level, Asfotase alfa therapy was associated with normalization of plasma levels of PPi and PLP [133].

2.5 PLPHP deficiency

2.5.1 Disease mechanism

PLPHP deficiency is the latest addition to B6EEs that is caused by recessive mutations in *PLPBP*, a gene previosly known as proline synthetase co-transcribed homolog (*PROSC*) [7]. The product of this gene, known as PLP homeostasis protein (PLPHP), belongs to a highly conserved family of proteins known to bind PLP. The function of these PLP-binding proteins in humans as well as other species is poorly understood. Their structures have remarkable similarity with a bacterial enzyme known as alanine racemase [134]. An insight into the function of this protein came from ananlysis of samples from PLPHP-deficient pateints which showed a widely deranged vitB6 vitamer profile. It has therefore been suggeseted that this protein plays an imprtant role in vitB6 homeostasis [7, 68]. However, the exact mechanism of how PLPHP dysfunction disrupts PLP homeostasis and leads to the observed epileptic encephalopathy is still unknown. Darin et al. [7] hypothesized that PLPHP is a PLP-carrier that protects the reactive cofactor from binding to other cellular molecules, shields it from degradative enzymes like phosphatases and securely delivers it to PLP-dependent enzymes.

2.5.2 Clinical features

The general clinical picture of PLPHP deficiency remarkably overlaps with that of ALDH7A1 deficiency and PNPO deficiency which is dominated by pharmaco-resistant seizures that respond to vitB6 treatment. Seizures typically manifest during the first week of life [7, 67, 68, 135] with possible prenatal onset in some cases [68] and a recent report of late onset at 14 months of age [136]. Johnstone et al. [68] reported two patients who presented with fatal mitochondrial encephalopathy and a patient with unique movement disorder who lacked epileptic seizures. Developmental delay, intellectual disability, acquired microcephaly and structural brain abnormalities are common co-morbidities observed in this form of B6EEs [7, 67, 68, 137–139]. Systemic features like metabolic acidosis, anemia and gastrointestinal problems have been also described in PLPHP-deficient pateints [7, 67].

2.5.3 Biochemical features and diagnostic biomarkers

Biochemical investigations performed in patient samples revealed amino acid and neurotransmitter abnormalities, reflecting the pleiotropic metabolic effects associated with altered PLP homeostasis. Among amino acids, elevated glycine in plasma and/or CSF was the most frequent alteration identified [7, 67, 68]. The enzyme that breaks down glycine, glycine cleavage system, requires PLP as a cofactor [140]. Abnormal monoamine neurotransmitter profile was detected in some patients, possibly due to suboptimal activity of the PLP-dependent enzyme AADC. Reported changes included low CSF levels of HVA (marker of low dopamine) and raised concentrations of 3-OMD, L-dopa, 5-HTP (CSF) and VLA (urine) indicating accumulation of AADC substrates [7, 67, 86]. Low PLP levels were detected in pre-treatment plasma [68] and CSF [7] samples from two patients. Johnstone et al. [68] described accumulation of high levels of PNP in patient fibroblasts and PLPHP-deficient HEK293 cells. There is currently no established biomarker for this disease.

2.5.4 Treatment and its outcome

Seizures typically respond well to vitB6 treatment (PN in majority of cases). In cases with inadequate response to PN, switching to PLP led to better seizure control [67]. About half of the cases required additional anti-seizure medications for optimal seizure control [7, 67]. The addition of folinic acid resulted in improved seizure control in one patient [68].

While seizures and secondary metabolic alterations are usually normalized with vitB6 therapy, a major fraction of patients still develop some form of neurodevelopmental disability. A recent review of 45 published PLPHP deficiency cases found that 65% of the patients suffered from intellectual disability [67]. The underlying pathophysiological mechanism is not well understood, and currently there is no effective treatment against the neurodevelopmental phenotype of this disorder.

3. Other vitB6-responsive conditions

The therapeutic effect of vitB6 supplementation have been also described in other disease conditions. The following section outlines some examples.

3.1 Hyperphosphatasia with mental retardation syndrome

Hyperphosphatasia with mental retardation (HPMR) syndrome (OMIM Phenotypic Series: PS239300) refers to a group of congenital disorders caused by defects in the biosynthetic pathway of glycosyl phosphatidylinositol (GPI) anchor. GPI-anchor is a glycolipid that is required for tethering of TNSALP and several other proteins (more than 150 in total) to the cell surface and at the blood–brain barrier (BBB) [6, 61]. Six subtypes of HPMR syndrome have been identified to date with variable phenotypic spectrum that extends from mild nonsyndromic intellectual disability (ID) to more complex forms with severe ID, seizures, increased serum alkaline phosphates and dysmorphic features [141–143]. Low serum PLP has been detected in some patients which may be ascribed to the elevated serum level of alkaline phosphate [144]. Seizures in some HPMR subtypes like PIGV deficiency [144] and PIGO deficiency [143] have been shown to respond to pyridoxine treatment.

3.2 PL kinase deficiency

PL kinase (PLK) is an important enzyme in the vitB6 salvage pathway (**Figure 2**). It is responsible for phosphorylating different vitameric compounds which is a pre-requisite step for their subsequent conversion to the active cofactor PLP (**Figure 2**). Biallelic mutations in the gene encoding PLK (*PDXK*) have been recently shown to cause an autosomal recessive disorder that is characterized by axonal peripheral polyneuropathy and optic atrophy [145]. Affected subjects had low plasma PLP and treatment with PLP supplementation was associated with biochemical and clinical improvements [145].

3.3 Molybdenum cofactor deficiency

Molybdenum cofactor (MoCoF) deficiency is a severe inherited metabolic disease that causes intractable seizures, developmental delay and structural brain defects. It is due to recessive mutations in either *MOCS1*, *MOCS2* or *GPHN*, all of which are important genes in the MoCoF biosynthetic pathway [146]. MoCoF deficiency impairs the activity of three MoCoF-dependent enzymes; sulfite oxidase, xanthine dehydrogenase and aldehyde oxidase [146, 147]. Patients with MoCoF deficiency excrete elevated levels of a number of metabolites, most notably sulfite and α-AASA [147] (the latter is the same metabolite that accumulates in PDE-ADLH7A1). *In vitro* experiments [147] demonstrated that sulfite inhibits the activity of ADLH7A1 which explains the accumulation of α-AASA in MoCoF deficiency. It is postulated that increased α-AASA, and consequently its cyclic form P6C, may lead to nonenzymatic trapping of PLP [148], in a mechanism analogous to that seen in PDE-ADLH7A1 (**Figure 6**). In line with this, Footitt et al. [149] described low CSF PLP levels in two MoCoF deficiency patients. Struys et al. [148] reported pyridoxine-responsive seizures in two patients with MoCoF deficiency due to *MOCS2* mutations.

3.4 Defects in PLP-dependent enzymes

In addition to its coenzymatic role, binding of PLP to its apo-enzymes may also be required for proper folding and correct subcellular targeting of these enzymes [6, 150]. Several inborn errors affecting PLP-dependent enzymes have been described to benefit from PN therapy. Examples are homocystinuria (cystathionine β-synthase deficiency), X-linked sideroblastic anemia (δ-aminolevulinate synthase deficiency), primary hyperoxaluria type I (alanine: glyoxylate aminotransferase (AGT) deficiency), ornithine aminotransferase deficiency and AADC deficiency [6, 9, 150]. The therapeutic effect of PN supplementation could to be attributed to the chaperone-like, stabilizing action of PLP on these mutated proteins [150]. In primary hyperoxaluria type I, it has been hypothesized that at high concentration, PLP promotes AGT dimerization and inhibit the accumulation of monomeric protein species which are mistargeted to the mitochondria [6, 150]. A recent addition to this category of PN-responsive disorders came from the discovery of *GOT2* mutations in patients who presented with a novel form of epileptic encephalopathy and serine deficiency [151]. *GOT2* encodes the PLP-dependent enzyme glutamate oxaloacetate transaminase (mitochondrial isoform). PN supplementation, either alone or in combination with serine, led to seizure control in these patients [151].

3.5 Other epileptic disorders

High-dose vitB6 treatment has been used for seizure control in several other epileptic disorders not related to vitB6 metabolism or its dependent enzymes; such as channelopathies [152–154] and West syndrome [155–157]. The specific mechanism of vitB6-repsosivness in these types of seizure disorders is not well recognized. Some authors [6, 154] suggested that vitB6 may have anticonvulsant effects because of the ability of PLP to block P2 purinoceptor 7 (P2X7 receptors), as demonstrated *in vitro* [158].

Author details

Hilal H. Al-Shekaili[1]*, Clara van Karnebeek[2,3] and Blair R. Leavitt[1]

1 Department of Medical Genetics, Centre for Molecular Medicine and Therapeutics, BC Children's Hospital Research Institute, University of British Columbia, Vancouver, BC, Canada

2 Department of Pediatrics, Amsterdam University Medical Centres, Amsterdam, The Netherlands

3 Department of Pediatrics, Radboud University Medical Centre, Nijmegen, The Netherlands

*Address all correspondence to: halshekaili@bcchr.ca

References

[1] di Salvo ML, Contestabile R, Safo MK. Vitamin B(6) salvage enzymes: mechanism, structure and regulation. Biochim Biophys Acta. 2011;1814(11): 1597-1608. doi:10.1016/j.bbapap.2010. 12.006

[2] ChemIDplus Lite - Chemical information with searchable synonyms, structures, and formulas. Accessed October 17, 2019. https:// chem.nlm.nih.gov/chemidplus/ chemidlite.jsp

[3] Mooney S, Leuendorf J-E, Hendrickson C, Hellmann H. Vitamin B6: a long known compound of surprising complexity. Mol Basel Switz. 2009;14(1):329-351. doi:10.3390/ molecules14010329

[4] Ghatge MS, Al Mughram M, Omar AM, Safo MK. Inborn errors in the vitamin B6 salvage enzymes associated with neonatal epileptic encephalopathy and other pathologies. Biochimie. 2021;183:18-29. doi:10.1016/j. biochi.2020.12.025

[5] Ghatge MS, Di Salvo ML, Contestabile R, et al. Molecular Defects of Vitamin B6 Metabolism Associated with Neonatal Epileptic Encephalopathy. In: Tanasescu R, ed. *Miscellanea on Encephalopathies - A Second Look*. InTech; 2012. doi:10.5772/29149

[6] Wilson MP, Plecko B, Mills PB, Clayton PT. Disorders affecting vitamin B6 metabolism. J Inherit Metab Dis. 2019;42(4):629-646. doi:10.1002/ jimd.12060

[7] Darin N, Reid E, Prunetti L, et al. Mutations in PROSC Disrupt Cellular Pyridoxal Phosphate Homeostasis and Cause Vitamin-B6-Dependent Epilepsy. Am J Hum Genet. 2016;99(6):1325-1337. doi:10.1016/j.ajhg.2016.10.011

[8] McCormick DB. Two interconnected B vitamins: riboflavin and pyridoxine. Physiol Rev. 1989;69(4):1170-1198. doi:10.1152/physrev.1989.69.4.1170

[9] Clayton PT. B6-responsive disorders: a model of vitamin dependency. J Inherit Metab Dis. 2006;29(2-3):317-326. doi:10.1007/s10545-005-0243-2

[10] Bohney JP, Fonda ML, Feldhoff RC. Identification of Lys190 as the primary binding site for pyridoxal 5′-phosphate in human serum albumin. FEBS Lett. 1992;298(2):266-268. doi:10.1016/0014-5793(92)80073-P

[11] Footitt EJ, Clayton PT, Mills K, et al. Measurement of plasma B6 vitamer profiles in children with inborn errors of vitamin B6 metabolism using an LC-MS/MS method. J Inherit Metab Dis. 2013;36(1):139-145. doi:10.1007/ s10545-012-9493-y

[12] Albersen M, Bosma M, Knoers NVVAM, et al. The Intestine Plays a Substantial Role in Human Vitamin B6 Metabolism: A Caco-2 Cell Model. Hamel FG, ed. *PLoS ONE*. 2013;8(1):e54113. doi:10.1371/journal. pone.0054113

[13] Sakurai T, Asakura T, Matsuda M. Transport and metabolism of pyridoxine and pyridoxal in mice. J Nutr Sci Vitaminol (Tokyo). 1987;33(1):11-19. doi:10.3177/jnsv.33.11

[14] Sakurai T, Asakura T, Matsuda M. Transport and metabolism of pyridoxine in the intestine of the mouse. J Nutr Sci Vitaminol (Tokyo). 1988;34(2):179-187. doi:10.3177/ jnsv.34.179

[15] Mukherjee T, Hanes J, Tews I, Ealick SE, Begley TP. Pyridoxal phosphate: Biosynthesis and catabolism. Biochim Biophys Acta BBA - Proteins

Proteomics. 2011;1814(11):1585-1596. doi:10.1016/j.bbapap.2011.06.018

[16] Gregory JF, Kirk JR. Determination of urinary 4-pyridoxic acid using high performance liquid chromatography. Am J Clin Nutr. 1979;32(4):879-883. doi:10.1093/ajcn/32.4.879

[17] Mahuren JD, Pauly TA, Coburn SP. Identification of 5-pyridoxic acid and 5-pyridoxic acid lactone as metabolites of vitamin B6 in humans. J Nutr Biochem. 1991;2(8):449-453. doi:10.1016/0955-2863(91)90115-L

[18] Edwards P, Liu PKS, Rose GA. Liquid chromatographic studies of vitamin B6 metabolism in man. Clin Chim Acta. 1990;190(1):67-80. doi:10.1016/0009-8981(90)90281-V

[19] Coburn SP, Mahuren JD. Identification of pyridoxine 3-sulfate, pyridoxal 3-sulfate, and N-methylpyridoxine as major urinary metabolites of vitamin B6 in domestic cats. J Biol Chem. 1987;262(6): 2642-2644.

[20] Metz TO, Alderson NL, Chachich ME, Thorpe SR, Baynes JW. Pyridoxamine traps intermediates in lipid peroxidation reactions in vivo: evidence on the role of lipids in chemical modification of protein and development of diabetic complications. J Biol Chem. 2003;278(43):42012-42019. doi:10.1074/jbc.M304292200

[21] Metz TO, Alderson NL, Thorpe SR, Baynes JW. Pyridoxamine, an inhibitor of advanced glycation and lipoxidation reactions: a novel therapy for treatment of diabetic complications. Arch Biochem Biophys. 2003;419(1):41-49. doi:10.1016/j.abb.2003.08.021

[22] Coburn SP, Mahuren JD. In vivo metabolism of 4′-deoxypyridoxine in rat and man. J Biol Chem. 1976;251(6): 1646-1652.

[23] Coburn SP, Mahuren JD, Wostmann BS, Snyder DL, Townsend DW. Role of Intestinal Microflora in the Metabolism of Vitamin B-6 and 4′-Deoxypyridoxine Examined Using Germfree Guinea Pigs and Rats. J Nutr. 1989;119(2):181-188. doi:10.1093/jn/119.2.181

[24] McCormick DB. Vitamin B6 transport and metabolism: Clues for delivery of bioactive compounds. In: Marino G, Sannia G, Bossa F, eds. *Biochemistry of Vitamin B6 and PQQ*. Advances in Life Sciences. Birkhäuser; 1994:311-317. doi:10.1007/978-3-0348-7393-2_50

[25] Said HM, Ortiz A, Ma TY. A carrier-mediated mechanism for pyridoxine uptake by human intestinal epithelial Caco-2 cells: regulation by a PKA-mediated pathway. Am J Physiol-Cell Physiol. 2003;285(5):C1219-C1225. doi:10.1152/ajpcell.00204.2003

[26] Said ZM, Subramanian VS, Vaziri ND, Said HM. Pyridoxine uptake by colonocytes: a specific and regulated carrier-mediated process. Am J Physiol-Cell Physiol. 2008;294(5):C1192-C1197. doi:10.1152/ajpcell.00015.2008

[27] Said HM, Ortiz A, Vaziri ND. Mechanism and regulation of vitamin B6 uptake by renal tubular epithelia: studies with cultured OK cells. Am J Physiol-Ren Physiol. 2002;282(3): F465-F471. doi:10.1152/ajprenal.00267.2001

[28] Bowman BB, McCormick DB. Pyridoxine uptake by rat renal proximal tubular cells. J Nutr. 1989;119(5):745-749. doi:10.1093/jn/119.5.745

[29] Pandey S, Garg P, Lim KT, et al. The efficiency of membrane transport of vitamin B6 coupled to poly(ester amine) gene transporter and transfection in cancer cells. Biomaterials. 2013;34(14):3716-3728. doi:10.1016/j.biomaterials.2013.01.098

[30] Said HM. Intestinal absorption of water-soluble vitamins in health and disease. Biochem J. 2011;437(3):357-372. doi:10.1042/BJ20110326

[31] Stolz J, Vielreicher M. Tpn1p, the Plasma Membrane Vitamin B6 Transporter of Saccharomyces cerevisiae. J Biol Chem. 2003;278(21): 18990-18996. doi:10.1074/jbc. M300949200

[32] Stolz J, Wöhrmann HJP, Vogl C. Amiloride Uptake and Toxicity in Fission Yeast Are Caused by the Pyridoxine Transporter Encoded by bsu1+ (car1+). Eukaryot Cell. 2005;4(2):319-326. doi:10.1128/EC.4.2.319-326.2005

[33] Szydlowski N, Bürkle L, Pourcel L, Moulin M, Stolz J, Fitzpatrick TB. Recycling of pyridoxine (vitamin B6) by PUP1 in Arabidopsis. Plant J. 2013;75(1):40-52. doi:10.1111/tpj.12195

[34] Lee Y-P, Kim D-W, Lee M-J, et al. Human brain pyridoxal-5′-phosphate phosphatase (PLPP): protein transduction of PEP-1-PLPP into PC12 cells. BMB Rep. 2008;41(5):408-413. doi:10.5483/BMBRep.2008.41.5.408

[35] Dakshinamurti S, Dakshinamurti K. Vitamin B6. In: *Handbook of Vitamins*. 4th ed. CRC Press; 2007. https://books.google.ca/books?id=6fYso23Mi5IC&lpg=PA322&vq=vitamin%20b6&pg=PP1#v=onepage&q&f=false

[36] Nichols TW, Gaiteri C. Morton's foot and pyridoxal 5′-phosphate deficiency: Genetically linked traits. Med Hypotheses. 2014;83(6):644-648. doi:10.1016/j.mehy.2014.09.003

[37] Plecko B, Struys EA, Jakobs C. Vitamin B6-Dependent and Responsive Disorders. In: Blau N, Duran M, Gibson KM, Dionisi Vici C, eds. *Physician's Guide to the Diagnosis, Treatment, and Follow-Up of Inherited Metabolic Diseases*. Springer; 2014:179-190. doi:10.1007/978-3-642-40337-8_11

[38] Sorolla MA, Rodríguez-Colman MJ, Tamarit J, et al. Protein oxidation in Huntington disease affects energy production and vitamin B6 metabolism. Free Radic Biol Med. 2010;49(4):612-621. doi:10.1016/j.freeradbiomed. 2010.05.016

[39] Locasale JW. Serine, glycine and the one-carbon cycle: cancer metabolism in full circle. Nat Rev Cancer. 2013;13(8): 572-583. doi:10.1038/nrc3557

[40] PubChem. PubChem. Accessed January 22, 2020. https://pubchem.ncbi.nlm.nih.gov/

[41] Ulvik A, Midttun Ø, Pedersen ER, Eussen SJ, Nygård O, Ueland PM. Evidence for increased catabolism of vitamin B-6 during systemic inflammation. Am J Clin Nutr. 2014;100(1):250-255. doi:10.3945/ajcn.114.083196

[42] Ducker GS, Rabinowitz JD. One-Carbon Metabolism in Health and Disease. Cell Metab. 2017;25(1):27-42. doi:10.1016/j.cmet.2016.08.009

[43] Surtees R, Mills PB, Clayton P. Inborn errors affecting vitamin B6 metabolism. Future Neurol. 2006;1(5):615-620. doi:10.2217/14796708.1.5.615

[44] Hyland K. Neurochemistry and defects of biogenic amine neurotransmitter metabolism. J Inherit Metab Dis. 1999;22(4):353-363. doi:10.1023/A:1005587719505

[45] Korevaar TIM, Grossman AB. Pheochromocytomas and paragangliomas: assessment of malignant potential. Endocrine. 2011;40(3):354-365. doi:10.1007/s12020-011-9545-3

[46] Luboshitzky R, Ophir U, Nave R, Epstein R, Shen-Orr Z, Herer P. The

effect of pyridoxine administration on melatonin secretion in normal men. Neuro Endocrinol Lett. 2002;23(3):213-217.

[47] Kanellis P, Gagliardi M, Banath JP, et al. A Screen for Suppressors of Gross Chromosomal Rearrangements Identifies a Conserved Role for PLP in Preventing DNA Lesions. PLOS Genet. 2007;3(8):e134. doi:10.1371/journal.pgen.0030134

[48] Allgood VE, Cidlowski JA. Vitamin B6 modulates transcriptional activation by multiple members of the steroid hormone receptor superfamily. J Biol Chem. 1992;267(6):3819-3824.

[49] Bilski P, Li MY, Ehrenshaft M, Daub ME, Chignell CF. Vitamin B6 (pyridoxine) and its derivatives are efficient singlet oxygen quenchers and potential fungal antioxidants. Photochem Photobiol. 2000;71(2):129-134. doi:10.1562/0031-8655(2000)071<0129:sipvbp>2.0.co;2

[50] Hellmann H, Mooney S. Vitamin B6: a molecule for human health? Mol Basel Switz. 2010;15(1):442-459. doi:10.3390/molecules15010442

[51] Christmann-Franck S, Fermandjian S, Mirambeau G, Der Garabedian PA. Molecular modelling studies on the interactions of human DNA topoisomerase IB with pyridoxal-compounds. Biochimie. 2007;89(4):468-473. doi:10.1016/j.biochi.2006.10.007

[52] Vermeersch JJ, Christmann-Franck S, Karabashyan LV, Fermandjian S, Mirambeau G, Der Garabedian PA. Pyridoxal 5′-phosphate inactivates DNA topoisomerase IB by modifying the lysine general acid. Nucleic Acids Res. 2004;32(18):5649-5657. doi:10.1093/nar/gkh897

[53] Ghatge MS, Karve SS, David TMS, et al. Inactive mutants of human pyridoxine 5′-phosphate oxidase: a possible role for a noncatalytic pyridoxal 5′-phosphate tight binding site. FEBS Open Bio. 2016;6(5):398-408. doi:https://doi.org/10.1002/2211-5463.12042

[54] Jeanclos E, Albersen M, Ramos RJJ, et al. Improved cognition, mild anxiety-like behavior and decreased motor performance in pyridoxal phosphatase-deficient mice. Biochim Biophys Acta BBA - Mol Basis Dis. 2019;1865(1):193-205. doi:10.1016/j.bbadis.2018.08.018

[55] Basura GJ, Hagland SP, Wiltse AM, Gospe SM. Clinical features and the management of pyridoxine-dependent and pyridoxine-responsive seizures: review of 63 North American cases submitted to a patient registry. Eur J Pediatr. 2009;168(6):697-704. doi:10.1007/s00431-008-0823-x

[56] Baumgartner-Sigl S, Haberlandt E, Mumm S, et al. Pyridoxine-responsive seizures as the first symptom of infantile hypophosphatasia caused by two novel missense mutations (c.677T>C, p.M226T; c.1112C>T, p.T371I) of the tissue-nonspecific alkaline phosphatase gene. Bone. 2007;40(6):1655-1661. doi:10.1016/j.bone.2007.01.020

[57] Mills PB, Surtees RAH, Champion MP, et al. Neonatal epileptic encephalopathy caused by mutations in the PNPO gene encoding pyridox(am)ine 5′-phosphate oxidase. Hum Mol Genet. 2005;14(8):1077-1086. doi:10.1093/hmg/ddi120

[58] Striano P, Battaglia S, Giordano L, et al. Two novel ALDH7A1 (antiquitin) splicing mutations associated with pyridoxine-dependent seizures. Epilepsia. 2009;50(4):933-936. doi:10.1111/j.1528-1167.2008.01741.x

[59] Walker V, Mills GA, Peters SA, Merton WL. Fits, pyridoxine, and hyperprolinaemia type II. Arch Dis

Child. 2000;82(3):236-237. doi:10.1136/adc.82.3.236

[60] Stockler S, Plecko B, Gospe SM, et al. Pyridoxine dependent epilepsy and antiquitin deficiency: clinical and molecular characteristics and recommendations for diagnosis, treatment and follow-up. Mol Genet Metab. 2011;104(1-2):48-60. doi:10.1016/j.ymgme.2011.05.014

[61] Knaus A, Awaya T, Helbig I, et al. Rare Noncoding Mutations Extend the Mutational Spectrum in the PGAP3 Subtype of Hyperphosphatasia with Mental Retardation Syndrome. Hum Mutat. 2016;37(8):737-744. doi:10.1002/humu.23006

[62] Mills PB, Struys E, Jakobs C, et al. Mutations in antiquitin in individuals with pyridoxine-dependent seizures. Nat Med. 2006;12(3):307-309. doi:10.1038/nm1366

[63] Farrant RD, Walker V, Mills GA, Mellor JM, Langley GJ. Pyridoxal Phosphate De-activation by Pyrroline-5-carboxylic Acid INCREASED RISK OF VITAMIN B6 DEFICIENCY AND SEIZURES IN HYPERPROLINEMIA TYPE II. J Biol Chem. 2001;276(18):15107-15116. doi:10.1074/jbc.M010860200

[64] Geraghty MT, Vaughn D, Nicholson AJ, et al. Mutations in the Δ1-Pyrroline 5-Carboxylate Dehydrogenase Gene Cause Type II Hyperprolinemia. Hum Mol Genet. 1998;7(9):1411-1415. doi:10.1093/hmg/7.9.1411

[65] Millán JL, Whyte MP. Alkaline Phosphatase and Hypophosphatasia. Calcif Tissue Int. 2016;98(4):398-416. doi:10.1007/s00223-015-0079-1

[66] Rockman-Greenberg C. Hypophosphatasia. Pediatr Endocrinol Rev PER. 2013;10 Suppl 2:380-388.

[67] Heath O, Pitt J, Mandelstam S, Kuschel C, Vasudevan A, Donoghue S. Early-onset vitamin B6-dependent epilepsy due to pathogenic PLPBP variants in a premature infant: A case report and review of the literature. JIMD Rep. 2021;58(1):3-11. doi:10.1002/jmd2.12183

[68] Johnstone DL, Al-Shekaili HH, Tarailo-Graovac M, et al. PLPHP deficiency: clinical, genetic, biochemical, and mechanistic insights. Brain J Neurol. 2019;142(3):542-559. doi:10.1093/brain/awy346

[69] Pena IA, Roussel Y, Daniel K, et al. Pyridoxine-Dependent Epilepsy in Zebrafish Caused by Aldh7a1 Deficiency. Genetics. 2017;207(4):1501-1518. doi:10.1534/genetics.117.300137

[70] Hallen A, Jamie JF, Cooper AJ. Lysine metabolism in mammalian brain: an update on the importance of recent discoveries. Amino Acids. 2013;45(6):1249-1272.

[71] Murty VSSY, Kishore MSS, Patel MR. A Rare Case of Pyridoxine-dependent Seizures in Infancy. J Clin Neonatol. 2013;2(1):39-41. doi:10.4103/2249-4847.109248

[72] Pena IA, MacKenzie A, Van Karnebeek CDM. Current knowledge for pyridoxine-dependent epilepsy: a 2016 update. Expert Rev Endocrinol Metab. 2017;12(1):5-20. doi:10.1080/17446651.2017.1273107

[73] Baxter. Pyridoxine-dependent and pyridoxine-responsive seizures. Dev Med Child Neurol. 2001;43(6):416-420. doi:10.1017/S0012162201000779

[74] Bejsovec M, Kulenda Z, Ponca E. Familial intrauterine convulsions in pyridoxine dependency. Arch Dis Child. 1967;42(222):201-207.

[75] Mills PB, Footitt EJ, Mills KA, et al. Genotypic and phenotypic spectrum of

pyridoxine-dependent epilepsy (ALDH7A1 deficiency). Brain J Neurol. 2010;133(Pt 7):2148-2159. doi:10.1093/brain/awq143

[76] Srinivasaraghavan R, Parameswaran N, Mathis D, Bürer C, Plecko B. Antiquitin Deficiency with Adolescent Onset Epilepsy: Molecular Diagnosis in a Mother of Affected Offsprings. Neuropediatrics. 2018;49(2):154-157. doi:10.1055/s-0037-1621721

[77] van Karnebeek CDM, Tiebout SA, Niermeijer J, et al. Pyridoxine-Dependent Epilepsy: An Expanding Clinical Spectrum. Pediatr Neurol. 2016;59:6-12. doi:10.1016/j.pediatrneurol.2015.12.013

[78] Friedman SD, Ishak GE, Poliachik SL, et al. Callosal alterations in pyridoxine-dependent epilepsy. Dev Med Child Neurol. 2014;56(11):1106-1110. doi:10.1111/dmcn.12511

[79] Toldo I, Bonardi CM, Bettella E, et al. Brain malformations associated to Aldh7a1 gene mutations: Report of a novel homozygous mutation and literature review. Eur J Paediatr Neurol EJPN Off J Eur Paediatr Neurol Soc. 2018;22(6):1042-1053. doi:10.1016/j.ejpn.2018.06.010

[80] Al Teneiji A, Bruun TUJ, Cordeiro D, et al. Phenotype, biochemical features, genotype and treatment outcome of pyridoxine-dependent epilepsy. Metab Brain Dis. 2017;32(2):443-451. doi:10.1007/s11011-016-9933-8

[81] Mercimek-Mahmutoglu S, Horvath GA, Coulter-Mackie M, et al. Profound neonatal hypoglycemia and lactic acidosis caused by pyridoxine-dependent epilepsy. Pediatrics. 2012;129(5):e1368-1372. doi:10.1542/peds.2011-0123

[82] Yusuf IH, Sandford V, Hildebrand GD. Congenital cataract in a child with pyridoxine-dependent epilepsy. J AAPOS Off Publ Am Assoc Pediatr Ophthalmol Strabismus. 2013;17(3):315-317. doi:10.1016/j.jaapos.2013.01.006

[83] Engelke UF, van Outersterp RE, Merx J, et al. Untargeted metabolomics and infrared ion spectroscopy identify biomarkers for pyridoxine-dependent epilepsy. *J Clin Invest*. Published online June 17, 2021:148272. doi:10.1172/JCI148272

[84] Wempe MF, Kumar A, Kumar V, et al. Identification of a novel biomarker for pyridoxine-dependent epilepsy: Implications for newborn screening. J Inherit Metab Dis. 2019;42(3):565-574. doi:10.1002/jimd.12059

[85] Yuzyuk T, Thomas A, Viau K, et al. Effect of dietary lysine restriction and arginine supplementation in two patients with pyridoxine-dependent epilepsy. Mol Genet Metab. 2016;118(3):167-172. doi:10.1016/j.ymgme.2016.04.015

[86] Stockler-Ipsiroglu S, Van Karnebeek CDM. Pyridoxine and Pyridoxal-Phosphate Dependent Epilepsies. In: Hoffmann GF, Blau N, eds. *Congenital Neurotransmitter Disorders: A Clinical Approach*. Neuroscience Research Progress. Nova Science Publishers; 2014:149-166. https://www.novapublishers.com/wp-content/uploads/2019/04/Pyridoxine-and-Pyridoxal-Phosphate-Dependent-Epilepsies.pdf

[87] Mills PB, Camuzeaux SSM, Footitt EJ, et al. Epilepsy due to PNPO mutations: genotype, environment and treatment affect presentation and outcome. Brain. 2014;137(5):1350-1360. doi:10.1093/brain/awu051

[88] Alghamdi M, Bashiri FA, Abdelhakim M, et al. Phenotypic and molecular spectrum of pyridoxamine-5′-phosphate oxidase deficiency:

A scoping review of 87 cases of pyridoxamine-5′-phosphate oxidase deficiency. Clin Genet. 2021;99(1):99-110. doi:10.1111/cge.13843

[89] Gallagher RC, Van Hove JLK, Scharer G, et al. Folinic acid-responsive seizures are identical to pyridoxine-dependent epilepsy. Ann Neurol. 2009;65(5):550-556. doi:10.1002/ana.21568

[90] Jansen LA, Hevner RF, Roden WH, Hahn SH, Jung S, Gospe SM. Glial localization of antiquitin: implications for pyridoxine-dependent epilepsy. Ann Neurol. 2014;75(1):22-32. doi:10.1002/ana.24027

[91] van Karnebeek CDM, Hartmann H, Jaggumantri S, et al. Lysine restricted diet for pyridoxine-dependent epilepsy: first evidence and future trials. Mol Genet Metab. 2012;107(3):335-344. doi:10.1016/j.ymgme.2012.09.006

[92] Mercimek-Mahmutoglu S, Cordeiro D, Cruz V, et al. Novel therapy for pyridoxine dependent epilepsy due to ALDH7A1 genetic defect: l-arginine supplementation alternative to lysine-restricted diet. Eur J Paediatr Neurol. 2014;18(6):741-746. doi:10.1016/j.ejpn.2014.07.001

[93] Coughlin CR, van Karnebeek CDM, Al-Hertani W, et al. Triple therapy with pyridoxine, arginine supplementation and dietary lysine restriction in pyridoxine-dependent epilepsy: Neurodevelopmental outcome. Mol Genet Metab. 2015;116(1-2):35-43. doi:10.1016/j.ymgme.2015.05.011

[94] Kava MP, Bryant L, Rowe P, Lewis B, Greed L, Balasubramaniam S. Beneficial outcome of early dietary lysine restriction as an adjunct to pyridoxine therapy in a child with pyridoxine dependant epilepsy due to Antiquitin deficiency. JIMD Rep. 2020;54(1):9-15. doi:10.1002/jmd2.12121

[95] Mahajnah M, Corderio D, Austin V, et al. A Prospective Case Study of the Safety and Efficacy of Lysine-Restricted Diet and Arginine Supplementation Therapy in a Patient With Pyridoxine-Dependent Epilepsy Caused by Mutations in ALDH7A1. Pediatr Neurol. 2016;60:60-65. doi:10.1016/j.pediatrneurol.2016.03.008

[96] Mercimek-Mahmutoglu S, Corderio D, Nagy L, et al. Lysine-restricted diet and mild cerebral serotonin deficiency in a patient with pyridoxine-dependent epilepsy caused by ALDH7A1 genetic defect. Mol Genet Metab Rep. 2014;1:124-128. doi:10.1016/j.ymgmr.2014.02.001

[97] Farmania R, Gupta A, Ankur K, Chetry S, Sharma S. Complexities of pyridoxine response in PNPO deficiency. Epilepsy Behav Rep. 2021;16. doi:10.1016/j.ebr.2021.100443

[98] Pavitt S, Karamian AS, Chattree G, Klotz J, Beres S. Atypical abnormal eye movements in PNPO-related epilepsy (1190). *Neurology*. 2020;94(15 Supplement). Accessed May 26, 2021. https://n.neurology.org/content/94/15_Supplement/1190

[99] van den Berge NT, Müller MA, Molkenboer JFM. A fetus with intrauterine convulsions: PNPO deficiency? Dutch J Obstet Gynecol. 2020;133(4):194-197.

[100] Ware TL, Earl J, Salomons GS, et al. Typical and atypical phenotypes of PNPO deficiency with elevated CSF and plasma pyridoxamine on treatment. Dev Med Child Neurol. 2014;56(5):498-502. doi:https://doi.org/10.1111/dmcn.12346

[101] Mathis D, Abela L, Albersen M, et al. The value of plasma vitamin B6 profiles in early onset epileptic encephalopathies. J Inherit Metab Dis. 2016;39(5):733-741. doi:10.1007/s10545-016-9955-8

[102] Mathis D, Beese K, Rüegg C, Plecko B, Hersberger M. LC-MS/MS method for the differential diagnosis of treatable early onset inherited metabolic epilepsies. J Inherit Metab Dis. 2020;43(5):1102-1111. doi:10.1002/jimd.12244

[103] Wilson MP, Footitt EJ, Papandreou A, et al. An LC-MS/MS-Based Method for the Quantification of Pyridox(am)ine 5'-Phosphate Oxidase Activity in Dried Blood Spots from Patients with Epilepsy. Anal Chem. 2017;89(17):8892-8900. doi:10.1021/acs.analchem.7b01358

[104] Hoffmann GF, Schmitt B, Windfuhr M, et al. Pyridoxal 5′-phosphate may be curative in early-onset epileptic encephalopathy. J Inherit Metab Dis. 2007;30(1):96-99. doi:10.1007/s10545-006-0508-4

[105] Khayat M, Korman SH, Frankel P, et al. PNPO deficiency: an under diagnosed inborn error of pyridoxine metabolism. Mol Genet Metab. 2008;94(4):431-434. doi:10.1016/j.ymgme.2008.04.008

[106] Plecko B, Paul K, Mills P, et al. Pyridoxine responsiveness in novel mutations of the PNPO gene. Neurology. 2014;82(16):1425-1433. doi:10.1212/WNL.0000000000000344

[107] Pearl PL, Gospe SM. Pyridoxine or pyridoxal-5′-phosphate for neonatal epilepsy: the distinction just got murkier. Neurology. 2014;82(16):1392-1394. doi:10.1212/WNL.0000000000000351

[108] Mohanlal S, Bindu PS, Sureshbabu S, Kumar S. Variable treatment response in a patient with pyridoxal N phosphate oxidase (PNPO) deficiency- understanding the paradox. Epilepsy Behav Rep. 2020;14. doi:10.1016/j.ebr.2020.100357

[109] Porri S, Fluss J, Plecko B, Paschke E, Korff CM, Kern I. Positive Outcome following Early Diagnosis and Treatment of Pyridoxal-5′-Phosphate Oxidase Deficiency: A Case Report. Neuropediatrics. 2014;45(1):64-68. doi:10.1055/s-0033-1353489

[110] Schmitt B, Baumgartner M, Mills PB, et al. Seizures and paroxysmal events: symptoms pointing to the diagnosis of pyridoxine-dependent epilepsy and pyridoxine phosphate oxidase deficiency. Dev Med Child Neurol. 2010;52(7):e133-e142. doi:10.1111/j.1469-8749.2010.03660.x

[111] Sudarsanam A, Singh H, Wilcken B, et al. Cirrhosis associated with pyridoxal 5′-phosphate treatment of pyridoxamine 5′-phosphate oxidase deficiency. JIMD Rep. 2014;17:67-70. doi:10.1007/8904_2014_338

[112] Hatch J, Coman D, Clayton P, et al. Normal Neurodevelopmental Outcomes in PNPO Deficiency: A Case Series and Literature Review. In: Morava E, Baumgartner M, Patterson M, Rahman S, Zschocke J, Peters V, eds. *JIMD Reports, Volume 26*. Vol 26. JIMD Reports. Springer Berlin Heidelberg; 2015:91-97. doi:10.1007/8904_2015_482

[113] Phang JM, Liu W. Proline metabolism and cancer. Front Biosci Landmark Ed. 2012;17:1835-1845. doi:10.2741/4022

[114] Kaur R, Paria P, Saini AG, Suthar R, Bhatia V, Attri SV. Metabolic epilepsy in hyperprolinemia type II due to a novel nonsense ALDH4A1 gene variant. *Metab Brain Dis*. Published online May 26, 2021. doi:10.1007/s11011-021-00757-w

[115] Pavone L, Mollica F, Levy HL. Asymptomatic type II hyperprolinaemia associated with hyperglycinaemia in

three sibs. Arch Dis Child. 1975;50(8):637-641. doi:10.1136/adc.50.8.637

[116] Motte J, Fisse AL, Grüter T, et al. Novel variants in a patient with late-onset hyperprolinemia type II: diagnostic key for status epilepticus and lactic acidosis. BMC Neurol. 2019;19(1):345. doi:10.1186/s12883-019-1583-0

[117] Emery FA, Goldie L, Stern J. Hyperprolinaemia Type 2. J Intellect Disabil Res. 1968;12(3-4):187-195. doi:10.1111/j.1365-2788.1968.tb00258.x

[118] Flynn MP, Martin MC, Moore PT, Stafford JA, Fleming GA, Phang JM. Type II hyperprolinaemia in a pedigree of Irish travellers (nomads). Arch Dis Child. 1989;64(12):1699-1707.

[119] van de Ven S, Gardeitchik T, Kouwenberg D, Kluijtmans L, Wevers R, Morava E. Long-term clinical outcome, therapy and mild mitochondrial dysfunction in hyperprolinemia. J Inherit Metab Dis. 2014;37(3):383-390. doi:10.1007/s10545-013-9660-9

[120] Onenli-Mungan N, Yüksel B, Elkay M, Topaloğlu AK, Baykal T, Ozer G. Type II hyperprolinemia: a case report. Turk J Pediatr. 2004;46(2):167-169.

[121] Walker V, Mills GA. N-(pyrrole-2-carboxyl) glycine a diagnostic marker of hyperprolinaemia type II: mass spectra of trimethylsilyl derivatives. Clin Chim Acta Int J Clin Chem. 2009;405(1-2):153-154. doi:10.1016/j.cca.2009.04.008

[122] Shapiro JR, Lewiecki EM. Hypophosphatasia in Adults: Clinical Assessment and Treatment Considerations. J Bone Miner Res Off J Am Soc Bone Miner Res. 2017;32(10):1977-1980. doi:10.1002/jbmr.3226

[123] Whyte MP. Hypophosphatasia — aetiology, nosology, pathogenesis, diagnosis and treatment. Nat Rev Endocrinol. 2016;12(4):233-246. doi:10.1038/nrendo.2016.14

[124] Simon S, Resch H, Klaushofer K, Roschger P, Zwerina J, Kocijan R. Hypophosphatasia: From Diagnosis to Treatment. Curr Rheumatol Rep. 2018;20(11):69. doi:10.1007/s11926-018-0778-5

[125] Whyte MP, Zhang F, Wenkert D, et al. Hypophosphatasia: Validation and expansion of the clinical nosology for children from 25years experience with 173 pediatric patients. Bone. 2015;75:229-239. doi:10.1016/j.bone.2015.02.022

[126] Whyte MP. Chapter 66 - Hypophosphatasia: Nature's window on alkaline phosphatase function in humans. In: Bilezikian JP, Martin TJ, Clemens TL, Rosen CJ, eds. *Principles of Bone Biology (Fourth Edition)*. Academic Press; 2020:1569-1599. doi:10.1016/B978-0-12-814841-9.00066-X

[127] Martos-Moreno GÁ, Calzada J, Couce ML, Argente J. Hypophosphatasia: Clinical manifestations, diagnostic recommendations and therapeutic options. *An Pediatría Engl Ed*. 2018;88(6):356.e1-356.e11. doi:10.1016/j.anpede.2017.06.006

[128] Colazo JM, Hu JR, Dahir KM, Simmons JH. Neurological symptoms in Hypophosphatasia. Osteoporos Int J Establ Result Coop Eur Found Osteoporos Natl Osteoporos Found USA. 2019;30(2):469-480. doi:10.1007/s00198-018-4691-6

[129] Nunes ML, Mugnol F, Bica I, Fiori RM. Pyridoxine-dependent

seizures associated with hypophosphatasia in a newborn. J Child Neurol. 2002;17(3):222-224. doi:10.1177/088307380201700314

[130] Shohat M, Rimoin DL, Gruber HE, Lachman RS. Perinatal lethal hypophosphatasia; clinical, radiologic and morphologic findings. Pediatr Radiol. 1991;21(6):421-427. doi:10.1007/BF02026677

[131] Akiyama T, Kubota T, Ozono K, et al. Pyridoxal 5′-phosphate and related metabolites in hypophosphatasia: Effects of enzyme replacement therapy. Mol Genet Metab. 2018;125(1-2):174-180. doi:10.1016/j.ymgme.2018.07.006

[132] Whyte MP, Greenberg CR, Salman NJ, et al. Enzyme-replacement therapy in life-threatening hypophosphatasia. N Engl J Med. 2012;366(10):904-913. doi:10.1056/NEJMoa1106173

[133] Kishnani PS, Rockman-Greenberg C, Rauch F, et al. Five-year efficacy and safety of asfotase alfa therapy for adults and adolescents with hypophosphatasia. Bone. 2019;121:149-162. doi:10.1016/j.bone.2018.12.011

[134] Ito T, Iimori J, Takayama S, et al. Conserved pyridoxal protein that regulates Ile and Val metabolism. J Bacteriol. 2013;195(24):5439-5449. doi:10.1128/JB.00593-13

[135] Plecko B, Zweier M, Begemann A, Mathis D, Schmitt B, Striano P. Confirmation of mutations in PROSC as a novel cause of vitamin B 6 -dependent epilepsy. J Med Genet. 2017;54(12):809-814.

[136] Espinoza AC, Wright MA, Candee MS, Trandafir C, Nelson GR. Child Neurology: Late-Onset Vitamin B6-Dependent Epilepsy Identified by Rapid Genome Sequencing. Neurology. 2021;96(19):911-914. doi:10.1212/WNL.0000000000011891

[137] Jensen KV, Frid M, Stödberg T, et al. Diagnostic pitfalls in vitamin B6-dependent epilepsy caused by mutations in the PLPBP gene. JIMD Rep. 2019;50(1):1-8. doi:10.1002/jmd2.12063

[138] Johannsen J, Bierhals T, Deindl P, et al. Excessive Seizure Clusters in an Otherwise Well-Controlled Epilepsy as a Possible Hallmark of Untreated Vitamin B6-Responsive Epilepsy due to a Homozygous PLPBP Missense Variant. J Pediatr Genet. 2019;08(4):222-225. doi:10.1055/s-0039-1685501

[139] Shiraku H, Nakashima M, Takeshita S, et al. PLPBP mutations cause variable phenotypes of developmental and epileptic encephalopathy. Epilepsia Open. 2018;3(4):495-502. doi:10.1002/epi4.12272

[140] Wilson MP, Plecko B, Mills PB, Clayton PT. Disorders affecting vitamin B $_6$ metabolism. *J Inherit Metab Dis*. Published online March 20, 2019:jimd.12060. doi:10.1002/jimd.12060

[141] Doğan ÖA, Demir GÜ, Kosukcu C, et al. Hyperphosphatasia with mental retardation syndrome type 4 In two siblings-expanding the phenotypic and mutational spectrum. Eur J Med Genet. 2019;62(6):103535. doi:10.1016/j.ejmg.2018.09.002

[142] Horn D, Krawitz P, Mannhardt A, Korenke GC, Meinecke P. Hyperphosphatasia-mental retardation syndrome due to PIGV mutations: Expanded clinical spectrum. Am J Med Genet A. 2011;155(8):1917-1922. doi:10.1002/ajmg.a.34102

[143] Kuki I, Takahashi Y, Okazaki S, et al. Vitamin B6-responsive epilepsy

due to inherited GPI deficiency. Neurology. 2013;81(16):1467-1469. doi:10.1212/WNL.0b013e3182a8411a

[144] Thompson MD, Killoran A, Percy ME, Nezarati M, Cole DEC, Hwang PA. Hyperphosphatasia with neurologic deficit: a pyridoxine-responsive seizure disorder? Pediatr Neurol. 2006;34(4):303-307. doi:10.1016/j.pediatrneurol.2005.08.020

[145] Chelban V, Wilson MP, Chardon JW, et al. PDXK mutations cause polyneuropathy responsive to pyridoxal 5′-phosphate supplementation. Ann Neurol. 2019;86(2):225-240. doi:https://doi.org/10.1002/ana.25524

[146] Atwal PS, Scaglia F. Molybdenum cofactor deficiency. Mol Genet Metab. 2016;117(1):1-4. doi:10.1016/j.ymgme.2015.11.010

[147] Mills PB, Footitt EJ, Ceyhan S, et al. Urinary AASA excretion is elevated in patients with molybdenum cofactor deficiency and isolated sulphite oxidase deficiency. J Inherit Metab Dis. 2012;35(6):1031-1036. doi:10.1007/s10545-012-9466-1

[148] Struys EA, Nota B, Bakkali A, Al Shahwan S, Salomons GS, Tabarki B. Pyridoxine-dependent epilepsy with elevated urinary α-amino adipic semialdehyde in molybdenum cofactor deficiency. Pediatrics. 2012;130(6):e1716-e1719. doi:10.1542/peds.2012-1094

[149] Footitt EJ, Heales SJ, Mills PB, Allen GF, Oppenheim M, Clayton PT. Pyridoxal 5′-phosphate in cerebrospinal fluid; factors affecting concentration. J Inherit Metab Dis. 2011;34(2):529-538.

[150] Cellini B, Montioli R, Oppici E, Astegno A, Voltattorni CB. The chaperone role of the pyridoxal 5′-phosphate and its implications for rare diseases involving B6-dependent enzymes. Clin Biochem. 2014;47(3):158-165. doi:10.1016/j.clinbiochem.2013.11.021

[151] van Karnebeek CDM, Ramos RJ, Wen X-Y, et al. Bi-allelic GOT2 Mutations Cause a Treatable Malate-Aspartate Shuttle-Related Encephalopathy. Am J Hum Genet. 2019;105(3):534-548. doi:10.1016/j.ajhg.2019.07.015

[152] Du X, Chen Y, Zhao Y, Luo W, Cen Z, Hao W. Dramatic response to pyridoxine in a girl with absence epilepsy with ataxia caused by a de novo CACNA1A mutation. Seizure. 2017;45:189-191. doi:10.1016/j.seizure.2016.12.020

[153] Mefford HC, Cook J, Gospe SM. Epilepsy due to 20q13.33 subtelomere deletion masquerading as pyridoxine-dependent epilepsy. Am J Med Genet A. 2012;158A(12):3190-3195. doi:10.1002/ajmg.a.35633

[154] Reid ES, Williams H, Stabej PLQ, et al. Seizures Due to a KCNQ2 Mutation: Treatment with Vitamin B6. JIMD Rep. 2015;27:79-84. doi:10.1007/8904_2015_460

[155] Ohtsuka Y, Ogino T, Asano T, Hattori J, Ohta H, Oka E. Long-term follow-up of vitamin B(6)-responsive West syndrome. Pediatr Neurol. 2000;23(3):202-206. doi:10.1016/s0887-8994(00)00185-5

[156] Pietz J, Benninger C, Schäfer H, Sontheimer D, Mittermaier G, Rating D. Treatment of infantile spasms with high-dosage vitamin B6. Epilepsia. 1993;34(4):757-763. doi:10.1111/j.1528-1157.1993.tb00458.x

[157] Wang H-S, Kuo M-F, Chou M-L, et al. Pyridoxal phosphate is better than pyridoxine for controlling idiopathic intractable epilepsy. Arch Dis Child.

2005;90(5):512-515. doi:10.1136/adc.2003.045963

[158] Thériault O, Poulin H, Thomas GR, Friesen AD, Al-Shaqha WA, Chahine M. Pyridoxal-5′-phosphate (MC-1), a vitamin B6 derivative, inhibits expressed P2X receptors. Can J Physiol Pharmacol. 2014;92(3):189-196. doi:10.1139/cjpp-2013-0404

Chapter 8

Several Dosage Forms Containing Vitamin B and Their Use in Therapy

Özlem Çoban

Abstract

Vitamin B plays a critical role in the synthesis of DNA and maintaining the normal functioning of tissues. Therefore, its deficiency may lead to mental problems such as depression, schizophrenia, dementia, and systemic problems such as megaloblastic anemia and peripheral neuropathy. Vitamin B deficiency may be based on nutrition, as well as the use of some drugs such as metformin and omeprazole suppress the absorption of B vitamins, which may lead to deficiency. Since B vitamin is water soluble, it cannot be stored in the body. For this reason, it should be taken continuously with food. However, in cases where the vitamin B taken with food is not sufficient for the body, it should be reinforced with drugs or dietary supplements from outside. Studies have shown that the absorption of Vitamin B is 50% higher in food supplements than in foods. It can also be used as a targeting agent in tumor therapy, due to its overexpression in some tumor cells. Due to these properties of Vitamin B, various dosage forms are being developed. In this chapter, vitamin B-containing dosage forms, their production techniques, and their use in therapy will be mentioned.

Keywords: Liposomes, emulsions, microparticles, nanoparticles, encapsulation, vitamin B, 3D printing, targeting, tumor, electrospinning

1. Introduction

Vitamins cannot be synthesized by the human body and therefore must be obtained through the diet. Vitamin B complex is also a vitamin taken in this way, and there are various derivatives such as thiamine (B1), riboflavin (B2), niacin or nicotinic acid (B3), pantothenic acid (B5), pyridoxine (B6), biotin (B7), folic acid (B9) and cobalamin (B12) [1, 2]. Foods containing Vitamin B can be: Lean pork, legumes and cereal grains (B1); milk, egg white, fish, roe, kidney and leafy vegetables (B2), especially yeast and liver; meat, fish or poultry, roasted coffee (B3); chicken, beef, potatoes, oat cereals, tomato products, liver, kidney, yeast, egg yolk, broccoli and whole grains (B5); fish and meat, seeds, non-citrus fruits such as bananas and watermelons (B6); liver, kidney, egg yolk, soybeans, nuts, spinach, mushrooms, lentils (B7); fruits and green-leafy vegetables, yeast, liver (B9); meat, fish, liver, dairy products (B12) [3].

B vitamins act as coenzymes for enzymes essential for cell function. Thanks to this their function, they take place in several physiological events including glucose, fatty acid, amino acid and homocysteine metabolism, tryptophan metabolism in the kynurenine pathway and synthesis and metabolism of various neurotransmitters and neurohormones such as serotonin, dopamine, adrenaline, acetylcholine, gamma-Aminobutyric acid (GABA), glutamate, D-serine, glycine, histamine and melatonin. Some also play a role in the regulation of intestinal and blood–brain barrier permeability [4]. For example, thiamine and riboflavin are essential in the oxidative decarboxylation of multienzyme branched chain ketoacid dehydrogenase complexes in the citric acid cycle and flavoenzymes of the respiratory chain, respectively, while niacin provides protons in nicotinamide adenine dinucleotide (NADH) synthesis and consequent oxidative phosphorylation. Pantothenic acid is required for coenzyme A (CoA) formation, α-ketoglutarate and pyruvate dehydrogenase complexes, and fatty acid oxidation. Biotin is a coenzyme of decarboxylases required for gluconeogenesis and fatty acid oxidation [5]. Pyridoxine plays a role in the catabolism of cysteine and, finally, folic acid and cobalamin have an essential role in the remethylation of methionine [6].

Due to these critical functions of the vitamin B family in the metabolic activities of our body and the key roles it plays in the functions of the central nervous system and psychopathology, some disorders occur in their deficiency. It has been shown by clinical data to they have important roles in various psychiatric diseases such as major depression, bipolar disorder, schizophrenia, autism, Alzheimer's and Parkinson's [4]. For example; in a study conducted in 3 884 elderly people with depressive symptoms, it was observed that folate values, especially cobalamin values, were lower compared to the control group. As a result of this study, although it was found that especially cobalamin deficiency was associated with depression, a positive relationship was also found between folate deficiency and depression in 6 different clinical studies. In this case, it can be said that folic acid and cobalamin deficiency may contribute to mental illness and neurological disorders in older adults [7]. The reason for the emergence of this situation due to folate and cobalamin deficiency may be that these vitamins help the metabolism of monoamine neurotransmitters such as norepinephrine, and methylation and monoamine metabolism are impaired in their deficiency [7, 8]. In another study of 7 387 Iranian adults were found that a strong association had between high biotin intake and a lower probability of depression, anxiety, and stress for the entire population, but especially for women, and pyridoxine had a protective effect against stress. In addition, there were observed that an inverse association had between cobalamin and anxiety symptoms in men, women with high pantothenic acid intake had a lower rate of depression, but there was no statistically significant linear relationship for riboflavin in both genders [9].

High levels of intake of folic acid or vitamin B complex may be of use in Parkinson's disease due to decreased serum homocysteine levels or neuroprotective effects, respectively [6]. Similarly, since homocysteine is effective in the atherosclerosis process and accordingly hyperhomocysteinemia poses a cardiovascular risk, cobalamine and pyridoxine, especially folic acid may have efficacy in the prevention of cardiovascular diseases [10]. In addition to their effects on nutrition, axonal transport, the excitability of neurons or synthesis of neurotransmitters, studies are available showing that B vitamins (thiamine, pyridoxine and cobalamin) are clinically useful in the treatment of certain painful conditions such as lumbago, sciatica, trigeminal neuralgia, and chronic pain due to diabetic polyneuropathy and rheumatoid arthritis [11].

In conducted another study, it was shown that the risk of non-alcoholic fatty liver disease (NAFLD) was lower in subjects who followed a diet rich in vitamin D, thiamine, riboflavin, cobalamin, niacin and zinc [12]. Cobalamin deficiency is quite common in patients with type 1 and type 2 diabetes. Depending on this deficiency, various clinical symptoms such as memory impairment, dementia, delirium, peripheral neuropathy, sub-acute combined degeneration of the spinal cord, megaloblastic anemia and pancytopenia are observed [13].

However, vitamin deficiency should not be attributed only to insufficient dietary intake of vitamins. For example, cobalamin deficiency can occur due to gastric atrophy, which causes absorption problems, and the use of gastric reflux medications, which lower acid levels in the stomach [7, 14].

As mentioned above, vitamin B has a protective effect in mental disorders such as schizophrenia, depression, neurological disorders such as Alzheimer's and Parkinson's, and metabolic disorders such as cardiovascular and diabetes. However, there is also the problem of absorption and stability of vitamin B. For this reason, these parameters should be taken into account when developing a formulation. In this section, the properties of various dosage forms containing vitamins belonging to the vitamin B family and the methods used during their preparation will be discussed.

2. Several dosage forms containing vitamin B and their use in therapy

Thiamine (vitamin B1) can improve immune system function and reduce the risk of type-2 diabetes, cardiovascular and kidney diseases, age-related, mental and neurodegenerative disorders and cancer. Therefore, its deficiency may affect the cardiovascular system, develop neuroinflammation, increase inflammation and cause atypical antibody responses [15]. Thiamine also acts as a carbonic anhydrase isoenzyme inhibitor [16, 17].

Riboflavin (vitamin B2) is the precursor of flavin adenine dinucleotide (FAD+) and makes folate coenzymes stabilize the C677T variant of 5,10-methylenetetrahydrofolate reductase (C677T MTHFR) by preventing the polymorphic enzyme from displacing the flavin cofactor [18]. In combination with UV light, riboflavin causes irreversible damage to nucleic acids such as DNA and RNA, rendering microbial pathogens unable to reproduce, and in this way may be effective against MERS-CoV virus as well as SARS-CoV-2 [19].

Niacin (vitamin B3, nicotinamide) serves as the building block of NAD and nicotinamide adenine dinucleotide phosphate (NADP), which are vital in chronic systemic inflammation [20]. In this respect, it serves as the sole substrate for poly (ADP-ribose) polymerase-1 (PARP), which is required for cell differentiation, DNA repair and expression, and apoptosis [18]. Since NAD+ acts as a coenzyme in various metabolic pathways, its high levels are required to treat a variety of pathophysiological conditions. For example, NAD+ has immunomodulatory properties, is released in the early stages of inflammation, and reduces serum levels of the proinflammatory cytokines IL-1β, IL-6, and TNF-α [7, 21, 22]. In addition, niacin reduces neutrophil infiltration and exerts anti-inflammatory effects in patients with ventilator-induced lung injury. It has been observed that niacin and nicotinamide prevent lung tissue damage in hamsters [23]. In addition, nicotinamide reduces viral replication and strengthens the body's defense mechanisms [24, 25].

Pantothenic acid (vitamin B5) has cholesterol and triglyceride lowering properties. It also speeds up wound healing, reduces inflammation and improves mental health [15]. However, there are limited studies showing its effects on the immune system [17].

Pyridoxine (vitamin B6, Pyridoxal 5′-phosphate) is required for the transsulfuration of vasculotoxic homocysteine and the activity of serine hydroxymethyl transferase (SHMT), which is the main entry point of one-carbon units into folate-dependent one-carbon metabolism [18]. Pyridoxal 5′-phosphate (PLP), the active form of pyridoxine, is an essential cofactor in various inflammatory pathways and its deficiency leads to immune dysregulation. Also, thirty years ago, PLP levels were shown to reduce abnormalities in platelet aggregation and blood clot formation. During inflammation, the use of PLP increases and accordingly, its amount in the body decreases. PLP levels are low in patients with type-2 diabetes and cardiovascular disease, and the elderly [26–29]. Recently, researchers at the University of Victoria reported that pyridoxine (as well as riboflavin and folic acid) is a potent anti-inflammatory that can deactivate macrophages and monocytes, inhibit antigen-presenting cells and T cells, and upregulate the immunosuppressive cytokine IL-10 [30].

Biotin (vitamin B7) acts as a coenzyme for three important carboxylation reactions involving the conversion of pyruvate to oxalacetate, acetyl-CoA to malonyl-CoA, and propionyl-CoA to methylmalonyl-CoA in humans. These enzymes are vital in numerous metabolic processes. For example, these conversions break down food into glucose, the primary source of carbohydrates for the brain and body. Recently, the roles of biotin in cell signaling and epigenetic regulation have been recognized [31].

Folic acid (vitamin B9, folate) is an essential vitamin for DNA and protein synthesis and adaptive immune response [32]. It plays several essential and direct roles in the synthesis, repair, and expression of DNA [18]. Furin is an enzyme associated with bacterial and viral infections and is a promising target for the treatment of infections. Recently, it has been noted that folic acid can inhibit furin and thus preventing SARS-CoV-2 spike protein binding, cell entry and virus transformation [32].

Cobalamin (vitamin B12) is essential for red blood cell synthesis, nervous system health, myelin synthesis, cellular growth and rapid synthesis of DNA. Its active forms are hydroxo-, adenosyl- and methylcobalamin. Cobalamin acts as a modulator of the gut microbiota and its low levels elevate methylmalonic acid and homocysteine, resulting in increased inflammation, reactive oxygen species (ROS) and oxidative stress [22]. Hyperhomocysteinemia causes endothelial dysfunction, activation of platelet and coagulation cascades, megaloblastic anemia, disruption of myelin sheath integrity and decreased immune responses. In addition, cobalamin deficiency can cause disorders in the respiratory, gastrointestinal, and central nervous systems [17].

2.1 Preparation and characterization of the dosage forms containing vitamin B

In this title, particular attention has been given to drug delivery systems in nano/micro size, from formulations prepared to contain various vitamins from the vitamin B family alone or in combination. Various nano/microsystems containing vitamin B are summarized in **Table 1** in terms of preparation methods and characterization results. Apart from this, alternative dosage forms were also evaluated since nano/micro drug carrier systems were not encountered or rarely encountered in some vitamins.

Edible hydrogels containing thiamine were prepared by hot extrusion-based three-dimensional (3D) printing and casting method. In the 3D printing method, products are produced as three-dimensional sized on special printers by the layering of the models created in the digital environment. First of all, hydrogels containing thiamine based on agar or kappa-carrageenan were prepared at 70°C and 80°C, respectively. In the preparation of these gels, 1 M NaOH was added to the medium to protect agar, kappa-carrageenan and thiamine from degradation, and the pH was maintained at 5.5. The prepared hydrogels were filled into the syringe of the printer while they were hot, the temperature of the environment where the plate was located was adjusted to 20°C, and various formulations were prepared by changing

Vitamin B Type	Molecular Stracture	Formülation Type	Preparation Method	Characterization				References
				Particle Size	Polydispersity Index	Zeta Potential	Encapsulation Efficiency	
Thiamine	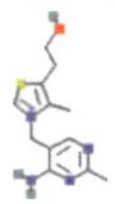	• Microparticle • Liposome	• Self-emulsion polymerization • High speed homogenization	• 681–1604 nm • 108–151 nm	• nd • 0.264–0.308	• (−)5.69 - (−) 16.90 mV • (−)21 - (−)34 mV	• 120–140 mg/g • 97%	• [33] • [34]
Riboflavin	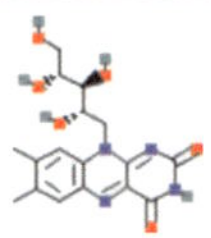	• Microparticle • Y/S emulsion • Liposome	• Emulsion-enzymatic gelation • High shear mixer/ Microfludization • Thin film	• 31.7–151.6 μm • 45–216 nm • 113–121 nm	• nd • nd • nd	• nd • (−)5 - (−) 7 mV • nd	• 56.5%–84.1% • nd • 25.86%–42.34%	• [35] • [36] • [37]
Niacin		• Dendrimer • Microsphere • Microsphere • Microemultion	• Simple esterification reaction • W/O/O double emulsion-solvent diffusion • Chemical denaturation • Dilution	• nd • 202–560 μm • 54–94 μm • 13.6–78.9 nm	• nd • nd • nd • nd	• nd • nd • nd • nd	• nd • 37–86% • 18.22%–70.42% • nd	• [38] • [39] • [40] • [41]
Pantothenic acid		• Liposome/ Microparticle • Nanofiber	• Proliposome/Encapsulator • Electrospinning	• 100–240 nm/240–300 μm • 625 nm	• nd/nd • nd	• nd/nd • nd	• 0.75 ± 0.02/0.60 ± 0.02 • nd	• [42] • [43]
Pyridoxine	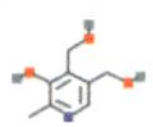	• Liposome • Microsphere • Nanoparticle	• Ethanol injection/Thin film / Modified thin film /Reverse phase evaporation • Spray drying • Emultion-solvent evaporation	• 176.70–260.30/164.80–256.40/177.80–223.20/153.60–235.00 nm • 4.5–4.8 μm • 585.7 nm	• 0.22–0.49/ 0.23–0.39/ 0.19–0.57/ 0.18–0.59 • nd • nd	• (+)4.7 – (+) 5.31 mV • (+)37 – (+)44 mV • (+) 8.98 mV	• 33.83%–34.94%/ 30.42%–34.31%/39.17%–43.69%/29.17%–40.12% • 83 ± 3.17% • nd	• [44] • [45] • [46]
Biyotin	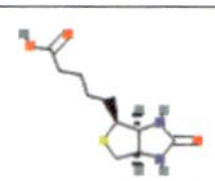	No biotin encapsulated nanoformulation was found.						

Vitamin B Type	Molecular Stracture	Formülation Type	Preparation Method	Characterization				References
				Particle Size	Polydispersity Index	Zeta Potential	Encapsulation Efficiency	
Folic acid	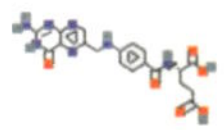	• Nanoemultion • Nanoparticle • Microparticle	• Double emultion-spray drying • Ionic gelation • Spray drying	• 32.5–90 nm • 209.5–479.4 nm • 5.4–267.5 μm	• nd • 39.9–120.4 • nd	• nd • nd • (−)6.7 - (−) 29.7 mV	• 86.60% • 43.7%–92.1% • ~100%	• [47] • [48] • [49]
Cobalamin		• Microstructure • Protein-lipid nanoparticle • Nanoemultion	• Electrospinning/spray drying • Emultion-solvent evaporation • solvent evaporation	• 0.25–1.25 μm, 0.63–20 μm/2.23–6.42 μm • 243–255 nm • 210.8–1947 nm	• nd • 0.17–0.19 • 0.21–0.85	• nd • (−)7.5 - (−)20 mV (for pH 6) • (−)9.67 - (−) 34.9 mV	• 71.0%–94.6% /61.3%–100.0% • 69–71% • 105.89%–110.65%	• [50] • [51] • [52]

Table 1.
Various nano/micro systems containing vitamin B, preparation methods and characterization results (The molecular stractures of the vitamins are taken from the PubChem database. nm; nanometer, μm; micrometer, nd; no data, mV; millivolt).

the parameters in terms of % filling, printing speed, flow percentage and layer height. Cast samples, on the other hand, were obtained by pouring the hydrogel solutions into the mold and leaving them to gel for 2 minutes (close to the printing time) at ambient temperature. When these two preparation methods were compared in terms of hardness, Young's modulus and release rate, it was observed that cast samples were harder and had higher Young's modulus than 3D ones, while their release rate was lower. Since it is based on layering, the microstructure of 3D products differs from cast samples. Printed gels are less resistant to external damage as they are a discontinuous network with several semi-fused small meshes, while cast gels can resist greater amounts of force since they are a single continuous mesh [53]. Thiamine-containing microparticles were prepared by the self-emulsion polymerization method. According to this method, first of all, empty polycitric acid (CA), polymaleic acid (MA) or polycitric acid-co-maleic acid (CA-MA) microparticles were synthesized at 25°C using a mechanical stirrer. For this, CA or MA, or CA/MA and 20 mL of deionized water were added to the reaction vessel, and then 10% mol of N,N′-methylenebisacrylamide (MBA), 0.08 mL of N,N,N′,N′-tetramethylethylenediamine (TEMED) and finally 3 mol% of ammonium persulfate (APS) aqueous solution were added to the reaction vessel, respectively, and finally mixed. The prepared poly(MA) or poly(CA) microparticles were washed with water and acetone and centrifuged for 20 minutes to remove the remaining chemicals without reacting. Thiamine was adsorbed to these prepared microparticles and loaded [33]. Finally, liposomes containing thiamine were prepared using lecithin, L-α-phosphatidylcholine and high-speed homogenization. Phosphatidylcholine was dissolved in chloroform and chloroform was removed with a fast evaporator to form a phosphatidylcholine layer. Thiamine solution was added to the layer and hydrated, and this mixture was homogenized at room temperature using a highspeed homogenizer. The nanoliposome solution was passed through a 0.2 μm syringe filter to collect small particles [34].

Studies as systems containing riboflavin include microparticles prepared using pea protein as carrier and transglutaminase as cross-linker. Emulsion-enzymatic gelation was preferred as the method. In this method, riboflavin was dissolved in the solution of pea proteins in phosphate buffer saline (PBS). Transglutaminase was added to this solution. The water phase thus formed was added to the preheated oil phase consisting of Mgliol under stirring. This mixture was left to incubate for a few more hours under stirring to allow gelation to occur. Microparticles formed over time were separated from the oil phase by centrifugation, washing with 2% Tween 80 solution, and filtration, respectively, and were lyophilized [35]. In addition, liposome formulations of riboflavin were prepared by the thin-film method using phosphatidylcholine and cholesterol. During the preparation, phosphatidylcholine and cholesterol were dissolved in methanol: chloroform mixture and the organic phase was removed in a rotary evaporator. The solution of riboflavin in PBS was added to the thin film and hydrated by rotation at ~40°C. The centrifugally formed liposomes were removed from the dispersion medium and extruded to reduce their size [37].

Microspheres were prepared by Water/Oil/Oil (W/O/O) double emulsion solvent diffusion method using ethyl cellulose for niacin. For this, niacin and ethyl cellulose were dissolved in an acetonitrile-dichloromethane mixture and deionized water was added to this mixture under a stirrer to prepare a Water/Oil (W/O) primer emulsion. Afterwards, this emulsion was added into liquid paraffin containing Span 80 under stirring. Pet ether was then added to the double emulsion to harden the microspheres. To obtain the resulting microspheres, filtration, washing with pet ether (to remove liquid paraffin) and then drying were performed, respectively. Size, encapsulation efficiency and release studies were performed on the obtained microspheres. As the polymer-drug ratio increased up to a certain value, the encapsulation efficiency increased but decreased after a certain value. The reason for this may be that the viscosity of the solution increases due to the increase in the polymer ratio, and accordingly the formation of large polymer/ solvent droplets, in this case, the curing of the microspheres is delayed and drug diffusion occurs during [39]. Niacin microemulsions were prepared by dilution method using the triangular phase diagram. Labrasol was used as a surfactant, Peceol as cosurfactant, isopropyl myristate as oil phase and water as water phase. Dimethylformamide (DMF), propylene glycol and lauric acid were preferred as penetration enhancers. In this method, the oil phase and the surfactant: cosurfactant mixture were mixed first and then titrated with water to form microemulsions. Particle size, rheological properties, conductivity and pH, refractive index, physical stability and in vitro skin penetration studies were performed on the formulations. In these studies, the highest permeation values were obtained with DMF in pH values ranging between 4.32 and 5.09 [41].

To protect pantothenic acid from environmental factors and improve its stability, liposomes, and alginate and alginate-pectin microparticles loaded liposomes have been developed. Liposomal formulations were prepared by the proliposome method. After dissolving Phospholipon 90 G (phosphatidylcholine) in ethanol, the vitamin was added to this solution. This mixture was raised to 60°C for a few minutes and then cooled to room temperature. Liposomes were obtained by adding citric acid solution dropwise to this mixture under stirring. Microparticles were prepared using an encapsulator with a 120 mm nozzle. For liposome-loaded microparticles, the pantothenic acid-loaded liposome suspension was mixed with 3% alginate or alginate-pectin mixture at a ratio of 1:1 (v/v), then these mixtures were passed through the encapsulator and sent to a solution containing 1.5% $CaCl_2$ (w/v) for hardening. The formulations prepared were evaluated in terms of morphology, encapsulation efficiency and release rate. Liposomes containing encapsulated pantothenic acid showed higher stability at moderate-acidic pH (4.0). However, microencapsulation of pantothenic acid-loaded liposomes using alginate or alginate pectin mixtures did not increase the encapsulation efficiency and retention of pantothenic acid [42]. Since pantothenic acid can be used in wound healing due to its fibroblast migration and proliferation effect, silk nanofibers containing pantothenic acid have been produced. Pantothenic acid and silk fibroin were dissolved in ultrapure water and nanofibers were produced by spraying this solution in an electrospinning device at a distance of 20 cm from the collector and a rate of 0.3 mL/h. The produced nanofibers were then exposed to 75% ethanol vapor. Morphological analyzes, cell viability and antioxidant activity were examined as characterization studies [43].

Unilamellar liposomes containing pyridoxine were prepared using ethanol injection, thin film, modified thin film and reverse phase evaporation methods. In the ethanol injection method, Lipoid S 100 was dissolved in ethanol and injected dropwise into an aqueous medium containing pyridoxine under stirring. In the thin film method, Lipoid S 100 was dissolved in chloroform and the organic solvent was evaporated in a rotary evaporator at low pressure at a temperature above the phase transition temperature of the lipid. The resulting dry lipid film was hydrated with an aqueous solution of pyridoxine. In the modified thin film method, a small amount of ethanol was used to dissolve the pyridoxine. Lipoid S 100 was dissolved in this ethanolic solution and the ethanol was evaporated at reduced pressure. The resulting lipid film was hydrated with deionized water. And finally, in the reverse phase evaporation method, Lipoid S 100 was dissolved in chloroform and pyridoxine was dissolved in 10% distilled water. Then, these solutions were mixed in a bath sonicator for 5 minutes to obtain an emulsion. The organic solvent was evaporated at reduced pressure and the remainder of the aqueous phase was added. Deionized water was used instead of buffer in all methods as the dispersion medium because when the buffer is used, the electrostatic interaction between pyridoxine and liposome decreases and accordingly the encapsulation efficiency decreases [44]. Within the scope of personalized treatment, the tablet formulation containing 4 different substances (thiamine, niacin, pyridoxine and caffeine) together and in which vitamins are released rapidly and caffeine is released in a sustained was prepared using a semi-solid pneumatic activated paste extrusion 3D printer at room temperature. Weight variation, moisture content, hardness, in vitro dissolution and active ingredient content analyzes were performed on 3D tablets [54].

The tablet dosage form in the form of a combination of biotin with various vitamins and minerals, although not alone, was prepared by the researchers. Prepared as a bilayer, this tablet contains biotin (with A, D, E, C and various B vitamins) in one layer and minerals in another layer. Since the active ingredients can be included in the form of premixes, the direct tableting method was used in the preparation of bilayer tablets. For this formulation, powder blends were evaluated in terms of angle of repose, bulk density, tapped density, Hausner ratio and Carr Index, while tablets were evaluated in terms of weight uniformity, diameter-thickness, friability, hardness, disintegration time and dissolution [55]. Unfortunately, no studies have been found on biotin loaded nano/microsystem formulations. However, there are studies on nano/microsystems where biotin is used as a ligand for active targeting. As an example of these; is the design of biotin-conjugated sunitinib-loaded nanostructures for use in lung cancer therapy. The reason why biotin is conjugated to the surface of the nanocarrier is that it is overexpressed by various cancer cells such as lung cancer [56]. Thus, it is envisaged that the nanosystem can only be delivered to tumor cells. In this study, after the preparation of biotin-sterylamine conjugate, nanostructured lipid carriers were prepared by emulsification-solvent evaporation method. In the conjugation process, the amine group of stearylamine and the activated carboxyl group of biotin were reacted in the medium of dicyclohexylcarbodiimide (DCC) and 4-N-hydroxysuccinimide (NHS) [56]. To increase oral penetration of insulin by targeting it to enterocytes, biotin-conjugated muco-inert nanocomplexes were prepared by Cui and coworkers. Although there are many ligands such as biotin, transferrin, and folic acid in enterocytes, the reason for choosing biotin in this study is that biotin receptors are widely located on the surface of enterocytes and that it has specific intracellular traffic and basal exocytosis through the transport of sodium-dependent multivitamins. In this study, after the chitosan-biotin copolymer was synthesized, insulin-loaded and hyaluronic acid (HA) coated chitosan-biotin nanocomplexes were prepared by self-assembly method (HA was used to increase mucus penetration) [57]. Apart from these studies, there are several studies in which biotin is used in surface modification; photoactive gold nanoparticles for improving the photothermal therapy of brain cancer, liposomes for use in targeted breast cancer treatment, and polymeric micelles to release paclitaxel [58–60].

Folic acid-containing nanoemulsion was prepared by Assadpour and coworkers by double emulsion method to protect folic acid from environmental and production conditions. According to this method, first of all, the W/O primary microemulsion was obtained. For this, the aqueous phase consisting of folic acid solution and Span 80 was added dropwise to the oil phase consisting of canola oil at 1 000 rpm under magnetic stirring. This primer emulsion was then added to the aqueous solution containing maltodextrin-whey protein, and the Water/Oil/Water (W/O/W) double emulsion was obtained mixing with a homogenizer at 12 000 rpm-10°C for 5 min, and then in the same homogenizer at 15 000 rpm-10°C for 8 min. In this way, nanoemulsions in which folic acid is encapsulated have been obtained [47]. Pamunuwa et al., on the other hand, encapsulated folic acid into alginate-pectin nanoparticles using the ionic gelation method to provide controlled release of folic acid. Accordingly, a small amount of Span 80 and then folic acid was added to the aqueous solution of the polymer adjusted to pH 5, and the mixture was stirred at 1 500 rpm for 30 min. $CaCl_2$ solution was added dropwise to this solution to ensure nanoparticle formation by gelation [48]. Apart from these, various nanosystems have been developed in which folic acid is conjugated to the surface, as in biotin. The general reason for using folic acid in these studies is the widespread presence of folic acid receptors in cancer cells. Examples of these studies are doxorubicin-loaded chitosan nanospheres for use in nuclear targeted cancer therapy, spiropyran and imidazole-loaded photoactive gold nanoparticles to increase the radiosensitivity of prostate cancer cells, and chitosan-coated trans-resveratrol-ferulic acid-loaded solid lipid nanoparticles to act on colon cancer cells [61–63].

Cobalamin loaded zein microstructures were prepared by Coelho et al. using electrospinning and spray drying methods. In the electrospinning method, 3 different microstructures were obtained as film, bead and fiber, while only microparticles were obtained in spray drying. A solution of zein in ethanol was used for both methods. However, while absolute ethanol is used in spray drying, aqueous alcohol is preferred in electrospinning. The formulation solutions were prepared by adding cobalamin to the solutions of zein in alcohol [50]. As a non-invasive supporting approach in cobalamin deficiency, cobalamin-loaded buccoadhesive films were prepared using solvent casting as a method, and poly(vinyl alcohol) (PVA) and chitosan as polymers. For this, glycerin or poly(ethylene glycol) 400 (PEG 400) as a plastisizer, propylene glycol as a penetration enhancer and maleic acid for chemical polymerization were added to the PVA: chitosan solution in various proportions in addition to cobalamin. After the prepared mixture was poured into the mold, it was kept at ambient temperature for 24–48 hours to form a film. Thickness, weight variation, drug content, moisture uptake and moisture content percentage, surface pH, mechanical properties, in vitro release and mucoadhesion tests were performed on the prepared formulations. In addition, in vivo pharmacokinetic studies on rabbits showed that the AUC value of bucoadhesive films was approximately 1.5 times higher than the market preparation administered intramuscularly [64]. As with other B vitamins, cobalamin has been used in the functionalization of various nanosystems. For example, calcium phosphate nanoparticles coated with cobalamin-chitosan conjugate and sodium alginate were prepared to increase receptor-mediated endocytosis and thus absorption of insulin from epithelial cells. For this, first of all, insulin-loaded calcium phosphate nanoparticles were prepared by microemulsion method, and then layer-by-layer coating process was performed [65]. In addition, solid lipid nanoparticles loaded with amphotericin B and coated with cobalamin-stearic acid conjugate were prepared to evaluate the anti-leishmanial effect of amphotericin B in vitro [66].

2.2 Vitamin B use in therapy

When cobalamin is metabolically significantly deficient, the methylation process, which is also essential in the conversion of dietary folate (methyl-tetrahydrofolate) to its active metabolic form (tetrahydrofolate), will be disrupted and the amount of intracellular and serum homocysteine will increase. The state of hyperhomocysteinemia has a potentially toxic effect on neuron and vascular endothelium. Also, as a cofactor, cobalamin helps convert methylmalonyl CoA to succinyl-CoA. Therefore, in its deficiency, this conversion decreases and an increase in serum methylmalonic acid (MMA) occurs. As a result, a defect develops in fatty acid synthesis in neuronal membranes [13]. Because cobalamin plays a vital role in the metabolism of fatty acids, which are necessary for the protection of nerve myelin [67]. Cobalamin is essential in the synthesis of monoamines or neurotransmitters such as serotonin and dopamine, and its deficiency disrupts this synthesis [13]. Similarly, pyridoxine plays a role in the production of serotonin and norepinephrine, chemicals that transmit signals in the brain, and in the formation of myelin [67].

Diabetes mellitus is one of the most common and serious metabolic disorders characterized by hyperglycemia, altered lipid, carbohydrate and protein metabolism, leading to oxidative stress and cell death in the brain, resulting in cognitive and behavioral disorders. The disease is associated with peripheral neuropathy and dysfunctions in the central nervous system. Patients have moderate changes in memory and cognitive functions, poor motor coordination and reduced motor activity. It is also associated with progressive end-organ damage in "diabetic encephalopathy" in the central nervous system. When the role of vitamin B complex in the prevention of neuronal death is examined, it has been observed that thiamine supplementation can improve and usually completely resolve the symptoms associated with Wernicke's encephalopathy (ophthalmoplegia, ataxia and confusion triad), folic acid improves memory in the elderly, and vitamin B complex protects Purkinje neurons, which is the main relay neurons of the cerebellum and play a very important role in motor coordination and learning, from degeneration and loss in diabetes [67–69].

There is substantial evidence to suggest that one-carbon metabolism is associated with cardiovascular disease, and particularly stroke. A one-carbon metabolism requires adequate folate along with cobalamin, pyridoxine, and riboflavin. The suboptimal state of any of these B vitamins and/or genetic polymorphisms on B-vitamin-dependent enzymes in one-carbon metabolism may cause disruption of one-carbon metabolic pathways that can lead to adverse phenotypes even if dietary folate and vitamin B intake are adequate. However, riboflavin supplementation may offer an effective personalized approach to managing hypertension in genetically at risk individuals. In fact, according to the findings from the 5-year Chinese Stroke Primary Prevention Study (CSPPT), which included 20 702 hypertensive patients, it was observed that intervention with folic acid did not reduce blood pressure but reduced the incidence of first stroke by 21% [70].

Because neuropathic pain that develops after nerve injury is severe and persistent, current drugs and non-drug treatments cannot provide significant pain relief in most patients. In a study investigating the analgesic effects of B vitamins (thiamine, pyridoxine and cobalamin) in rats with neuropathic pain due to spinal ganglion compression (CCD) or loose ligation of the sciatic nerve (CCI), the results showed that intraperitoneal injection of thiamine/pyridoxine or cobalamin significantly reduced thermal hyperalgesia; when used in combination synergistically inhibited thermal hyperalgesia; when administered repeatedly, it provided prolonged inhibition of thermal hyperalgesia and did not affect mechanical hyperalgesia or normal pain sensation, but showed similar effects on CCD and CCI-induced hyperalgesia. In conclusion, the effects of B vitamins on pain and hyperalgesia following primary sensory neuron injury have been demonstrated, and the possible clinical use of B vitamins has been suggested in the treatment of neuropathic pain conditions following injury, inflammation, degeneration or other disorders in the nervous system for humans [11].

Because vitamin B plays a crucial role in cell function, energy metabolism and immune function, it helps activate both the hereditary and adaptive immune response, reduces proinflammatory cytokine levels, improves respiratory function, maintains endothelial integrity and prevents hypercoagulability. Because of these effects, it can be used in the prevention or reduction of COVID 19 symptoms, or in the treatment of SARS-CoV-2 infection. For this reason, the vitamin B status of COVID-19 patients can be evaluated and their existing treatments can be supplemented [17].

It is known that behavioral symptoms mimicking schizophrenia and long-term changes in brain function are observed in animals exposed to maternal deprivation (MD) during early development, and the positive effect of vitamin B complex on schizophrenia. In an experiment in Wistar rats, a decrease in long and short-term memory was observed in addition to anxiety-like behaviors in subjects exposed to maternal deprivation. However, it was determined that rats receiving vitamin B complex showed a decrease in behavioral decline, histomorphological deterioration and oxidative stress caused by MD due to the enhancement of endogenous antioxidant defense, and thus nootropic behavior and reduced anxiety. In conclusion, it has been confirmed that the vitamin B complex is neuroprotective against neuropathological changes caused by maternal deprivation [71].

Vitamin B has effects not only on humans but also on animal and plant health. For example, B vitamins mainly have an indirect effect on the immune function of the gut and their interaction with the bacterial cells and network in the gut. However, impaired redox balance and uncontrolled inflammation can often occur in pigs, especially due to rapid genetic improvement and intensive production methods. Therefore, antioxidant vitamins, such as vitamin B, can be used for the control of reactive oxygen species and peroxides and, accordingly, improved enteric immunity and inflammation control [72]. As in humans, pigs have a decrease in plasma homocysteine due to folic acid and/or cobalamin supplements (although not as much as in humans) [73, 74]. To evaluate the effect of high homocysteinemia on piglets (birth to 8 weeks of age), two piglet populations with high or low homocysteinemia were used in one study. The low homocysteine group was formed by giving sows folate and cobalamin during pregnancy and lactation, and by intramuscular injection of cobalamin into piglets during lactation. As an indicator of cell-mediated immunity, the proliferative response of lymphocyte to mitogen activation was investigated. However, as a positive correlation between growth rate and feed conversion (gain:feed) and plasma homocysteine was observed in piglets, young "high performance" piglets producing high levels of plasma homocysteine may appear immunologically suppressed [75] and are may be less prone to develop an optimal adaptive immune response to antigenic challenges. However, the disease resistance of high-performing piglets is also high in the clinic. This may be due to homocysteine being an important metabolite that plays a central role in the interface between transsulfuration and trans- and re-methylation, linking dietary intake of at least three B vitamins and regulation of endogenous antioxidation activity, innate and adaptive immune responses [72]. Although ultraviolet-B radiation is essential for plant growth at low doses, it can have serious adverse effects on plants at high doses. Reactive oxygen species (ROS) can form in some plants in response to UV-B, and pyridoxine is known to be a quencher of ROS [76].

3. Conclusions

Vitamin B is a family of vitamins that are water-soluble and cannot be accumulated and synthesized in the body, is essential for vital functions in the body, therefore it must be taken with food. It mostly functions as a coenzyme and is also responsible for DNA repair. It has antioxidant activity. It has various activities in metabolic disease, inflammatory conditions, and mental disorders such as schizophrenia, dementia, Alzheimer's. In this respect, its deficiency may trigger these conditions in people with a genetic predisposition. Also, although it can be taken with food, some gastrointestinal conditions and medications can reduce its absorption. Therefore, external reinforcement is recommended. However, effective formulations are needed due to their low in vitro stability and limited in vivo bioavailability. In this respect, it is aimed to develop nano/microsystems that are known to be more effective rather than conventional dosage forms such as tablets and capsules. However, when we look at the literature, there are not enough studies on these drug delivery systems containing vitamin B. Therefore, researchers can be advised to develop dosage forms containing vitamin B with high bioavailability and in vitro stability.

Acknowledgements

The work was not funded by any institute or person.

Conflict of interest

The authors declare no conflict of interest.

Author details

Özlem Çoban[1,2]

1 Department of Pharmaceutical Technology, Faculty of Pharmacy, Karadeniz Technical University, Trabzon, Turkey

2 Drug and Pharmaceutical Technology Application and Research Center, Karadeniz Technical University, Trabzon, Turkey

*Address all correspondence to: ozlemcoban@ktu.edu.tr

References

[1] Chantereau G, Sharma M, Abednejad A, Vilela C, Costa EM, Veiga M, Antunes F, Pintado MM, Sèbe G, Coma V, Freire MG, Freire CSR, Silvestre AJD. Bacterial nanocellulose membranes loaded with vitamin B-based ionic liquids for dermal care applications. Journal of Molecular Liquids. 2020;302:112547. DOI: 10.1016/j.molliq.2020.112547

[2] Kennedy DO. B vitamins and the brain: Mechanisms, dose and efficacy—A review. Nutrients. 2016;8(2):68. DOI: 10.3390/nu8020068

[3] Bonetti F, Brombo G, Zuliani G. The role of B group vitamins and choline in cognition and brain aging. Nutrition and Functional Foods for Healthy Aging. 2017;139-158. DOI: 10.1016/b978-0-12-805376-8.00015-0

[4] Rudzki L, Stone T, Maes M, Misiak B, Samochowiec J, Szulc A. Gut microbiota-derived vitamins–underrated powers of a multipotent ally in psychiatric health and disease. Progress in Neuro-Psychopharmacology and Biological Psychiatry. 2021;107:110240. DOI: 10.1016/j.pnpbp.2020.110240

[5] Depeint F, Bruce WR, Shangari N, Mehta R, O'Brien PJ. Mitochondrial function and toxicity: role of the B vitamin family on mitochondrial energy metabolism. Chemico-Biological Interactions. 2006;163(1-2):94-112. DOI: 10.1016/j.cbi.2006.04.014

[6] Haghdoost-Yazdi H, Fraidouni N, Faraji A, Jahanihashemi H, Sarookhani M. High intake of folic acid or complex of B vitamins provides anti-Parkinsonism effect: no role for serum level of homocysteine. Behavioural Brain Research. 2012;233(2): 375-381. DOI: 10.1016/j.bbr.2012.05.011

[7] Mikkelsen K, Stojanovska L, Apostolopoulos V. The effects of vitamin B in depression. Curr Med Chem. 2016;23(38):4317-4337. DOI: 10.2174/0929867323666160920110810

[8] Bottiglieri T, Laundy M, Crellin R, Toone BK, Carney MW, Reynolds EH. Homocysteine, folate, methylation, and monoamine metabolism in depression. Journal of Neurology, Neurosurgery & Psychiatry. 2000;69(2):228-232. DOI: 10.1136/jnnp.69.2.228

[9] Mahdavifar B, Hosseinzadeh M, Salehi-Abargouei A, Mirzaei M, Vafa M. Dietary intake of B vitamins and their association with depression, anxiety, and stress symptoms: A cross-sectional, population-based survey. Journal of Affective Disorders. 2021;288:92-98. DOI: 10.1016/j.jad.2021.03.055

[10] Ntaios G, Savopoulos C, Grekas D, Hatzitolios A. The controversial role of B-vitamins in cardiovascular risk: an update. Archives of Cardiovascular Diseases. 2009;102(12):847-854. DOI: 10.1016/j.acvd.2009.07.002

[11] Wang ZB, Gan Q, Rupert RL, Zeng YM, Song XJ. Thiamine, pyridoxine, cyanocobalamin and their combination inhibit thermal, but not mechanical hyperalgesia in rats with primary sensory neuron injury. Pain. 2005;114(1-2):266-277. DOI: 10.1016/j.pain.2004.12.027

[12] Vahid F, Hekmatdoost A, Mirmajidi S, Doaei S, Rahmani D, Faghfoori Z. Association between index of nutritional quality and nonalcoholic fatty liver disease: the role of vitamin D and B group. Am J Med Sci. 2019;358(3): 212-218. DOI: 10.1016/j.amjms.2019.06.008

[13] Kibirige D, Mwebaze R. Vitamin B12 deficiency among patients with diabetes mellitus: is routine screening and

supplementation justified?. Journal of Diabetes & Metabolic Disorders. 2013; 12(1):1-6. DOI: 10.1186/2251-6581-12-17

[14] Marcuard SP, Albernaz L, Khazanie PG. Omeprazole therapy causes malabsorption of cyanocobalamin (vitamin B12). Annals of Internal Medicine. 1994;120(3): 211-215. DOI: 10.7326/0003-4819-120-3-199402010-00006

[15] Mikkelsen K., Apostolopoulos V. Vitamin B1, B2, B3, B5, and B6 and the immune system. In: Mahmoudi M, Rezaei N, editors. Nutrition and Immunity. Cham: Springer; 2019. p. 115-125. DOI: 10.1007/978-3-030-16073-9_7

[16] Özdemir ZÖ, Şentürk M, Ekinci D. Inhibition of mammalian carbonic anhydrase isoforms I, II and VI with thiamine and thiamine-like molecules. Journal of Enzyme Inhibition and Medicinal Chemistry. 2013;28(2):316-319. DOI: 10.3109/14756366.2011.637200

[17] Editorial. Be well: A potential role for vitamin B in COVID-19. Maturitas. 2021;144:108-111. DOI: 10.1016/j.maturitas.2020.08.007

[18] Lucock M, Yates Z, Glanville T, Leeming R, Simpson N, Daskalakis I. A critical role for B-vitamin nutrition in human developmental and evolutionary biology. Nutrition Research. 2003;23 (11):1463-1475. DOI: 10.1016/S0271-5317(03)00156-8

[19] Ragan I, Hartson L, Pidcoke H, Bowen R,Goodrich R. Pathogen reduction of SARS-CoV-2 virus in plasma and whole blood using riboflavin and UV light. PLoSONE, 2020;15(5): e0233947. DOI: 10.1371/journal.pone.0233947

[20] Boergeling Y, Ludwig S. Targeting a metabolic pathway to fight the flu. The FEBS Journal. 2017:284(2):218-221. DOI: 10.1111/febs.13997

[21] Mikkelsen K, Apostolopoulos V. B Vitamins and ageing. In: Harris J, Korolchuk V, editors. Biochemistry and Cell Biology of Ageing: Part I Biomedical Science. Subcellular Biochemistry. Vol 90. Singapore:Springer; 2018. p. 451-470. DOI: 10.1007/978-981-13-2835-0_15

[22] Mikkelsen K, Stojanovska L, Prakash M, Apostolopoulos V. The effects of vitamin B on the immune/cytokine network and their involvement in depression. Maturitas. 2017;96:58-71. DOI: 10.1016/j.maturitas.2016.11.012

[23] Nagai A, Matsumiya H, Hayashi M, Yasui S, Okamoto H, Konno K. Effects of nicotinamide and niacin on bleomycin-induced acute injury and subsequent fibrosis in hamster lungs. Experimental Lung Research. 1994;20 (4):263-281. DOI: 10.3109/01902149409064387

[24] Zhang L, Liu Y. Potential interventions for novel coronavirus in China: A systematic review. Journal of Medical Virology. 2020;92(5):479-490. DOI: 10.1002/jmv.25707

[25] Mehmel M, Jovanović N, Spitz U. Nicotinamide riboside—the current state of research and therapeutic uses. Nutrients. 2020;12(6):1616. DOI: 10.3390/nu12061616

[26] Merigliano C, Mascolo E, Burla R, Saggio I, Vernì F. The relationship between vitamin B6, diabetes and cancer. Frontiers in Genetics. 2018;9: 388. DOI: 10.3389/fgene.2018.00388

[27] Lengyel CO, Whiting SJ, Zello GA. Nutrient inadequacies among elderly residents of long-term care facilities. Can J Diet Pract Res. 2008;69(2):82-88. DOI: 10.3148/69.2.2008.82

[28] Nix WA, Zirwes R, Bangert V, Kaiser RP, Schilling M, Hostalek U, Obeid R. Vitamin B status in patients with type 2 diabetes mellitus with and

without incipient nephropathy. Diabetes Research and Clinical Practice. 2015;107(1):157-165. DOI: 10.1016/j.diabres.2014.09.058

[29] Van Wyk V, Luus HG, du P. Heyns A. The in vivo effect in humans of pyridoxal-5′-phosphate on platelet function and blood coagulation. Thrombosis Research. 1992;66(6):657-668. DOI: 10.1016/0049-3848(92)90042-9

[30] Mikkelsen K, Prakash MD, Kuol N, Nurgali K, Stojanovska L, Apostolopoulos V. Anti-tumor effects of vitamin B2, B6 and B9 in promonocytic lymphoma cells. International Journal of Molecular Sciences. 2019;20(15):3763. DOI: 10.3390/ijms20153763

[31] Bistas KG, Tadi P. Biotin [Internet]. 2021. Available from: https://www.ncbi.nlm.nih.gov/books/NBK554493/

[32] Sheybani Z, Dokoohaki MH, Negahdaripour M, Dehdashti M, Zolghadr H, Moghadami M, Masoompour SM, Zolghadr AR. The role of folic acid in the management of respiratory disease caused by COVID-19. ChemRxiv. Cambridge: Cambridge Open Engage. DOI: 10.26434/chemrxiv.12034980.v1

[33] Alpaslan D, Dudu TE, Kubilay Ş, Aktaş N. Synthesis and characterization of biocompatible poly (maleic acid-co-citric acid) microparticles as a smart carrier for thiamine. Polymer Bulletin. 2020;1-16. DOI: 10.1007/s00289-020-03405-y

[34] Fathima SJ, Fathima I, Abhishek V, Khanum F. Phosphatidylcholine, an edible carrier for nanoencapsulation of unstable thiamine. Food Chemistry. 2016;197, 562-570. DOI: 10.1016/j.foodchem.2015.11.005

[35] Djoullah A, Saurel R. Controlled release of riboflavin encapsulated in pea protein microparticles prepared by emulsion-enzymatic gelation process. Journal of Food Engineering. 2021;292:110276. DOI: 10.1016/j.jfoodeng.2020.110276

[36] Uluata S, McClements DJ, Decker EA. Riboflavin-induced oxidation in fish oil-in-water emulsions: Impact of particle size and optical transparency. Food Chemistry. 2016;213:457-461. DOI: 10.1016/j.foodchem.2016.06.103

[37] Ahmad I, Arsalan A, Ali SA, Sheraz MA, Ahmed S, Anwar Z, Munir I, Shah MR. Formulation and stabilization of riboflavin in liposomal preparations. Journal of Photochemistry and Photobiology B: Biology. 2015;153:358-366. DOI: 10.1016/j.jphotobiol.2015.10.017

[38] Sharma A, Khatchadourian A, Khanna K, Sharma R, Kakkar A, Maysinger D. Multivalent niacin nanoconjugates for delivery to cytoplasmic lipid droplets. Biomaterials. 2011;32(5):1419-1429. DOI: 10.1016/j.biomaterials.2010.10.025

[39] Maravajhala V, Dasari N, Sepur A, Joginapalli S. Design and evaluation of niacin microspheres. Indian Journal of Pharmaceutical Sciences. 2009;71(6):663-669. DOI: 10.4103/0250-474X.59549

[40] Vidyavathi M, Papishetty S. Design and evaluation of niacin microspheres with a biopolymer. Trends in Biomaterials & Artificial Organs. 2013;27(2):81-87.

[41] Tashtoush BM, Bennamani AN, AL-Taani, BM. Preparation and characterization of microemulsion formulations of nicotinic acid and its prodrugs for transdermal delivery, Pharmaceutical Development and Technology. 2013;18(4):834-843. DOI: 10.3109/10837450.2012.727003

[42] Ota A, Istenič K, Skrt M, Šegatin N, Žnidaršič N, Kogej K, Ulrih NP.

Encapsulation of pantothenic acid into liposomes and into alginate or alginate–pectin microparticles loaded with liposomes. Journal of Food Engineering. 2018;229:21-31. DOI: 10.1016/j.jfoodeng.2017.06.036

[43] Fan L, Cai Z, Zhang K, Han F, Li J, He C, Mo X, Wang X, Wang, H. Green electrospun pantothenic acid/silk fibroin composite nanofibers: Fabrication, characterization and biological activity. Colloids and Surfaces B: Biointerfaces. 2014;117:14-20. DOI: 10.1016/j.colsurfb.2013.12.030

[44] Abd-El-Azim H, Ramadan A, Nafee N, Khalafallah N. Entrapment efficiency of pyridoxine hydrochloride in unilamellar liposomes: experimental versus model-generated data, Journal of Liposome Research. 2018;28(2):112-116. DOI: 10.1080/08982104.2016.1275679

[45] Chatterjee NS, Anandan R, Navitha M, Asha KK, Kumar KA, Mathew S, Ravishankar CN. Development of thiamine and pyridoxine loaded ferulic acid-grafted chitosan microspheres for dietary supplementation. Journal of Food Science and Technology. 2016;53(1):551-560. DOI: 10.1007/s13197-015-2044-4

[46] Vigneshwaran R, Sankar M, Vigneshkumar S. Synthesis and characterization of gellan gum-pyridoxine nanoparticle for treatment of localized peripheral neuropathy. International Journal of Engineering and Advanced Technology (IJEAT). 2019;9 (1S4):576-580. DOI:10.35940/ijeat.A1127.1291S4194

[47] Assadpour E, Maghsoudlou Y, Jafari SM, Ghorbani M, Aalami M. Optimization of folic acid nano-emulsification and encapsulation by maltodextrin-whey protein double emulsions. International Journal of Biological Macromolecules. 2016;86: 197-207. DOI: 10.1016/j.ijbiomac.2016.01.064

[48] Pamunuwa G, Anjalee N, Kukulewa D, Edirisinghe C, Shakoor F, Karunaratne DN. Tailoring of release properties of folic acid encapsulated nanoparticles via changing alginate and pectin composition in the matrix. Carbohydrate Polymer Technologies and Applications. 2020;1:100008. DOI: 10.1016/j.carpta.2020.100008

[49] Estevinho BN, Lazar R, Blaga A, Rocha F. Preliminary evaluation and studies on the preparation, characterization and in vitro release studies of different biopolymer microparticles for controlled release of folic acid. Powder Technology. 2020; 369:279-288. DOI: 10.1016/j.powtec.2020.05.048

[50] Coelho SC, Laget S, Benaut P, Rocha F, Estevinho BN. A new approach to the production of zein microstructures with vitamin B12, by electrospinning and spray drying techniques. Powder Technology. 2021; 392:47-57. DOI: 10.1016/j.powtec.2021.06.056

[51] Liu G, Yang J, Wang Y, Liu X, Chen L. Protein-lipid composite nanoparticles for the oral delivery of vitamin B12: Impact of protein succinylation on nanoparticle physicochemical and biological properties. Food Hydrocolloids. 2019;92: 189-197. DOI: 10.1016/j.foodhyd.2018.12.020

[52] Çoban Ö, Yıldırım S, Bakır T. Alpha-lipoic acid and cyanocobalamin co-loaded nanoemulsions: Development, characterization, and evaluation of stability. Journal of Pharmaceutical Innovation. 2021;1-11. DOI: 10.1007/s12247-020-09531-4

[53] Kamlow MA, Vadodaria S, Gholamipour-Shirazi A, Spyropoulos F, Mills T. 3D printing of edible hydrogels containing thiamine and their comparison to cast gels. Food Hydrocolloids. 2021;116:106550. DOI: 10.1016/j.foodhyd.2020.106550

[54] Goh WJ, Tan SX, Pastorin G, Ho PCL, Hu J, Lim SH. 3D printing of four-in-one oral polypill with multiple release profiles for personalized delivery of caffeine and vitamin B analogues. International Journal of Pharmaceutics. 2021;598:120360. DOI: 10.1016/j.ijpharm.2021.120360

[55] Rajakumari R, Oluwafemi OS, Thomas S, Kalarikkal N. Dietary supplements containing vitamins and minerals: formulation, optimization and evaluation. Powder Technology. 2018; 336:481-492. DOI: 10.1016/j.powtec.2018.06.026

[56] Taymouri S, Alem M, Varshosaz J, Rostami M, Akbari V, Firoozpour L. Biotin decorated sunitinib loaded nanostructured lipid carriers for tumor targeted chemotherapy of lung cancer. Journal of Drug Delivery Science and Technology. 2019;50:237-247. DOI: 10.1016/j.jddst.2019.01.024

[57] Cui Z, Qin L, Guo S, Cheng H, Zhang X, Guan J, Mao S. Design of biotin decorated enterocyte targeting muco-inert nanocomplexes for enhanced oral insulin delivery. Carbohydrate Polymers. 2021;261:117873. DOI: 10.1016/j.carbpol.2021.117873

[58] He Y, Gao Q, Lv C, Liu L. Improved photothermal therapy of brain cancer cells and photogeneration of reactive oxygen species by biotin conjugated gold photoactive nanoparticles. Journal of Photochemistry and Photobiology B: Biology. 2021;215:112102. DOI: 10.1016/j.jphotobiol.2020.112102

[59] Tang B, Peng Y, Yue Q, Pu Y, Li R, Zhao Y, Hai L, Guo L, Wu Y. Design, preparation and evaluation of different branched biotin modified liposomes for targeting breast cancer. European Journal of Medicinal Chemistry. 2020;193:112204. DOI: 10.1016/j.ejmech.2020.112204

[60] Wang Y, van Steenbergen MJ, Beztsinna N, Shi Y, Lammers T, van Nostrum CF, Hennink WE. Biotin-decorated all-HPMA polymeric micelles for paclitaxel delivery. Journal of Controlled Release. 2020;328:970-984. DOI: 10.1016/j.jconrel.2020.09.013

[61] Ren X, He L, Tian X, Zhang P, Chen Z, Mei X. pH and folic acid dual responsive polysaccharide nanospheres used for nuclear targeted cancer chemotherapy. Colloids and Surfaces B: Biointerfaces. 2019;178:445-451. DOI: 10.1016/j.colsurfb.2019.03.028

[62] Changizi O, Khoei S, Mahdavian A, Shirvalilou S, Mahdavi SR, Rad JK. Enhanced radiosensitivity of LNCaP prostate cancer cell line by goldphotoactive nanoparticles modified with folic acid. Photodiagnosis and Photodynamic Therapy. 2020;29: 101602. DOI: 10.1016/j.pdpdt.2019.101602

[63] Kumar CS, Thangam R, Mary SA, Kannan PR, Arun G, Madhan B. Targeted delivery and apoptosis induction of trans-resveratrol-ferulic acid loaded chitosan coated folic acid conjugate solid lipid nanoparticles in colon cancer cells. Carbohydrate Polymers. 2020;231:115682. DOI: 10.1016/j.carbpol.2019.115682

[64] Mohamad SA, Sarhan HA, Abdelkader H, Mansour HF. Vitamin B12-loaded buccoadhesive films as a non-invasive supplement in vitamin B12 deficiency: In vitro evaluation and in vivo comparative study with intramuscular injection. Journal of Pharmaceutical Sciences. 2017;106(7): 1849-1858. DOI: 10.1016/j.xphs.2017.03.040.

[65] Verma A, Sharma S, Gupta PK, Singh A, Teja BV, Dwivedi P, Gupta GK, Trivedi R, Mishra PR. Vitamin B12 functionalized layer by layer calcium phosphate nanoparticles: A mucoadhesive and pH responsive carrier for improved oral delivery of insulin. Acta Biomaterialia. 2016;31:

288-300. DOI: 10.1016/j.actbio.2015.12.017

[66] Singh A, Yadagiri G, Parvez S, Singh OP, Verma A, Sundar S, Mudavath SL. Formulation, characterization and in vitro anti-leishmanial evaluation of amphotericin B loaded solid lipid nanoparticles coated with vitamin B12-stearic acid conjugate. Materials Science and Engineering: C. 2020;117:111279. DOI: 10.1016/j.msec.2020.111279

[67] Eltony SA. Histological study on the protective role of vitamin B complex on the cerebellum of diabetic rat. Tissue and Cell. 2016;48(4):283-296. DOI: 10.1016/j.tice.2016.06.009

[68] Galvin R, Bråthen G, Ivashynka A, Hillbom M, Tanasescu R, Leone MA. EFNS guidelines for diagnosis, therapy and prevention of Wernicke encephalopathy. European Journal of Neurology. 2010;17(12):1408-1418. DOI: 10.1111/j.1468-1331.2010.03153.x

[69] Reynolds EH. Folic acid, ageing, depression, and dementia. BMJ. 2002; 324(7352):1512-1515. DOI: 10.1136/bmj.324.7352.1512

[70] Psara E, Pentieva K, Ward M, McNulty H. Critical review of nutrition, blood pressure and risk of hypertension through the lifecycle: do B vitamins play a role?. Biochimie. 2020;173:76-90. DOI: 10.1016/j.biochi.2020.03.016

[71] Omotoso GO, Abdulsalam FA, Mutholib NY, Bature AI, Gbadamosi IT. Cortico-hippocampal morphology and behavioural indices improved in maternal deprivation model of schizophrenia following vitamin B complex supplementation. Neurology, Psychiatry and Brain Research. 2020; 38: 74-82. DOI: 10.1016/j.npbr.2020.10.007

[72] Lauridsen C, Matte JJ, Lessard M, Celi P, Litta G. Role of vitamins for gastro-intestinal functionality and health of pigs. Animal Feed Science and Technology. 2021;273:114823. DOI: 10.1016/j.anifeedsci.2021.114823

[73] Simard F, Guay F, Girard CL, Giguère A, Laforest JP, Matte JJ. Effects of concentrations of cyanocobalamin in the gestation diet on some criteria of vitamin B-12 metabolism in first-parity sows. J. Anim. Sci. 2007;85(12): 3294-3302. DOI: 10.2527/jas.2006-523.

[74] Giguère A, Girard CL, Matte JJ. Methionine, folic acid and vitamin B12 in growing-finishing pigs: Impact on growth performance and meat quality. Archives of Animal Nutrition. 2008;62 (3):193-206. DOI: 10.1080/17450390802027494

[75] Audet I, Girard CL, Lessard M, Lo Verso L, Beaudoin F, Matte JJ. Homocysteine metabolism, growth performance, and immune responses in suckling and weanling piglets. Journal of Animal Science. 2015;93(1):147-157. DOI:10.2527/jas.2014-7872

[76] Ristilä M, Strid H, Eriksson LA, Strid Å, Sävenstrand H. The role of the pyridoxine (vitamin B6) biosynthesis enzyme PDX1 in ultraviolet-B radiation responses in plants. Plant Physiology and Biochemistry. 2011;49(3):284-292. DOI: 10.1016/j.plaphy.2011.01.003

Chapter 9

Vitamins D and B_{12}, Altered Synaptic Plasticity and Extracellular Matrix

Marcela Bermudez Echeverry, Silvia Honda Takada, Bruna Petrucelli Arruda, Debora Sterzeck Cardoso, Pamela Pinheiro Martins, Juliane Midori Ikebara, Aline V. Sousa-Santos and Victor R.C. Torres da Silva

Abstract

Brain plasticity is regulated through dynamic interactions between perineuronal nets, matrix metalloproteases (MMPs) and the extracellular matrix (ECM). Several studies have identified a crucial role for vitamins D and B_{12} in brain development and a deficiency in these vitamins may contribute to the emergence of cognitive deficits, as well as the onset of both autism spectrum disorder and schizophrenia. However, the mechanisms underlying the interplay between ECM, MMPs, vitamins and these neuropsychiatric conditions are poorly understood. In this chapter, we seek to understand how the risk of neurodegeneration in vulnerable individuals and the aetiology of specific neuropsychiatric disorders are affected by vitamin D and B_{12} deficiency, in conjunction with low levels of the antioxidant glutathione, impaired GABAergic inhibition, and alterations in the permanent ECM.

Keywords: vitamin deficiency, perineuronal nets, matrix metalloproteases, parvalbumin interneurons, GABA, neurodevelopment

1. Introduction

A proteoglycan-rich matrix, the perineuronal net (PNN) is a dense structure within the extracellular matrix (ECM), whose synapses form through gaps around many neuronal bodies and dendrites at a late stage in brain development. PNNs are formed at the end of a critical period of neurodevelopment, following the transformation of the central nervous system (CNS) from an environment conducive to neuronal growth and motility to one that is more restrictive, in response to several sensory inputs from both neurons and glia driving increased neuroplasticity [1]. The main components of the PNN matrix include several chondroitin sulfate proteoglycans (CSPGs), such as hyaluronan, link proteins, and tenascin-R and -C.

During mammalian development, hyaluronan binds to members of the lectican family originally produced in neurons, including versican V0 and V1 and neurocan [2], whereas aggrecan seems to be expressed by astroglial cells in the juvenile matrix [3]. Other lecticans include versican 2, brevican, phosphacan, tenascin-R and the link proteins HAPLN2/Bral1 and HAPLN4/Bral2 are only observed in more mature matrix environments approximately 2 weeks after birth [4–6], in contrast to the composition of the juvenile matrix. Following this period, shifts in brevican expression occur at the end of myelination, leading to white-matter precursor changes from an oligodendroglial to an astrocytic lineage [7], and resulting in a compact extracellular matrix forming the PPN [8].

PNNs have been observed 2–5 weeks after birth around parvalbumin (PV^+)-expressing GABAergic interneurons in pyramidal cortex, and around large motor neurons of the brainstem and spinal cord. This period coincides with the end of experience-dependent refinement of the synaptic network [8], but marked by a still critical period of matrix turnover and proteoglycan degradation by ADAMTS metalloendopeptidases and matrix metalloproteinases (MMPs) [8, 9]. PNN formations can also be observed in several distinct areas of the CNS, such as other regions of the cerebral cortex, the hippocampus (HPC), thalamus, and cerebellum [8].

PNNs in the adult CNS secrete hyaluronan through the action of membrane-bound HA synthase, an enzyme linked to the action of link proteins, lecticans, tenascin-R and chondroitin sulphate proteoglycans (CSPGs), creating supramolecular aggregates on the surface of neurons [1]. Other relevant glycoproteins besides CSPGs include Reelin, mainly secreted by Cajal–Retzius cells and involved in the control of neuronal migration and the establishment of cell aggregation and dendrite formation during the embryonic and early postnatal stages of development [10]. In adulthood, Reelin signalling is involved in the modulation of synaptic function and binds to very-low-density lipoprotein receptors and apolipoprotein E receptor 2 [11]. Increased clustering of Reelin receptors leads to a build-up of DAB1 proteins on the neuron membrane, greater activation of Src/SFK family kinases, and tyrosine phosphorylation of N-methyl-D-aspartate receptors (NMDARs), resulting in a net increase of receptor activity (**Figure 1**) [12]. Reelin insufficiency may lead to alterations in NMDAR clustering and LTP, such as in dysfunctional GABA-ergic transmission in the cerebral cortex and hippocampus observed among the morphofunctional signalling changes in schizophrenia (SZ) [12].

In fact, alterations of GABAergic signalling within a prenatal stress period have been identified as important factors in the development of SZ [13], autism spectrum disorder (ASD) [14], and epilepsy [15], often leading to an altered density of GABAergic cells and aberrant oscillatory activity. However, one functional model of brain development has proposed that prenatal stress involves DNA methylation, possibly inducing methylation of the gene responsible for Reelin promoter, with the consequent down-expression of Reelin resulting in abnormalities within the neuronal architecture of the prefrontal cortex, a reduction in dendritic complexity and a decreased number of GABAergic neurons, leading to altered developmental neuronal connectivity [13].

Animal studies have demonstrated that DNA methylation in the BDNF gene controls its expression during forebrain development in mice [16]. Furthermore, binding of BDNF and nerve growth factor (NGF) neurotrophins to their respective receptors (TrkB/A) triggers the PI-3kinase/AKT pathway, with activation of the mammalian target of rapamycin (mTOR) [17] and Akt-dependent inhibition of the serine/threonine kinase Gsk3β, resulting in decreased transcription of pro-inflammatory cytokines (IL-1β, TNF-α, and IL-6) [18]. Since Reelin/lipoprotein receptors do not contain a cytoplasmic kinase domain, the core Reelin signalling pathway seems to be associated with tyrosine kinase receptor (RTK or Trk) activity [19], the most likely coreceptor candidate [20]. As such, Reelin signalling from the ECM in collaboration with TrkB/A receptor activation, leads to increased phosphorylation of

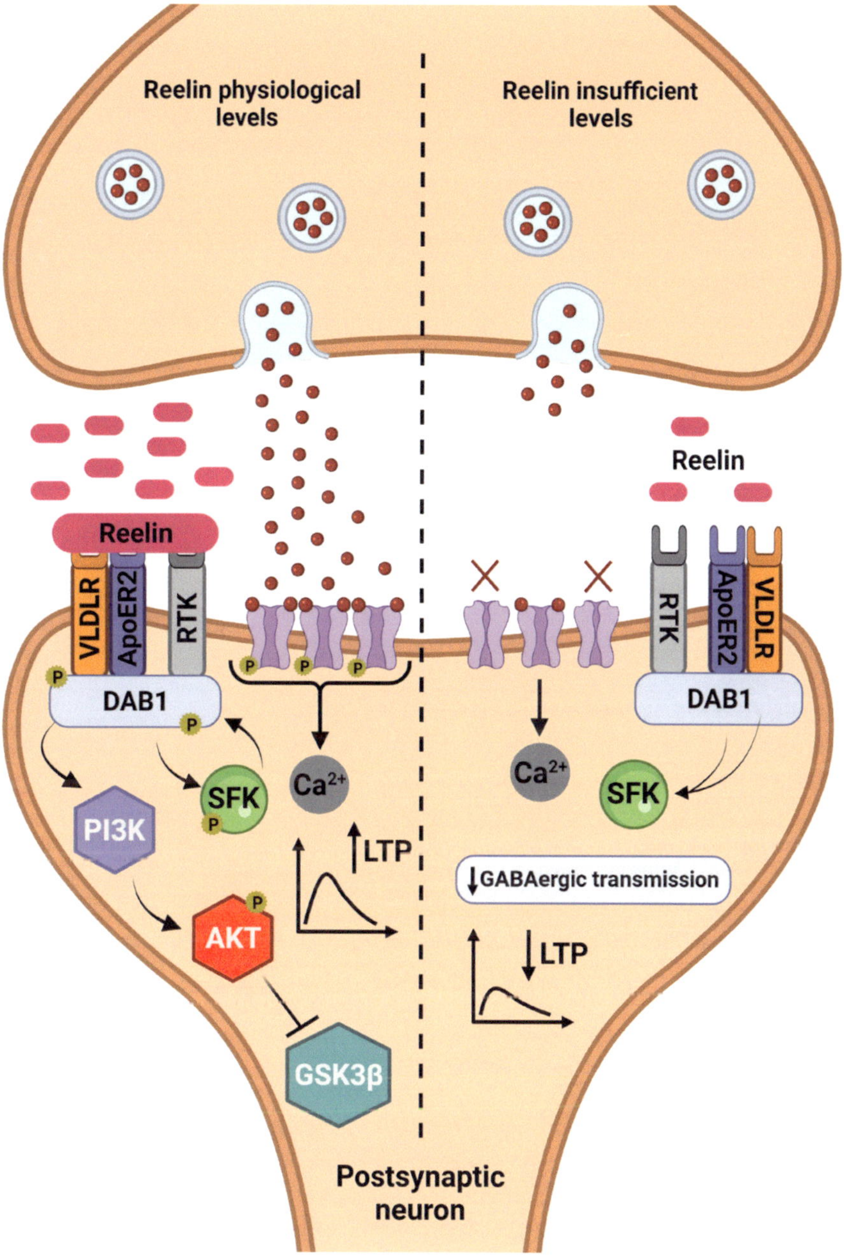

Figure 1.
Effect of Reelin concentrations in the ECM on NMDA signalling. Reelin activates adaptor protein disabled 1 (DAB1) by binding with its very-low-density lipoprotein receptor (VLDLR) and apolipoprotein E receptor type 2 (APOER2). DAB1 is phosphorylated by Src family kinases (SFK) at different sites on the protein - this phosphorylation occurs mainly through the action of a co-receptor, tyrosine kinase receptor (RTK or Trk), implicated in a variety of cellular processes including growth, differentiation, and regulation of energy metabolism in the neuron. DAB1 phosphorylation leads to inhibition of the serine/threonine kinase Gsk3β via protein kinase B (Akt), where a decrease in AKT phosphorylation levels with subsequent high levels of GSK-3β phosphorylation, has been observed in lymphocytes and brains of individuals with schizophrenia (SZ). Clustering of Reelin receptors via SKF activation also leads to greater tyrosine phosphorylation of N-methyl-D-aspartate receptors (NMDARs), resulting in a net increase of receptor activity following the induction of long-term potentiation via $Ca2^{+}$ regulation. This signalling cascade appears to be an essential process for neurobiological regulation during neurodevelopment (modified from [12]). Created with BioRender.

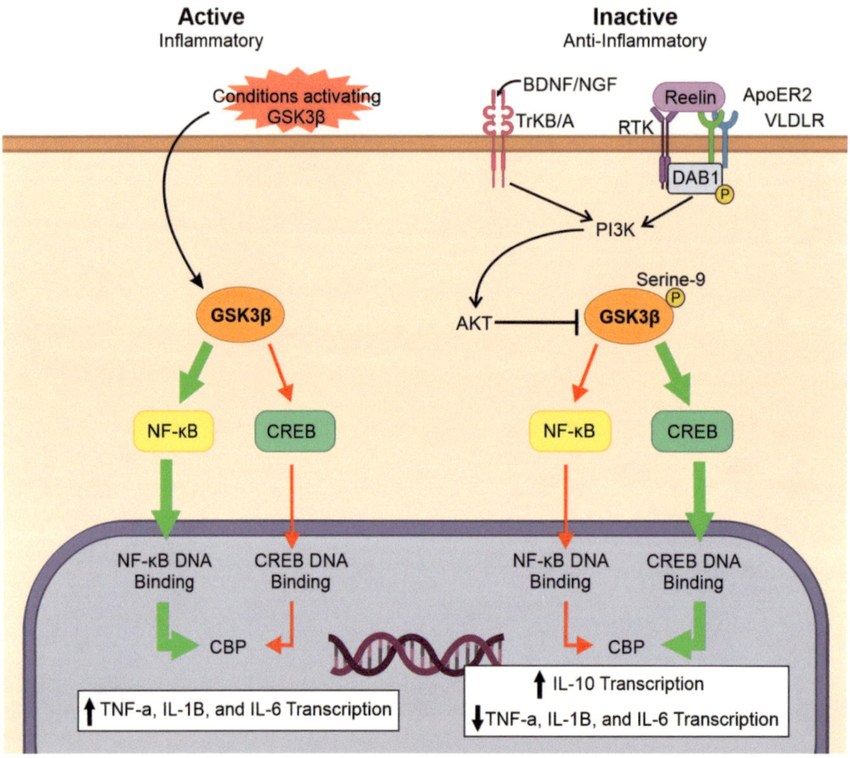

Figure 2.
Hypothetical signalling cascades for GSK3β modulation of the expression of pro-inflammatory and anti-inflammatory cytokines in glial cells. Receptor crosstalk between receptors TrkB/A and Reelin in the ECM (see ***Figure 1****), increase the phosphorylation of serine-9 GSK3β leading to GSK3β inhibition, an increase in the translocation of CREB from the cytoplasm to the nucleus, and an increase in the transcription of anti-inflammatory cytokine (IL-10). In fact, GSK3β can modulate the expression of both pro-inflammatory and anti-inflammatory cytokines. GSK3β activation in glial cells triggers the NF-κB pathway and the translocation of NF-κB from the cytoplasm to the nucleus, leading to an increase in the transcription of pro-inflammatory cytokines ((IL-1β, TNF-α, and IL-6), via the action of CREB-binding protein (CBP). Inhibition of GSK3β results in an increase of CREB translocation from the cytoplasm to the nucleus. Phosphorylated CREB binds specifically to the nuclear CBPs at transcriptional sites, resulting in increased transcription of anti-inflammatory cytokines such as IL-10 (modified from [21]).*

serine-9 GSK3β, with the inhibition of GSK3β in glial cells and leukocytes resulting in more CREB being translocated from the cytoplasm to the nucleus, and an increase in the transcription of anti-inflammatory cytokines (IL-10) (**Figure 2**) [21].

PNN development and the maturation of PV^+ inhibitory cells, as well as processes such as myelination, mark the end of the critical period of human neurodevelopment [22]. Disruption or delay to the formation of the PNN results in the resumption or extension of the time window for neuroplasticity in the brain [23], wherein the nervous system is more sensitive to epigenetic, physical, biochemical, environmental, and nutritional factors. The effects of nutrition on individuals during gestational and early development have been extensively researched, leading many researchers to conclude that nutritional factors such as vitamins, folate and iodine can cause long-lasting impacts in neurodevelopment [24, 25]. As the foetus' and newborn's acquisition of vitamins like B_{12} and D, depends to a great extent on maternal diet, such research has increasingly focussed on the impact of the mothers' vitamin deficiency on their offspring's brain development during the foetal and exclusive breastfeeding stages.

S-adenosyl methionine (SAM) is a universal methyl donor for some of the main methylation reactions. Vitamin B_{12} is an important cofactor in the one-carbon cycle and is involved in the formation of SAM. Vitamin B_{12} supplements have been shown to improve pregnancy outcomes and reduce the risk of neurodevelopmental disorders in the developing child [26]. In rats, dose-dependent vitamin B_{12} supplementation was able to maintain the levels of docosahexaenoic acid (DHA) and BDNF in the hippocampus and cortex in pups at birth, and BDNF in the hippocampus at 3 months of age [17]. In addition, the combination of omega-3 fatty acid and vitamin B_{12} administration maintained spatial memory performance in neonates [17]. Experimental evidence suggests that DHA, together with greater levels of physical exercise, increases activated forms of CREB and synapsin I, reducing oxidative stress in the hippocampus [18].

Vitamin D deficiency may also reduce the integrity of PNNs and synaptic plasticity in neuropsychiatric disorders through the modulation of MMPs. Vitamin D deficiency has been associated with vulnerability to SZ [27], as well as ASD [28] and attention deficit and hyperactivity disorder (ADHD) [29], the two most common neurodevelopmental disorders. As mentioned earlier, ADAMTS and MMPs are two families of endogenous zinc-dependent proteases, secreted as inactive proenzymes that cleave ECM components. Alterations in the genes that encode MMP-16 and MMP-9 have been observed in patients with SZ [30]. High levels of MMP-9 can support the proteolytic cleavage of ECM with permissive synaptic plasticity but also lead to abnormal aggrecan degradation, abnormal development and neural excitability [30]. Chronic stress and neurological trauma can enhance MMP-9 levels in the brain [31, 32], and consequently raise the risk of SZ. A plausible proposal has been made that vitamin D deficiency leads to PNN degradation in patients with SZ [27]. In fact, vitamin D

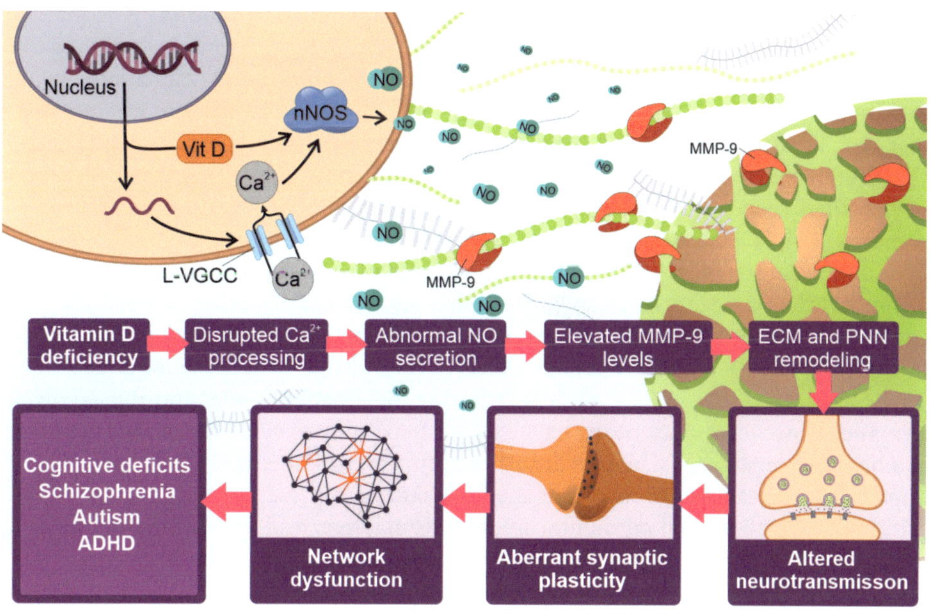

Figure 3.
Vitamin D deficiency and PNN formation during neurodevelopment. The figure above shows a neuron enveloped by a PNN. Vitamin D deficiency may induce a deficit in ECM organisation over the course of neurodevelopment, leading to the PNN loss and network-wide dysfunction in GABAergic, glutamatergic, and dopaminergic neurotransmission. Vitamin D deficiency is also linked to altered transcription of calcium channels (L-VGCC) potentially increasing the level of calcium input into the neuron, and to altered neuronal nitric oxide synthase (nNOS) activity, resulting in an increase of nitric oxide (NO) secretion into the extracellular space and elevated levels of MMP-9. This enhanced MMP-9 expression induces increased aggrecan synthesis, resulting in disruptions to the network of several neurotransmission systems important for normal cognitive function (spatial learning deficits), and SZ, autism and ADHD. Abbreviations: L-VGCC: L-type voltage-gated calcium channel; ECM: Extracellular matrix; MMP-9: Matrix metalloproteinase-9; nNOS: Neuronal nitric oxide synthase; NO: Nitric oxide; PNN: Perineuronal net; SZ: Schizophrenia; ADHD: Attention deficit and hyperactivity disorder (modified from [34]).

deficiency is associated with increased MMP-9 production [33] and calcium activity on the neuronal membrane, leading to increased nitric oxide (NO) formation and higher MMP-9 levels, and further appears to modulate its endogenous inhibitor TIMP1 (tissue inhibitor of MMP) [34]. Aggrecan-rich PNNs undergo restructuring leading to the occurrence of more synaptic anomalies and greater network dysfunction in GABAergic, glutamatergic, and dopaminergic neurotransmission, as evidenced by some forms of SZ (**Figure 3**) [34]. Cognitive deficits, such as spatial learning deficits, have been observed in adult mice with vitamin D deficiencies, with reduced density of PNNs and neural networks within the hippocampus [35].

2. Vitamin B_{12} in neurodevelopment

Vitamin B_{12} is a member of the cobalamin family and can usually be obtained in sufficient quantities from meat, eggs, and dairy products. It is critical for the production of red blood cells, and the growth and maintenance of the nervous system.

Major causes of vitamin B_{12} deficiency include autoimmune pernicious anaemia, gastrectomy, ileal resection, pancreatic insufficiency, and malabsorption syndromes [36]. The human body is incapable of endogenous B_{12} production so it must be obtained from external sources in the individual's diet [37, 38]. Currently, a nutritional B_{12} deficiency is common in vegans, a fast-growing eating trend in many Western countries [39]. Vitamin B_{12} deficiency is also common in developing countries, with more widespread occurrence over more widespread sections of society beginning in early life and persisting throughout adulthood [40].

Pregnant women require a greater amount of vitamin B_{12} (2.5 g/day) compared to the general adult population (2.4 g/day), to adequately meet the nutritional needs of the foetus via absorption through the placenta [41]. The lack of sufficient vitamin B_{12} ingestion by the mother during pregnancy results in its deficiency in breast milk and the foetal bloodstream [42, 43]. Indeed, several studies have linked maternal deficiency in vitamin B_{12} concentration to developmental complications, including spontaneous abortion [44], low birth weight [45, 46], intrauterine growth restriction [45], and neural tube defects [47]. Lack of sufficient vitamin B_{12} in the mother's diet is reflected in similarly low concentrations of B_{12} in the bloodstream of breastfed babies during the period of exclusive breastfeeding, when the baby is dependent on breast milk for vitamin absorption, despite the relatively short window of this development period [48].

As in adults, vitamin B_{12} is mainly stored in the newborn's liver and is used on demand. However, as newborns have limited hepatic reserves, even if they are born at a healthy weight and size, symptoms resulting from vitamin B_{12} deficiency may appear from as early as 2 months of age [49]. Such symptoms imply a clinical pattern including abnormal pigmentation, hypotonia, liver and spleen enlargement, anorexia and growth failure associated with poor brain growth, which were first described by Jadhav and colleagues in 1962 [50].

Cellular deficiency of vitamin B_{12} results in slower proliferation and a faster differentiation of neuroblastoma cells [51], which point to its pivotal role in cell division and differentiation. In addition, deficiency of vitamin B_{12} in rodents is associated with selective brain damage, vascular and cognitive impairment [52], long-lasting functional disabilities in exploratory behaviour, learning and memory functions, and a mild decrease in hippocampal neurogenesis [53]. Moreover, vitamin B_{12} seems to play an important role in myelination as its deficiency leads to demyelination and may cause severe retardation of myelination during prenatal development [54].

Different mechanisms have been proposed to account for the effects of vitamin B_{12} deficiency on general neural function, most especially in relation to neurodevelopment. The most well-understood mechanisms are related to the metabolic reactions in which vitamin B_{12} is the exclusive cofactor for mammalian cells [55], and the importance of the coenzyme forms of vitamin B_{12}, methylcobalamin (MeCbl) and adenosylcobalamin (AdoCbl) to the methionine and methyl groups that are used for DNA synthesis, epigenetic and cellular division.

When vitamin B_{12} is in MeCbl form it is a cofactor for methionine synthetase, which promotes the methylation of homocysteine (HCY) to methionine required in cases of a reduced methionine status of dietary ingestion. The synthesised form of methionine is subsequently condensed into SAM, which is finally demethylated into S-adenosylhomocysteine (SAH) and is a methyl donor for the conversion of phosphatidylethanolamine to phosphatidylcholine [56]. The altered SAM:SAH ratio may be the result of B_{12} deficiency as SAH and HCY levels increase while SAM levels decrease. This decreased SAM:SAH ratio may impair the methylation that is necessary for the synthesis of proteins, lipids, and neurotransmitters [57, 58] and leads to inhibition of DNA synthesis and cell division, since folate is not being recycled [59]. These results show that without methionine, the myelin and neurotransmitters considered essential for neurodevelopment cannot be produced. Methionine

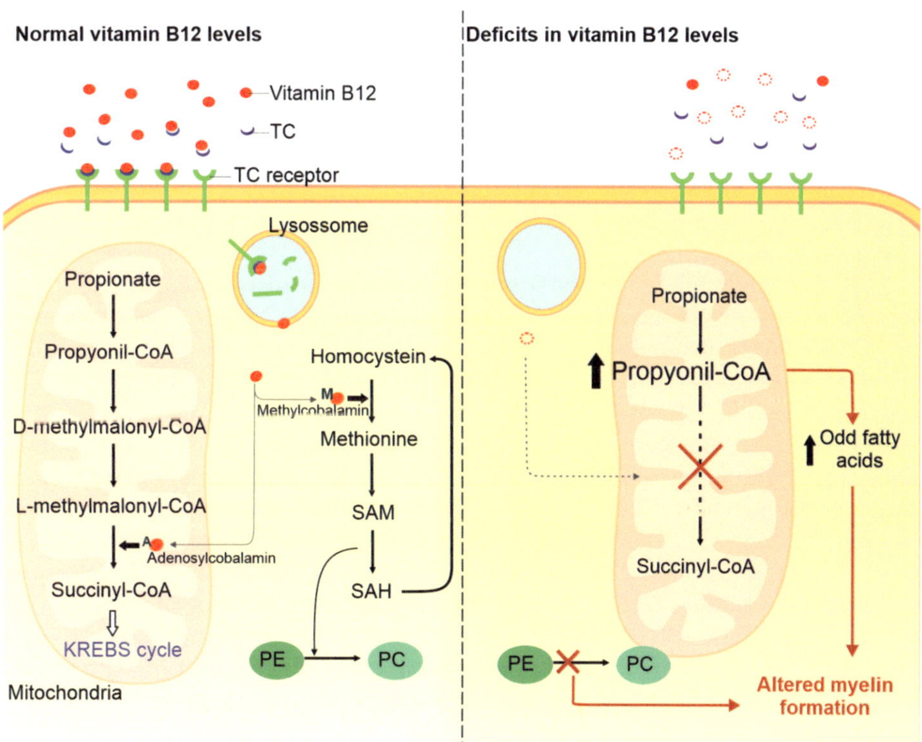

Figure 4.
Cellular processing of vitamin B in normal and deficiency conditions. Transcobalamin (TC) binds to vitamin B12 with high affinity and transports it by TC receptor-mediated endocytosis. The TC undergoes degradation into the lysosome, liberating the vitamin B12, that can be in two forms, methylcobalamin (MeCbl) and adenosylcobalamin (AdoCbl). When the vitamin B is in MeCbl form, it participates in HCY cascade as a cofactor, promoting the methylation of HCY to methionine that is subsequently condensed into SAM, which is finally demethylated into S-adenosylhomocysteine (SAH) and it a methyl donor for the conversion of phosphatidylethanolamine (PE) to phosphatidylcholine. The propionate (propionic acid) reacts with coenzyme A yielding propionyl CoA that is converted first to D-methylmalonyl-CoA and then to L-methylmalonyl-CoA and finally to Succinyl-CoA with the participation of AdoCbl form of vitamin B12 in this step. The deficiency of vitamin B12 leads to an inappropriate conversion of methylmalonyl CoA to succinyl CoA resulting in excessive production of the propionyl CoA, and consequently an increase of odd-chain fatty acid synthesis, and the inefficient conversion of PE to PC may impair propionyl CoA production by the mitochondria, resulting in an altered myelin formation.

is synthesised from homocysteine, increased levels of which are observed in many neurodegenerative diseases, and which are functionally linked to brain injury and cognitive impairment [60].

When vitamin B_{12} is found in adenosylcobalamin (AdoCbl) form, it is required as the cofactor of the enzyme L-methylmalonyl-CoA mutase (EC 5.4.99.2) in the mitochondria and promotes the conversion of methylmalonyl CoA to succinyl CoA. This pathway is important for the mitochondria to reuse propionyl CoA and the energy obtained through the Krebs cycle. In other instances, this pathway may be recruited in response to increased levels of SAM in which the elimination of HCY is necessary [61]. The inappropriate conversion of methylmalonyl CoA to succinyl CoA leads to excessive production of the precursor propionyl CoA, resulting in odd-chain fatty acid synthesis and subsequent incorporation of these abnormal fatty acids into the nerve sheaths, which leads to altered myelin formation, reduced quantities of ethanolamine, phospholipids, and sphingomyelin (**Figure 4**) [56]. In addition, the inefficient conversion of phosphatidylethanolamine to phosphatidylcholine may impair propionyl CoA production by the mitochondria and the myelination, or lead to demyelination [62]. Myelination, together with synaptic refinement and the physiological maturation of inhibitory neural networks, are key processes of brain development that seem to coincide with the appearance of PNNs. In fact, alterations in these processes are well described in disorders such as SZ and ASD [63, 64].

3. Vitamin D and neurodevelopment

Vitamin D is known as the "neglected neurosteroid", a term proposed by McGrath and colleagues in 2001 [65], and exerts a great variety of effects during brain development.

In 2005, a study conducted the first description of the distribution of 1,25-dihydroxyvitamin D3 receptors (VDR) and 1α-hydroxylase (1α-OHase), the enzyme responsible for the formation of the active vitamin D, in the human brain. They are found both in neurons and glial cells and show a region- and layer-specific pattern of expression in the brain. VDR receptors are absent in the macrocellular cells within the nucleus basalis of Meynert (NBM) and Purkinje cells in the cerebellum, while both VDRs and 1α-OHase have been identified in the substantia nigra and the hypothalamus [66]. VDRs can also be found in the developing brain during the critical period of cell proliferation, in the temporal lobe, cingulate, thalamus, cerebellum, amygdala and hippocampus [67].

Similarly to vitamin B_{12} deficiency, vitamin D deficiency during pregnancy appears to have serious consequences for foetal health. In the literature, abortions within the first trimester of gestation have been reported in association with the mothers' lack of sufficient vitamin D concentration [68–70]. Moreover, as foetus are dependent on access to the mother's supply of vitamin D, in conjunction with the large distribution of vitamin D receptors in the brain, there is a high probability that the vitamin D status of pregnant mothers may affect child neurocognitive development [67]. Among other sequelae, researchers have found cognitive and neural deficits in foetus and young infants due to maternal vitamin D deficiency during pregnancy, such as suboptimal neurocognitive development [71] and delays in the development of gross and fine motor function, problem-solving and communication [72]. However, a recent study analysed high-dosage vitamin D supplementation in the third trimester of pregnancy and its effects on child brain development and failed to report any improvements in the neurobiological mechanisms underlying developmental behaviour, such as motor milestones and cognitive and language development [73]. Few studies of the direct effects of vitamin D on brain development have been conducted to date in humans and many of them are inconclusive, probably because of differences in the timing of vitamin D exposure, the types of assessment used, and the age at which the assessment occurred.

Studies in rats have identified similar critical periods to vitamin B_{12} during neurodevelopment in which maternal vitamin D deficiency can result in neurodevelopmental alterations, such as abnormal cell proliferation and decreased expression of neuronal structure genes in the brain of mice offspring [74]. VDRs are temporally regulated in the rat brain during development, with the first VDR expression occurring in the mesencephalon on day 12 [75]. VDR expression continues across different areas throughout the course of development [75, 76] and is directly correlated to the onset of natural cell elimination [77]. Because of its key role in regulating developmental processes throughout the brain, altered levels of vitamin D or VDR dysfunction can result in long-lasting behavioural disruption in animal models [67]. Consistent with its role in PNN formation, altered vitamin D levels and VDR signalling during development have been associated with neuropsychiatric disorders such as SZ [78] and autism [79, 80].

The mechanisms by which vitamin D influences brain development are diverse. There is evidence that vitamin D modulates neurotrophic factors such as NGF and glial cell line-derived neurotrophic factor (GDNF), suggesting that it may have a neuroprotective function [81, 82]. Other neuroprotective functions of vitamin D include the reduction of the neurotoxicity of glutamate and reactive oxygen species (ROS), the upregulation of antioxidant molecules [83, 84], and the suppression of macrophage activity, leading to decreased neuroinflammation activity [85]. When used as an immune suppressant, vitamin D has been reported to suppress the concentration of proinflammatory cytokines in the brain [86, 87], to reduce blood–brain barrier disruption and macrophage/microglia activation in induced autoimmune encephalomyelitis [88], and to attenuate proinflammatory processes and up-regulate anti-inflammatory processes such as the M2 microglia phenotype, in a mouse model of Parkinson's disease [89].

Vitamin D is known to reduce calcium levels in the brain, preventing cell death via excitotoxicity, and to downregulate or modulate L-type voltage-gated calcium channels (L-VGCCs) [90] (**Figure 5**). L-VGCCs are expressed in great quantity in the developing brain and have a critical role in synaptic plasticity and in regulating basal and burst firing activity in dopaminergic neurons within the ventral tegmental area [91]. Interestingly, disruption of L-VGCC function in hippocampal PV^+ interneurons during development leads to significant morphological changes, such as reduced cell number and a decrease in dendritic arbour complexity [92, 93].

The association between vitamin D and L-VGCC function in the dopaminergic system and PV^+ interneurons strengthen the hypothesis that vitamin D (dys) regulation plays a role in the onset of SZ [94]. Both epidemiological research and rodent models have shown a correlation between vitamin D deficiency and SZ. Alterations in the dopaminergic system have been frequently reported in response to vitamin D deficiency, consistent with increased VDR expression in brain areas primarily innervated by dopaminergic projections [95].

Associations have also been made between vitamin D receptors, calcium channels, PV^+ interneurons and PNNs. L-VGCCs regulate PV^+ expression and interneuron development [92] which are significantly disrupted in SZ. Curiously, the appearance of PNNs coincides with the synaptic refinement, myelination and maturation of inhibitory networks, and there is therefore strong evidence that PNNs have a pivotal role in the pathogenesis of SZ [63].

Following this same reasoning, vitamin D deficiency is considered a potential candidate for the development of several key alterations in ASD pathophysiology, most of them present mid to early development. John Cannell has been the main

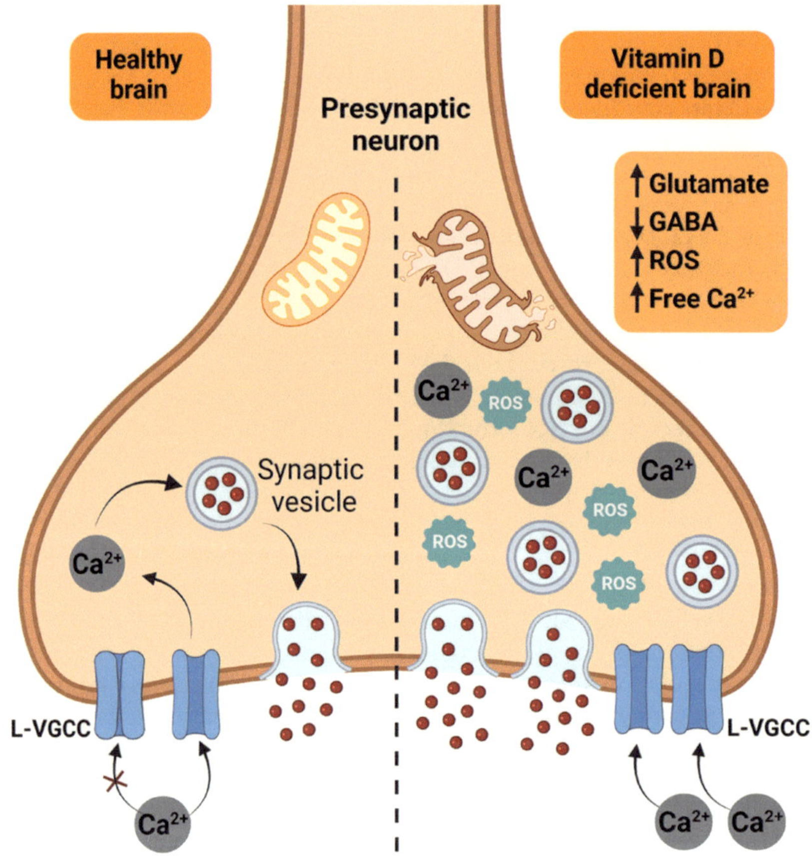

Figure 5.
Effects of vitamin D in presynaptic transmission Vitamin D plays an important role in several neurodevelopmental processes, such as neurotransmission and calcium signalling, preventing cell damage, especially mitochondrial dysfunction. As mentioned above, vitamin D acts by regulating L-VGCC in order that, for example, prevent the accumulation of intracellular calcium and the concomitant activation of pathways that can lead to cell death. With vitamin D deficiency, L-VGCC becomes functionality altered, leading to increased intracellular calcium, mitochondrial dysfunction, and the consequent generation of free radicals such as reactive oxygen species (ROS) and inhibitory/excitatory synaptic imbalance, illustrated by the increase in glutamate release and GABA neurotransmitter decrease, resulting in excitotoxicity and neuroinflammation (modified from [91]). Created with BioRender.

proponent of this hypothesis [96] and since its inception a number of studies have contributed to this line of research. The pathophysiology of ASD is multifactorial with solid evidence for both a genetic background and an environmental component. Some epidemiological findings have raised the additional hypothesis of specific genes which are environmentally responsive to the ASD, as epidemiological observations have pointed to changes in genotype expression. Cannell and others have cited neurosteroid pathway genes as good examples of environmentally responsive genes, since alterations in neurosteroid concentration may affect the genetic expression of the neural proteins regulated by steroids.

4. The link between vitamins B_{12} and D, PNNs and neurodevelopmental disorders: a mechanistic hypothesis

As already mentioned before, PNNs appear during postnatal development and surround cortical inhibitory GABA neurons expressing PV^+, which control the output of mainly cortical and hippocampal neurons and are necessary for fast rhythmic neuronal synchrony during information processing in cognitive tasks [97, 98]. PNNs enveloping PV^+-inhibitory interneurons are known to be vital for cognition [99]. These cortical PV^+-inhibitory interneurons express specific NMDA receptor (NMDAR) subunits such as NR2A, and are targets of the NMDA receptor antagonist MK-801. MK-801 is the main component for studying an important hypothesis for our understanding of the aetiology of SZ, the cortical hypofunction of NMDA receptors (i.e. the glutamatergic hypothesis) [100]. This hypothesis is compatible with the concurrent descriptions provided by the neurodevelopmental hypothesis, with respect to the disruptions to the central nervous system over the course of development [101].

An animal model using MK-801 treatment revealed a critical postnatal period, specifically from 7 to 14 days after birth (P 7–14), resulting in SZ-like behaviour during adulthood, with a significant reduction in PV^+-expressing cells in the PFC but not in the hippocampus, for mice treated with MK-801 from P 7–14 compared to matched controls [102]. In addition, mice in the treated group showed changes in performance on cognitive, social and behavioural tasks, while electrophysiological recordings from brain slices within the PFC showed significantly reduced frequency of up-states in MK-801-treated mice, but increased gamma activity in MK-801-treated mice compared to saline-treated mice [102]. Recent memory reactivation has been shown to occur in the presence of slow oscillatory up-states, contributing to memory consolidation [103]. This electrophysiological profile is altered in humans with SZ, with increased delta oscillations and irregular gamma rhythm [104, 105]. Moreover, PNN removal from visual cortex during a critical postnatal period alters the balance between excitatory and inhibitory PV^+ spiking activity, inducing greater potentiation of gamma activity and restoring the neural network to an immature or juvenile ECM state, suggesting that the maturation of GABAergic fast spiking and PV^+-inhibitory neurons suppresses the spontaneous activity of excitatory neurons [106].

Together with these electrophysiological findings, the metabolic requirements of PNN function have also been explored in fast-spiking cells, as well as their intrinsic vulnerability. In fact, PNNs appear to promote interneuron maturation and network stability and may also protect neurons against iron sequestration and oxidative stress [107, 108]. Oxidative stress has been observed in the PFC of SZ patients in conjunction with decreased levels of glutathione (GSH), an endogenous antioxidant and redox regulator [109]. This outcome can in turn be aggravated by the overexpression of truncated DISC1 that is associated with SZ in humans, causing mitochondrial dysfunction with decreased mitochondrial NADH dehydrogenase activity, diminished cellular ATP contents, and overactivity of mitochondrial $Ca2^+$ mechanisms [110].

As mentioned previously with regards to the role of GSH, there are two metabolically active forms of vitamin B_{12}, AdoCbl, and MeCbl: essential as a cofactor for the reaction necessary folate-dependent methylation of HCY by methionine synthase (MS) in the cytoplasm. As MS levels determine the ratio of methyl donor SAM to the endogenous methylation inhibitor SAH, the MeCbl reaction can influence SAM-dependent methylation reactions mainly through methylation of DNA and histones [111], an effect previously described with in relation to BDNF and epigenetic control. Zhang et al. 2016 analysed the postmortem human frontal cortex of autistic and schizophrenic individuals and 80-year-old individuals and found that the MeCbl form of vitamin B_{12} decreases with age, as well as to age of onset of ASD and SZ, as compared to age-matched controls. Additionally, they also observed an abnormally lower total cobalamin Cbl and MeCbl concentration in ASD and SZ subjects, leading the authors to propose a "Redox/Methylation Hypothesis of Autism" in light of the impaired GSH-dependent synthesis observed in the brains of autistic individuals [111]. In the same study, the authors further found that certain brain regions, as cerebellum and temporal cortex, in ASD, showed synthesis of reduced and oxidised glutathione (GSH/GSSG) (stable redox status), while other regions maintained SAM/SAH production (stable methylation status) [111], suggesting regional differences for metabolic disorders.

Concomitantly, several neurobiological changes found in ASD subjects or ASD animal models are consistent with an account of vitamin D deficiency. The main neurobiological processes mediated by vitamin D in neurodevelopment include neural cell proliferation [74], GABA, glutamate and serotonin neurotransmission [92], calcium signalling, mitochondrial regulation, oxidative stress and neuroinflammation (for review see [112]). Most of these processes are altered in ASD, resulting in the reduction of brain volume and changes in the number of glial and neuron cells, excitatory/inhibitory imbalance, disrupted calcium signalling, increased oxidative stress, the overactivation of microglia, and immune system dysregulation [113].

One of the best-studied hypotheses about the pathophysiology of ASD is the occurrence of changes in brain connectivity during development, which posits a reduced number of connections between distal brain regions and an increase in connections between proximal brain regions [114]. However, this hypothesis shows some inconsistencies in the light of several findings, such as reports of hyperconnectivity over long axonal fibres in autism or even mixed patterns of hypo- and hyper-connectivity [115–118].

Another hypothesis regarding the pathophysiology of ASD is one of excitatory-inhibitory imbalance, caused both by the probable hyperactivity of excitatory cells and a reduction in the number, activity or even delay of inhibitory interneurons in the maturation process [119–121].

PV^+-expressing interneurons are the main regulators of inhibitory/excitatory balance and orchestrate the coordinative function of brain microcircuitry via their fast-spiking inhibitory inputs onto pyramidal neurons [97, 122]. As such, the healthy maturation of PV^+-interneurons is crucial for the establishment of optimal neural and behavioural development. Disruption to PV^+ expression in PFC caused by early-life adversity leads to altered social interactions [123] and anxiety-related behaviour in rodents [124, 125]. Both forms of altered behaviour can be found in patients diagnosed with ASD [126–128].

As mentioned previously, the maturation of PV^+ interneurons are mostly regulated by PNNs [129, 130]. Despite the protective actions of the PNN-PV^+, the interneurons are very susceptible to oxidative stress [131, 132], and as such vitamin D deficiency represents a threat to the integrity of PNN function and consequently to the development of the GABAergic system.

Lastly, it is important to highlight the dysregulation of the immune system that is observed in ASD. ASD patients suffer from chronic systemic inflammation with a disbalance in cytokine expression, leading to increased production of proinflammatory cytokines. Vitamin D has been shown to contain immunomodulatory properties and may be an alternative treatment for slowing or minimising behavioural alterations in ASD (for review see [133]). Chronic systemic inflammation in early life may disrupt PV^+ interneuron maturation and consequently lead to changes in PNNs. Taken together, these findings reveal the important role of vitamin D in maintaining the integrity of PNNs and the efficiency of synaptic transmission as a whole.

5. Conclusion

The ECM is one component of the tetrapartite synapse, together with the network of glial cells. The PNN is a dense ECM structure that enwraps inhibitory fast-spiking parvalbumin (PV^+) interneurons, serving both as a protective barrier and to regulate synaptic plasticity. The destruction of those PNNs during the critical postnatal period of brain development alters the balance between excitatory and inhibitory PV^+ spiking activity, inducing a greater potentiation of gamma-band activity, and reverting the firing pattern of the neural network to a so-called immature or juvenile ECM state. Studies have shown that a vitamin D deficiency in early development may lead to a reduction in the PNN integrity and synaptic plasticity through modulation of MMPs, contributing to increased risk of the onset of neuropsychiatric disorders, as SZ, ASD, and ADHD. Also, in preclinical studies, dose-dependent vitamin B_{12} supplements were sufficient to maintain the levels of DHA and BDNF in the hippocampus and cortex in neonates, and BDNF levels in the hippocampus at 3 months of age, considered a sensitive window or critical period during neurodevelopment. In addition, the ingestion and metabolism of vitamin B_{12} methylcobalamin (MeCbl) variants can influence S- adenosyl methionine and SAM-dependent methylation reactions, mainly through the methylation of DNA and histones, and the MeCbl deficiency can result in impairment of GSH-dependent synthesis, inducing oxidative stress in the PFC of schizophrenic patients, or with autism diagnosis. Thus, adequate levels of vitamin B_{12} and D appear to contribute to maintaining the integrity of PNNs, consequently lead to PV^+ inter-neuron maturation.

Acknowledgements

The authors wish to thank IntechOpen for their support and payment of the Open Access Publishing Fee.

Conflict of interest

The authors declare no conflict of interest.

Author details

Marcela Bermudez Echeverry[1,2]*, Silvia Honda Takada[1], Bruna Petrucelli Arruda[1], Debora Sterzeck Cardoso[1], Pamela Pinheiro Martins[1], Juliane Midori Ikebara[1], Aline V. Sousa-Santos[1] and Victor R.C. Torres da Silva[1]

1 Universidade Federal do ABC (UFABC), Centro da Matemática, Computação e Cognição, São Bernardo do Campo, SP, Brazil

2 Neuroscience Laboratory, School of Medicine, University of Santander (UDES), Bucaramanga, Santander, Colombia

*Address all correspondence to: marcela.echeverry@ufabc.edu.br

References

[1] Tsien, R.Y., *Very long-term memories may be stored in the pattern of holes in the perineuronal net.* Proc Natl Acad Sci U S A, 2013. **110**(30): p. 12456-12461.

[2] Engel, M., et al., *Chondroitin sulfate proteoglycans in the developing central nervous system. I. cellular sites of synthesis of neurocan and phosphacan.* J Comp Neurol, 1996. **366**(1): p. 34-43.

[3] Domowicz, M.S., et al., *Aggrecan is expressed by embryonic brain glia and regulates astrocyte development.* Dev Biol, 2008. **315**(1): p. 114-124.

[4] Bekku, Y., et al., *Molecular cloning of Bral2, a novel brain-specific link protein, and immunohistochemical colocalization with brevican in perineuronal nets.* Mol Cell Neurosci, 2003. **24**(1): p. 148-159.

[5] Hirakawa, S., et al., *The brain link protein-1 (BRAL1): cDNA cloning, genomic structure, and characterization as a novel link protein expressed in adult brain.* Biochem Biophys Res Commun, 2000. **276**(3): p. 982-989.

[6] Meyer-Puttlitz, B., et al., *Chondroitin sulfate and chondroitin/keratan sulfate proteoglycans of nervous tissue: developmental changes of neurocan and phosphacan.* J Neurochem, 1995. **65**(5): p. 2327-2337.

[7] Ogawa, H., et al., *Lp3/Hapln3, a novel link protein that co-localizes with versican and is coordinately up-regulated by platelet-derived growth factor in arterial smooth muscle cells.* Matrix Biol, 2004. **23**(5): p. 287-298.

[8] Zimmermann, D.R. and M.T. Dours-Zimmermann, *Extracellular matrix of the central nervous system: from neglect to challenge.* Histochem Cell Biol, 2008. **130**(4): p. 635-653.

[9] Flannery, C.R., *MMPs and ADAMTSs: functional studies.* Front Biosci, 2006. **11**: p. 544-569.

[10] Kohno, T., et al., *Importance of Reelin C-terminal region in the development and maintenance of the postnatal cerebral cortex and its regulation by specific proteolysis.* J Neurosci, 2015. **35**(11): p. 4776-4787.

[11] Ishii, K., K.I. Kubo, and K. Nakajima, *Reelin and Neuropsychiatric Disorders.* Front Cell Neurosci, 2016. **10**: p. 229.

[12] Dityatev, A., M. Schachner, and P. Sonderegger, *The dual role of the extracellular matrix in synaptic plasticity and homeostasis.* Nat Rev Neurosci, 2010. **11**(11): p. 735-746.

[13] Negron-Oyarzo, I., et al., *Schizophrenia and reelin: a model based on prenatal stress to study epigenetics, brain development and behavior.* Biol Res, 2016. **49**: p. 16.

[14] Pizzarelli, R. and E. Cherubini, *Alterations of GABAergic signaling in autism spectrum disorders.* Neural Plast, 2011. **2011**: p. 297153.

[15] Trevino, M., C. Vivar, and R. Gutierrez, *Beta/gamma oscillatory activity in the CA3 hippocampal area is depressed by aberrant GABAergic transmission from the dentate gyrus after seizures.* J Neurosci, 2007. **27**(1): p. 251-259.

[16] Dennis, K.E. and P. Levitt, *Regional expression of brain derived neurotrophic factor (BDNF) is correlated with dynamic patterns of promoter methylation in the developing mouse forebrain.* Brain Res Mol Brain Res, 2005. **140**(1-2): p. 1-9.

[17] Rathod, R., et al., *Maternal omega-3 fatty acid supplementation on vitamin B12 rich diet improves brain omega-3 fatty acids, neurotrophins and cognition in the Wistar rat offspring.* Brain Dev, 2014. **36**(10): p. 853-863.

[18] Wu, A., Z. Ying, and F. Gomez-Pinilla, *Docosahexaenoic acid dietary supplementation enhances the effects of exercise on synaptic plasticity and cognition.* Neuroscience, 2008. **155**(3): p. 751-759.

[19] Cooper, J.A. and B.W. Howell, *Lipoprotein receptors: signaling functions in the brain?* Cell, 1999. **97**(6): p. 671-674.

[20] Bock, H.H. and P. May, *Canonical and Non-canonical Reelin Signaling.* Front Cell Neurosci, 2016. **10**: p. 166.

[21] Maixner, D.W. and H.R. Weng, *The Role of Glycogen Synthase Kinase 3 Beta in Neuroinflammation and Pain.* J Pharm Pharmacol (Los Angel), 2013. **1**(1): p. 001.

[22] Takesian, A.E. and T.K. Hensch, *Balancing plasticity/stability across brain development.* Prog Brain Res, 2013. **207**: p. 3-34.

[23] Beurdeley, M., et al., *Otx2 binding to perineuronal nets persistently regulates plasticity in the mature visual cortex.* J Neurosci, 2012. **32**(27): p. 9429-9437.

[24] Colombo, J., K.M. Gustafson, and S.E. Carlson, *Critical and Sensitive Periods in Development and Nutrition.* Ann Nutr Metab, 2019. 75 **Suppl 1**: p. 34-42.

[25] Scott, J.A., *The first 1000 days: A critical period of nutritional opportunity and vulnerability.* Nutr Diet, 2020. **77**(3): p. 295-297.

[26] Pawlak, R., et al., *How prevalent is vitamin B(12) deficiency among vegetarians?* Nutr Rev, 2013. **71**(2): p. 110-117.

[27] Chiang, M., R. Natarajan, and X. Fan, *Vitamin D in schizophrenia: a clinical review.* Evid Based Ment Health, 2016. **19**(1): p. 6-9.

[28] Fernell, E., et al., *Autism spectrum disorder and low vitamin D at birth: a sibling control study.* Mol Autism, 2015. **6**: p. 3.

[29] Sharif, M.R., et al., *The Relationship between Serum Vitamin D Level and Attention Deficit Hyperactivity Disorder.* Iran J Child Neurol, 2015. **9**(4): p. 48-53.

[30] Wen, T.H., et al., *Genetic Reduction of Matrix Metalloproteinase-9 Promotes Formation of Perineuronal Nets Around Parvalbumin-Expressing Interneurons and Normalizes Auditory Cortex Responses in Developing Fmr1 Knock-Out Mice.* Cereb Cortex, 2018. **28**(11): p. 3951-3964.

[31] Meyer-Lindenberg, A. and H. Tost, *Neural mechanisms of social risk for psychiatric disorders.* Nat Neurosci, 2012. **15**(5): p. 663-668.

[32] Zhang, H., et al., *Matrix metalloproteinases and neurotrauma: evolving roles in injury and reparative processes.* Neuroscientist, 2010. **16**(2): p. 156-170.

[33] Kim, S.H., et al., *Vitamin D Inhibits Expression and Activity of Matrix Metalloproteinase in Human Lung Fibroblasts (HFL-1) Cells.* Tuberc Respir Dis (Seoul), 2014. 77(2): p. 73-80.

[34] Mayne, P.E. and T.H.J. Burne, *Vitamin D in Synaptic Plasticity, Cognitive Function, and Neuropsychiatric Illness.* Trends Neurosci, 2019. **42**(4): p. 293-306.

[35] Al-Amin, M.M., et al., *Adult vitamin D deficiency disrupts hippocampal-dependent learning and structural brain connectivity in BALB/c mice.* Brain Struct Funct, 2019. **224**(3): p. 1315-1329.

[36] Watanabe, F., et al., *Biologically active vitamin B12 compounds in foods for preventing deficiency among vegetarians and elderly subjects.* J Agric Food Chem, 2013. **61**(28): p. 6769-6775.

[37] Anonymous, *Dietary reference values for food energy and nutrients for the*

United Kingdom. Report of the Panel on Dietary Reference Values of the Committee on Medical Aspects of Food Policy. Rep Health Soc Subj (Lond), 1991. **41**: p. 1-210.

[38] FAO/WHO, *Evaluation of certain veterinary drug residues in food. Thirty-second report of the Joint FAO/WHO Expert Committee on Food Additives.* World Health Organ Tech Rep Ser, 1988. **763**: p. 1-40.

[39] Simcikas, S.J.A.C.E., *Is the percentage of vegetarians and vegans in the US increasing.* 2018.

[40] Allen, L.H., *Causes of vitamin B12 and folate deficiency.* Food Nutr Bull, 2008. **29**(2 Suppl): p. S20-34; discussion S35-7.

[41] Graber, S.E., et al., *Placental transport of vitamin B12 in the pregnant rat.* J Clin Invest, 1971. **50**(5): p. 1000-1004.

[42] Baker, S.J. and H.J. Sinn, *Megaloblastic anaemia of infancy; report of a case.* Med J Aust, 1952. **1**(22): p. 750-752.

[43] Deegan, K.L., et al., *Breast milk vitamin B-12 concentrations in Guatemalan women are correlated with maternal but not infant vitamin B-12 status at 12 months postpartum.* J Nutr, 2012. **142**(1): p. 112-116.

[44] Kharb, S., et al., *Prospective study on role of folic acid and vitamin B12 in early pregnancy and spontaneous abortion.* 2018. **2**(4): p. 265-268.

[45] Muthayya, S., et al., *Low maternal vitamin B12 status is associated with intrauterine growth retardation in urban South Indians.* Eur J Clin Nutr, 2006. **60**(6): p. 791-801.

[46] Yajnik, C.S., et al., *Maternal total homocysteine concentration and neonatal size in India.* Asia Pac J Clin Nutr, 2005. **14**(2): p. 179-181.

[47] Molloy, A.M., et al., *Maternal vitamin B12 status and risk of neural tube defects in a population with high neural tube defect prevalence and no folic Acid fortification.* Pediatrics, 2009. **123**(3): p. 917-923.

[48] Rosenblatt, D.S. and V.M. Whitehead, *Cobalamin and folate deficiency: acquired and hereditary disorders in children.* Semin Hematol, 1999. **36**(1): p. 19-34.

[49] Michaud, J.L., et al., *Nutritional vitamin B12 deficiency: two cases detected by routine newborn urinary screening.* Eur J Pediatr, 1992. **151**(3): p. 218-220.

[50] Jadhav, M., et al., *Vitamin B12 deficiency in Indian infants. A clinical syndrome.* Lancet, 1962. **2**(7262): p. 903-907.

[51] Battaglia-Hsu, S.F., et al., *Vitamin B12 deficiency reduces proliferation and promotes differentiation of neuroblastoma cells and up-regulates PP2A, proNGF, and TACE.* Proc Natl Acad Sci U S A, 2009. **106**(51): p. 21930-21935.

[52] Troen, A.M., et al., *B-vitamin deficiency causes hyperhomocysteinemia and vascular cognitive impairment in mice.* Proc Natl Acad Sci U S A, 2008. **105**(34): p. 12474-12479.

[53] Daval, J.L., S. Blaise, and J.L. Gueant, *Vitamin B deficiency causes neural cell loss and cognitive impairment in the developing rat.* Proc Natl Acad Sci U S A, 2009. **106**(1): p. E1; author reply E2.

[54] Lovblad, K., et al., *Retardation of myelination due to dietary vitamin B12 deficiency: cranial MRI findings.* Pediatr Radiol, 1997. **27**(2): p. 155-158.

[55] Pons, L., et al., *Transcytosis and coenzymatic conversion of [(57)Co] cobalamin bound to either endogenous transcobalamin II or exogenous intrinsic factor in caco-2 cells.* Cell Physiol Biochem, 2000. **10**(3): p. 135-148.

[56] Stollhoff, K. and F.J. Schulte, *Vitamin B12 and brain development.* Eur J Pediatr, 1987. **146**(2): p. 201-205.

[57] Briddon, A., *Homocysteine in the context of cobalamin metabolism and deficiency states.* Amino Acids, 2003. **24**(1-2): p. 1-12.

[58] Weir, D.G., et al., *Correlation of the ratio of S-adenosyl-L-methionine to S-adenosyl-L-homocysteine in the brain and cerebrospinal fluid of the pig: implications for the determination of this methylation ratio in human brain.* Clin Sci (Lond), 1992. **82**(1): p. 93-97.

[59] Guerra-Shinohara, E.M., et al., *Low ratio of S-adenosylmethionine to S-adenosylhomocysteine is associated with vitamin deficiency in Brazilian pregnant women and newborns.* Am J Clin Nutr, 2004. **80**(5): p. 1312-1321.

[60] Lipton, S.A., et al., *Neurotoxicity associated with dual actions of homocysteine at the N-methyl-D-aspartate receptor.* Proc Natl Acad Sci U S A, 1997. **94**(11): p. 5923-5928.

[61] Hoffer, L.J., *Homocysteine remethylation and trans-sulfuration.* Metabolism, 2004. **53**(11): p. 1480-1483.

[62] Grattan-Smith, P.J., et al., *The neurological syndrome of infantile cobalamin deficiency: developmental regression and involuntary movements.* Mov Disord, 1997. **12**(1): p. 39-46.

[63] Bitanihirwe, B.K. and T.U. Woo, *Perineuronal nets and schizophrenia: the importance of neuronal coatings.* Neurosci Biobehav Rev, 2014. **45**: p. 85-99.

[64] Singh, V.K., et al., *Antibodies to myelin basic protein in children with autistic behavior.* Brain Behav Immun, 1993. **7**(1): p. 97-103.

[65] McGrath, J.J., et al., *Vitamin D insufficiency in south-east Queensland.* Med J Aust, 2001. **174**(3): p. 150-151.

[66] Eyles, D.W., et al., *Distribution of the vitamin D receptor and 1 alpha-hydroxylase in human brain.* J Chem Neuroanat, 2005. **29**(1): p. 21-30.

[67] Eyles, D.W., T.H. Burne, and J.J. McGrath, *Vitamin D, effects on brain development, adult brain function and the links between low levels of vitamin D and neuropsychiatric disease.* Front Neuroendocrinol, 2013. **34**(1): p. 47-64.

[68] Urrutia, R.P. and J.M. Thorp, *Vitamin D in pregnancy: current concepts.* Curr Opin Obstet Gynecol, 2012. **24**(2): p. 57-64.

[69] Andersen, L.B., et al., *Vitamin D depletion aggravates hypertension and target-organ damage.* J Am Heart Assoc, 2015. **4**(2).

[70] Li, N., et al., *Women with recurrent spontaneous abortion have decreased 25(OH) vitamin D and VDR at the fetal-maternal interface.* Braz J Med Biol Res, 2017. **50**(11): p. e6527.

[71] Darling, A.L., et al., *Association between maternal vitamin D status in pregnancy and neurodevelopmental outcomes in childhood: results from the Avon Longitudinal Study of Parents and Children (ALSPAC).* Br J Nutr, 2017. **117**(12): p. 1682-1692.

[72] Dhamayanti, M., et al., *Association of maternal vitamin D deficiency and infants' neurodevelopmental status: A cohort study on vitamin D and its impact during pregnancy and childhood in Indonesia.* J Paediatr Child Health, 2020. **56**(1): p. 16-21.

[73] Sass, L., et al., *High-Dose Vitamin D Supplementation in Pregnancy and Neurodevelopment in Childhood: A Prespecified Secondary Analysis of a Randomized Clinical Trial.* JAMA Netw Open, 2020. **3**(12): p. e2026018.

[74] Feron, F., et al., *Developmental Vitamin D3 deficiency alters the adult rat*

brain. Brain Res Bull, 2005. **65**(2): p. 141-148.

[75] Veenstra, T.D., et al., *Zinc-induced conformational changes in the DNA-binding domain of the vitamin D receptor determined by electrospray ionization mass spectrometry.* J Am Soc Mass Spectrom, 1998. **9**(1): p. 8-14.

[76] Cui, X., et al., *Maternal vitamin D depletion alters neurogenesis in the developing rat brain.* Int J Dev Neurosci, 2007. **25**(4): p. 227-232.

[77] Ko, P., et al., *Maternal vitamin D3 deprivation and the regulation of apoptosis and cell cycle during rat brain development.* Brain Res Dev Brain Res, 2004. **153**(1): p. 61-68.

[78] Schoenrock, S.A. and L.M. Tarantino, *Developmental vitamin D deficiency and schizophrenia: the role of animal models.* Genes Brain Behav, 2016. **15**(1): p. 45-61.

[79] Currenti, S.A., *Understanding and determining the etiology of autism.* Cell Mol Neurobiol, 2010. **30**(2): p. 161-171.

[80] Schmidt, R.J., et al., *Selected vitamin D metabolic gene variants and risk for autism spectrum disorder in the CHARGE Study.* Early Hum Dev, 2015. **91**(8): p. 483-489.

[81] Kalueff, A.V. and P. Tuohimaa, *Neurosteroid hormone vitamin D and its utility in clinical nutrition.* Curr Opin Clin Nutr Metab Care, 2007. **10**(1): p. 12-19.

[82] Brown, J., et al., *1,25-dihydroxyvitamin D3 induces nerve growth factor, promotes neurite outgrowth and inhibits mitosis in embryonic rat hippocampal neurons.* Neurosci Lett, 2003. **343**(2): p. 139-143.

[83] Garcion, E., et al., *1,25-dihydroxyvitamin D3 regulates the synthesis of gamma-glutamyl transpeptidase and glutathione levels in rat primary astrocytes.* J Neurochem, 1999. **73**(2): p. 859-866.

[84] Ibi, M., et al., *Protective effects of 1 alpha,25-(OH)(2)D(3) against the neurotoxicity of glutamate and reactive oxygen species in mesencephalic culture.* Neuropharmacology, 2001. **40**(6): p. 761-771.

[85] Garcion, E., et al., *Expression of inducible nitric oxide synthase during rat brain inflammation: regulation by 1,25-dihydroxyvitamin D3.* Glia, 1998. **22**(3): p. 282-294.

[86] Moore, D.D., et al., *International Union of Pharmacology. LXII. The NR1H and NR1I receptors: constitutive androstane receptor, pregnene X receptor, farnesoid X receptor alpha, farnesoid X receptor beta, liver X receptor alpha, liver X receptor beta, and vitamin D receptor.* Pharmacol Rev, 2006. **58**(4): p. 742-759.

[87] van Etten, E. and C. Mathieu, *Immunoregulation by 1,25-dihydroxyvitamin D3: basic concepts.* J Steroid Biochem Mol Biol, 2005. **97**(1-2): p. 93-101.

[88] de Oliveira, L.R.C., et al., *Calcitriol Prevents Neuroinflammation and Reduces Blood-Brain Barrier Disruption and Local Macrophage/Microglia Activation.* Front Pharmacol, 2020. **11**: p. 161.

[89] Cianciulli, A., et al., *Folic Acid Is Able to Polarize the Inflammatory Response in LPS Activated Microglia by Regulating Multiple Signaling Pathways.* Mediators Inflamm, 2016. **2016**: p. 5240127.

[90] Zanatta, L., et al., *1alpha,25-dihydroxyvitamin D(3) mechanism of action: modulation of L-type calcium channels leading to calcium uptake and intermediate filament phosphorylation in cerebral cortex of young rats.* Biochim Biophys Acta, 2012. **1823**(10): p. 1708-1719.

[91] Liu, Y., et al., *Cav1.2 and Cav1.3 L-type calcium channels regulate dopaminergic firing activity in the mouse ventral tegmental area.* J Neurophysiol, 2014. **112**(5): p. 1119-1130.

[92] Jiang, M. and J.W. Swann, *A role for L-type calcium channels in the maturation of parvalbumin-containing hippocampal interneurons.* Neuroscience, 2005. **135**(3): p. 839-850.

[93] Kasatkina, L.A., et al., *Vitamin D deficiency induces the excitation/inhibition brain imbalance and the proinflammatory shift.* Int J Biochem Cell Biol, 2020. **119**: p. 105665.

[94] Chung, G.E., et al., *The serum vitamin D level is inversely correlated with nonalcoholic fatty liver disease.* Clin Mol Hepatol, 2016. **22**(1): p. 146-151.

[95] Cui, X., et al., *The vitamin D receptor in dopamine neurons; its presence in human substantia nigra and its ontogenesis in rat midbrain.* Neuroscience, 2013. **236**: p. 77-87.

[96] Cannell, J.J., *Autism and vitamin D.* Med Hypotheses, 2008. **70**(4): p. 750-759.

[97] Sohal, V.S., et al., *Parvalbumin neurons and gamma rhythms enhance cortical circuit performance.* Nature, 2009. **459**(7247): p. 698-702.

[98] Whittington, M.A., et al., *Multiple origins of the cortical gamma rhythm.* Dev Neurobiol, 2011. **71**(1): p. 92-106.

[99] Murray, A.J., et al., *Parvalbumin-positive interneurons of the prefrontal cortex support working memory and cognitive flexibility.* Sci Rep, 2015. **5**: p. 16778.

[100] Nakazawa, K. and K. Sapkota, *The origin of NMDA receptor hypofunction in schizophrenia.* Pharmacol Ther, 2020. **205**: p. 107426.

[101] Owen, M.J., et al., *Neurodevelopmental hypothesis of schizophrenia.* Br J Psychiatry, 2011. **198**(3): p. 173-175.

[102] Plataki, M.E., et al., *Effect of Neonatal Treatment With the NMDA Receptor Antagonist, MK-801, During Different Temporal Windows of Postnatal Period in Adult Prefrontal Cortical and Hippocampal Function.* Front Behav Neurosci, 2021. **15**: p. 689193.

[103] Tatsuno, M., et al., *Memory reactivation in rat medial prefrontal cortex occurs in a subtype of cortical UP state during slow-wave sleep.* Philos Trans R Soc Lond B Biol Sci, 2020. **375**(1799): p. 20190227.

[104] Narayanan, B., et al., *Multivariate genetic determinants of EEG oscillations in schizophrenia and psychotic bipolar disorder from the BSNIP study.* Transl Psychiatry, 2015. **5**: p. e588.

[105] Uhlhaas, P.J. and W. Singer, *Oscillations and neuronal dynamics in schizophrenia: the search for basic symptoms and translational opportunities.* Biol Psychiatry, 2015. **77**(12): p. 1001-1009.

[106] Lensjo, K.K., et al., *Removal of Perineuronal Nets Unlocks Juvenile Plasticity Through Network Mechanisms of Decreased Inhibition and Increased Gamma Activity.* J Neurosci, 2017. **37**(5): p. 1269-1283.

[107] Morawski, M., et al., *Perineuronal nets potentially protect against oxidative stress.* Exp Neurol, 2004. **188**(2): p. 309-315.

[108] Suttkus, A., et al., *Neuroprotection against iron-induced cell death by perineuronal nets - an in vivo analysis of oxidative stress.* Am J Neurodegener Dis, 2012. **1**(2): p. 122-129.

[109] Do, K.Q., et al., *Schizophrenia: glutathione deficit in cerebrospinal fluid*

and prefrontal cortex in vivo. Eur J Neurosci, 2000. **12**(10): p. 3721-3728.

[110] Park, Y.U., et al., *Disrupted-in-schizophrenia 1 (DISC1) plays essential roles in mitochondria in collaboration with Mitofilin.* Proc Natl Acad Sci U S A, 2010. **107**(41): p. 17785-17790.

[111] Zhang, Y., et al., *Decreased Brain Levels of Vitamin B12 in Aging, Autism and Schizophrenia.* PLoS One, 2016. **11**(1): p. e0146797.

[112] Berridge, M.J., *Vitamin D, reactive oxygen species and calcium signalling in ageing and disease.* Philos Trans R Soc Lond B Biol Sci, 2016. **371**(1700).

[113] Siracusano, M., et al., *Vitamin D Deficiency and Autism Spectrum Disorder.* Curr Pharm Des, 2020. **26**(21): p. 2460-2474.

[114] Belmonte, M.K., et al., *Autism and abnormal development of brain connectivity.* J Neurosci, 2004. **24**(42): p. 9228-9231.

[115] Courchesne, E. and K. Pierce, *Brain overgrowth in autism during a critical time in development: implications for frontal pyramidal neuron and interneuron development and connectivity.* Int J Dev Neurosci, 2005. **23**(2-3): p. 153-170.

[116] Maximo, J.O., E.J. Cadena, and R.K. Kana, *The implications of brain connectivity in the neuropsychology of autism.* Neuropsychol Rev, 2014. **24**(1): p. 16-31.

[117] Mizuno, A., et al., *Partially enhanced thalamocortical functional connectivity in autism.* Brain Res, 2006. **1104**(1): p. 160-174.

[118] Shih, P., et al., *Atypical network connectivity for imitation in autism spectrum disorder.* Neuropsychologia, 2010. **48**(10): p. 2931-2939.

[119] Markicevic, M., et al., *Cortical Excitation:Inhibition Imbalance Causes Abnormal Brain Network Dynamics as Observed in Neurodevelopmental Disorders.* Cereb Cortex, 2020. **30**(9): p. 4922-4937.

[120] Lacaille, H., et al., *Impaired Interneuron Development in a Novel Model of Neonatal Brain Injury.* eNeuro, 2019. **6**(1).

[121] Trakoshis, S., et al., *Intrinsic excitation-inhibition imbalance affects medial prefrontal cortex differently in autistic men versus women.* Elife, 2020. **9**.

[122] Ferguson, B.R. and W.J. Gao, *PV Interneurons: Critical Regulators of E/I Balance for Prefrontal Cortex-Dependent Behavior and Psychiatric Disorders.* Front Neural Circuits, 2018. **12**: p. 37.

[123] Holland, F.H., et al., *Early life stress disrupts social behavior and prefrontal cortex parvalbumin interneurons at an earlier time-point in females than in males.* Neurosci Lett, 2014. **566**: p. 131-136.

[124] Lee, M.J., et al., *Investigation of anxiety and depressive disorders and psychiatric medication use before and after cancer diagnosis.* Psychooncology, 2021. **30**(6): p. 919-927.

[125] Page, C.E., et al., *Prefrontal parvalbumin cells are sensitive to stress and mediate anxiety-related behaviors in female mice.* Sci Rep, 2019. **9**(1): p. 19772.

[126] Gillis, R.F. and G.A. Rouleau, *The ongoing dissection of the genetic architecture of autistic spectrum disorder.* Mol Autism, 2011. **2**(1): p. 12.

[127] Coffey, B.J., et al., *Anxiety disorders and tic severity in juveniles with Tourette's disorder.* J Am Acad Child Adolesc Psychiatry, 2000. **39**(5): p. 562-568.

[128] Howlin, K.J., et al., *Evidence for electroneutral sodium chloride transport in rat proximal convoluted tubule.* Am J

Physiol, 1986. **250**(4 Pt 2): p. F644-F648.

[129] Hartig, W., K. Brauer, and G. Bruckner, *Wisteria floribunda agglutinin-labelled nets surround parvalbumin-containing neurons.* Neuroreport, 1992. **3**(10): p. 869-872.

[130] Enwright, J.F., et al., *Reduced Labeling of Parvalbumin Neurons and Perineuronal Nets in the Dorsolateral Prefrontal Cortex of Subjects with Schizophrenia.* Neuropsychopharmacology, 2016. **41**(9): p. 2206-2214.

[131] Cabungcal, J.H., et al., *Perineuronal nets protect fast-spiking interneurons against oxidative stress.* Proc Natl Acad Sci U S A, 2013. **110**(22): p. 9130-9135.

[132] Brenhouse, H.C. and J.M. Schwarz, *Immunoadolescence: Neuroimmune development and adolescent behavior.* Neurosci Biobehav Rev, 2016. **70**: p. 288-299.

[133] Vegelin, M., G. Teodorowicz, and H. Savelkoul, *Vitamin D and Autism Spectrum Disorder*. 2021.

Section 2

Novel Applications of B-Complex Vitamins

Chapter 10

Thiamin (B1) and Its Application in Patients with Critical Condition

Bastian Lubis, Putri Amelia, Aznan Lelo,
Muhammad Akil and Vincent Viandy

Abstract

Thiamin is an essential water-soluble nutrient that is naturally available in some foods and available as a supplement. This nutrient plays a vital role in metabolism, cell growth and development. The recommended daily intake of thiamin for adults is around 1.1–1.2 mg/day. Several studies have described that thiamin deficiency is commonly seen in critically ill patients, mainly sepsis. Thiamin deficiency reduces pyruvate access to the Krebs cycle, therefore, increases the production of lactate. The administration of thiamin in critically ill patients has been linked to better outcomes and depletion of mortality rate.

Keywords: thiamin, vitamin B1, sepsis, septic shock

1. Introduction

Thiamin is the first vitamin B discovered; thus, it is known as vitamin B1. This micronutrient is also known as aneurine. Thiamin is an essential micronutrient that is water-soluble and involved in aerobic metabolism. Humans' daily requirement of thiamin is highly dependent on food intake due to their inability to synthesise it endogenously. Some bacteria in the human intestine can produce thiamin. However, the amount is limited [1].

The chemical name for B1 is 3-[(4-amino-2-methyl-5-pyrimidine)methyl]-(2-hydroxy ethyl)-4-methylthiazolium. Thiamin originated from pyrimidine and thiazole ring substitution, combined with the methylene bridge (**Figure 1**) [2].

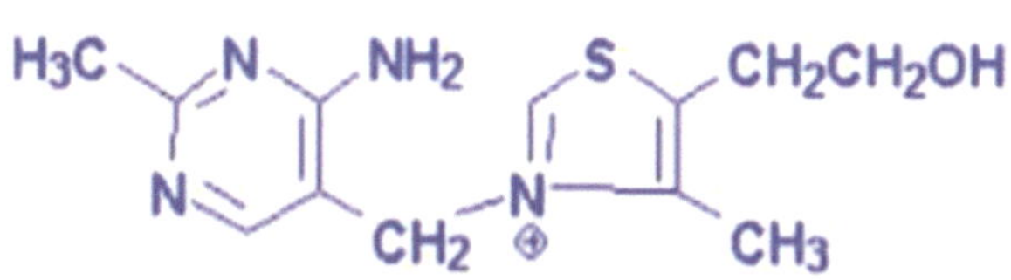

Figure 1.
Thiamin structure [2].

2. Pharmacokinetics and pharmacodynamics of thiamin

Thiamin works as a cofactor in citric acid cycles. Vitamin B1 reacts with adenosine triphosphate (ATP) to form an active form called thiamin pyrophosphate.

Thiamin is an essential cofactor for enzymes pyruvate dehydrogenase, alpha-ketoglutarate dehydrogenase, and transketolase. Pyruvate dehydrogenase enzymes are the main entrance to the Krebs cycle, catalysing oxidative decarboxylation of pyruvate to form acetyl-coenzyme A (acetyl-CoA). Without this enzyme, pyruvate would be converted to lactate. Alpha-ketoglutarate dehydrogenase catalyses the oxidative decarboxylation of alpha-ketoglutarate to succinyl-CoA to complete the Krebs cycle. Transketolase is an enzyme necessary for the pentose phosphate pathway and the production of nicotinamide adenine dinucleotide phosphate (NADPH). Thiamin is required in each of these three steps (**Figure 2**) [3, 4].

The mechanism of thiamin absorption in the body is still controversial. Some researchers argue that it is only absorbed from active transport mechanisms in the proximal small intestine. However, recent studies show that thiamin is also absorbed by passive diffusion [5]. Thiamin absorption by the intestine is mediated by a transport system and absorbed by cells in the liver, heart, and other various tissues from the blood, except neural fibres. In the nervous system, thiamin is transported from circulation blood towards cerebrospinal fluid across the blood–brain barrier [2, 6]. Vitamin B1 is rapidly absorbed and transformed through a

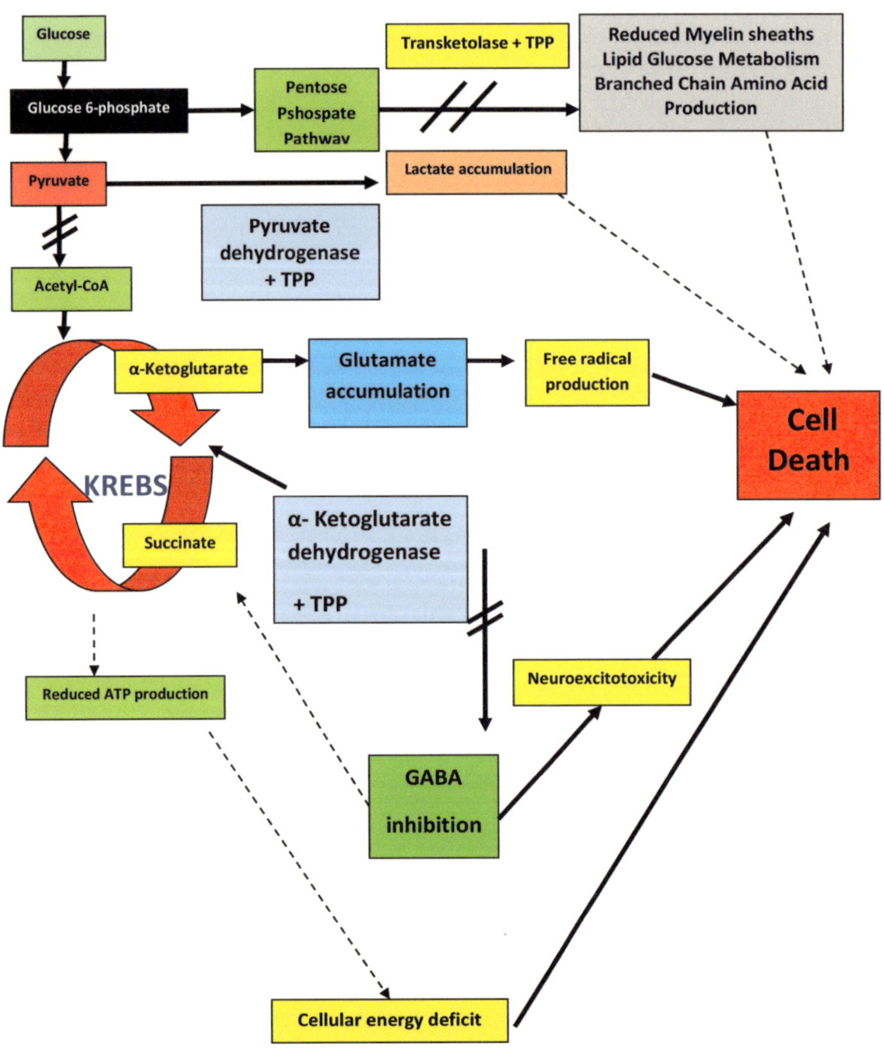

Figure 2.
Pathogenesis of cell death in deficiency thiamin [2, 16].

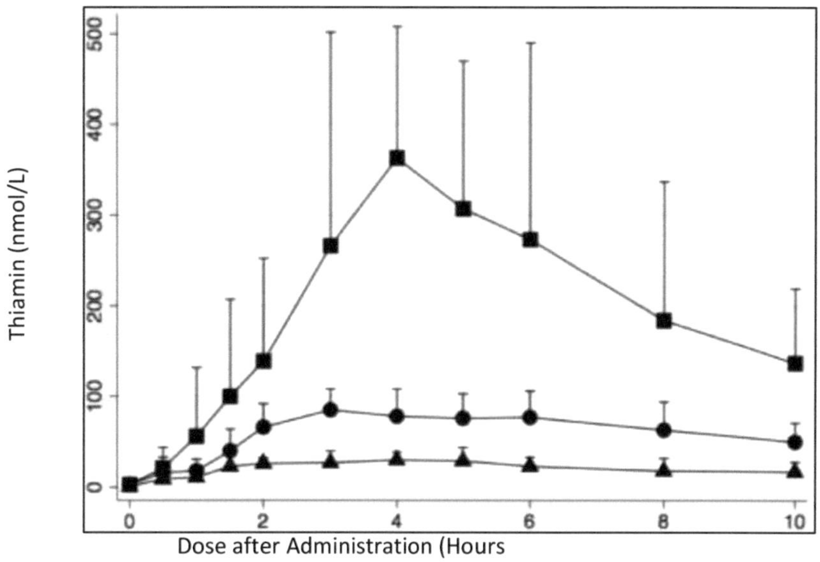

Figure 3.
Oral Thiamin concentration in blood [5].

phosphorylation process into an active coenzyme, thiamin pyrophosphate. Vitamin B1 is absorbed in the jejunum at low concentrations, involving the phosphorylation process through an active transport system. At high concentrations, absorption of vitamin B1 occurs by passive diffusion. The relative bioavailability of vitamin B1 is about 5.3%. A study by Smithline et al. (2012) shows oral thiamin concentration reaching the peak at 4 hours after consumption (**Figure 3**) [5].

Thiamin is widely distributed to almost all body tissues, including breast milk. Thiamin is not stored in the body. Thiamin transport occurs through the blood, both in erythrocytes and plasma. About 90–94% of vitamin B1 is bounded to protein [2].

Thiamin metabolism occurs in the liver and produces active metabolites, thiamin pyrophosphate, thiamin monophosphate, and thiamin triphosphate. Thiamin diphosphate is the primary active metabolite, which acts as a coenzyme in carbohydrate metabolism through transketolase reaction [2, 7].

Thiamin half-life ranges from 9 to 18 days on daily consumption, and the elimination or dephosphorylation process occurs in kidneys. This half-life appears to be variable and highly dose-dependent. One study showed that the half-life of thiamin is only about 6 hours at high doses (500-1500 mg). For intravenous administration, peak levels reached within 2–6 hours depending on doses [5]. If there is an excess of free-form vitamin B1, it will be excreted in the urine. In regular doses, it is secreted in the urine in unchanged form [2].

3. Sources

Thiamin cannot be produced indigenously in the human body. Therefore, we rely on dietary intake [8]. The sources of thiamin include fortified flours, whole grain cereals, meat (pork, beef or poultry), eggs, dried beans, soybeans and nuts. Nevertheless, polished rice, fats, processed flours, dairy products and vegetables are not reliable sources to satisfy the daily requirements of thiamin [9]. Significant losses of thiamin happen when the food is cooked or undergone other heating processes. Polyphenolic compounds in tea and coffee may inactivate thiamin;

therefore, their consumption must be in moderation. Similarly, uncooked fish and shellfish contain thiaminases that inactivate and break down thiamin [9, 10].

4. Intake and novel applications

4.1 Recommended daily intake of thiamin

Recommended daily thiamin intake in healthy adult men is 1.2 mg/day, while adult women are 1.1 mg/day. For children aged 1–8 years, recommended intake of vitamin B1 ranges from 0.5 to 0.6 mg/day and for ages 9–13 years, it starts from 0.9 mg/day (**Table 1**) [5]. Thiamin is water-soluble across the placenta. Its requirements increase during pregnancy. Pregnancy also increases the risk of thiamin deficiency when prolonged nausea and vomiting (including hyperemesis gravidarum). The dose of parenteral nutritional supplements is 6 mg/day. The parenteral form can be given by intramuscular (IM) or intravenous (IV) injection. Administration of thiamin intravenously can be given as much as 100 mg over 5 minutes [11].

4.2 Side effects of thiamin

Adverse reactions to thiamin administration have been reported as reactions at the injection site, but their frequency is unknown. Other side effects can be diaphoresis, pruritus, skin sclerosis (at the injection site after IM administration), urticaria, nausea, bleeding (in the digestive tract). For hypersensitivity side effects, reported anaphylaxis (after IV administration), angioedema, hypersensitivity reactions (following IV administration). Side effects of intravenous thiamin are rarely reported. A prospective study by Wrenn in 989 samples given thiamin 100 mg IV found adverse reactions in the form of minor reactions, which were transient local irritation in 1.1% and pruritus in 0.0093% of patients. Thiamin hydrochloride can be given intravenously without problems. An intradermal test dose before administration is not required unless the patient had a previous allergic reaction [12]. Local side effects for larger doses can be minimised by slow administration into a larger and more proximal vein. Thiamin should be administered before parenteral glucose solutions to prevent Thiamin deposition as a symptom of acute thiamin deficiency in malnourished patients [11].

Patients by age	mg per day
Infant and children	
Newborn to 6 months	0.3
6 months–1 year	0.4
1–3	0.7
4–6 years	0.9
7–10 years	1.0
Teens and adults	
Men	1.2–1.5
Women	1.0–1.1
Pregnant women	1.5
Breastfeeding women	1.6

Table 1.
Recommended dietary allowance *(RDA) of thiamin [2].*

Alcohol consumption can interfere with intestinal absorption of vitamin B1, and chronic alcoholism leads to *Wernicke-Korsakoff syndrome* (WKS) [2]. Intoxication may occur with ingestion of more than 3000 mg thiamin in the long term. Based on animal research, thiamin lethal dose/LD50 are 8224 mg/kg, while the LD50 in rats are 3710 mg/kg [13]. Sporadic anaphylactic reactions have been reported. Some researchers suggest that intravenous thiamin should be administered in a resuscitation facility. However, due to the life-threatening nature of WKS, EFNS (European Federation of Neurological Society) guidelines recommend starting treatment immediately, even in the absence of facilities for resuscitation [14].

4.3 Thiamin and brain metabolism

Thiamin has an essential role in brain metabolism. Nerve cells use glucose as the primary fuel in producing energy. Glucose reaches brain tissue by diffusion across the blood–brain barrier. Around 30% of glucose absorbed by the brain undergoes complete oxidation through the Krebs cycle [15]. Various mechanisms contribute to selective brain lesions observed in WKS and thiamin deficiency. Recent evidence of early microglial activation and increased production of free radicals suggests that oxidative stress processes play a vital role in brain cell death associated with thiamin deficiency. Recent studies in animal models of WKS demonstrated changes in thiamin-dependent enzymes in the brain and suggested that changes in these enzyme activities may result in neuronal death, characteristic of this syndrome [16].

Thiamin is needed as an enzyme cofactor essential for brain metabolism, and around 80% of total thiamin is in the neural tissues [10, 17]. In addition to its co-enzymatic function in metabolism, thiamin also has a structural role [18, 19]. Thiamin affects membrane structure and function, including axoplasmic, mitochondrial, and synaptosomal membranes, which act against agent-induced cytotoxicity and improve membrane location [20, 21]. Thiamin also intervenes in synaptic transmission and plays a role in cell differentiation, synapse formation, axonal growth, and myogenogenesis [2].

4.4 Thiamin as a novel treatment for sepsis

Thiamin deficiency is common in critically ill patients and correlated with increased mortality in some cases. In addition, its levels are depleted throughout the illness, and administration of thiamin during critical illness can improve organ dysfunction [22]. Predisposing factors to thiamin deficiency result from several associated problems associated with nutritional disorders and other accompanying diseases. Several conditions can reduce thiamin levels, such as impaired carbohydrate metabolism, increased metabolic requirements for parenteral or enteral nutrition, diuretics, and haemodiafiltration. Several studies have found the presence of thiamin deficiency in critically ill patients. Thiamin deficiency is associated with poor prognostic outcomes [23].

A cohort study in Australia with 129 patients found no association between plasma thiamin concentrations with systemic inflammation and mortality in critically ill patients. In addition, it also shows that level of thiamin intake in patients who had not received its supplementation before ICU admission did not differ between patients who died and those who survived, 264 compared 268 nanomol/L (normal value: 190–400 nanomol/L). Besides that, only a weak correlation was found between thiamin levels and disease severity index. A study had investigated the correlation of thiamin, APACHE II, SOFA score, maximum SOFA score, SOFA delta (DSOFA), and CRP. However, the correlation is not statistically significant [24].

Thiamin deficiency can also be observed in septic shock patients, occurring in 8.5–72% depending on the cutoff value used to determine thiamin deficiency [4, 22, 25, 26]. Lack of thiamin reduces pyruvate access to the Krebs cycle, increasing lactate production as it converts metabolism to anaerobic [23].

A prospective observational study examined the association between thiamin levels and lactic acidosis in 30 septic shock patients and found no correlation between these variables. However, after excluding patients with abnormal liver function, a significant negative correlation was found between thiamin concentrations and lactic acidosis ($r = -0.53$, $P = 0.01$). This finding implies a potential relationship between thiamin levels and lactic acidosis in septic shock patients with normal liver function. Thus, by reducing pyruvate dehydrogenase complex activity, thiamin deficiency contributes to an increase in lactic acid in septic patients [26].

Parenteral administration of thiamin 250 mg once daily for 3–5 consecutive days is recommended to treat thiamin deficiency. Slow intravenous administration of thiamin diluted in isotonic NaCl or 5% dextrose is also safe. However, there is no consensus on the optimal daily dose of thiamin, its formulation, and duration of treatment [14]. The half-life of unphosphorylated thiamin blood is 96 hours. Therefore, two or three daily doses can achieve better concentrations in the brain than a single daily dose. In patients who do not consume alcohol, a daily intravenous dose of 100 or 200 mg is sufficient to meet thiamin requirements. However, alcoholic patients with WKS may require doses as high as 500 mg three times daily [14].

A clinical trial of *ascorbic acid and Thiamin effect in septic shock* (ATESS) conducted in South Korea compare outcomes of a combination of ascorbic acid (IV 50 mg/kg, maximum dose per dose of 3 g) and thiamin (200 mg) every 12 hours for two days with placebo groups on 111 subjects. The results showed no significant differences in SOFA scores and organ function but found an increase in serum levels of vitamin C and thiamin [27].

This finding is in contrast to another clinical study in 94 patients who received a combination of 1500 mg of vitamin C IV q6, 200 mg of thiamin IV q12 for four days or until discharge from ICU, and 50 mg of hydrocortisone IV q6 (with the optional alternative of 50 mg bolus, followed by continuous infusion of 200 mg in 24 hours) for four days. Thiamine administration had significantly reduced the progression of organ dysfunction and mortality in patients with severe sepsis and septic shock. However, the research design had weaknesses include small study size, pre-and post-study design, single-centre, absence of blinding, and presence of three simultaneous interventions limiting the generalizability of the conclusion. Although it can help invent new hypotheses in future, this study is still not strong enough to produce a change in clinical practice [22].

5. Conclusion

Thiamin plays a vital role in cell metabolism. The administration of thiamin supplementation should be considered adjunctive therapy in critically ill patients as it may improve their outcomes. Further research should be developed to determine the optimal dosage and timing to achieve the maximum effect.

Author details

Bastian Lubis[1,2*], Putri Amelia[3,4], Aznan Lelo[5], Muhammad Akil[1] and Vincent Viandy[1]

1 Department of Anesthesiology, Faculty of Medicine, Universitas Sumatera Utara, Medan, Indonesia

2 Functional Medical Unit of Anesthesiology, H. Adam Malik Central Hospital, Medan, Indonesia

3 Department of Child Health, Faculty of Medicine, Universitas Sumatera Utara, Medan, Indonesia

4 Functional Medical Unit of Child Health, H. Adam Malik Central Hospital, Medan, Indonesia

5 Department of Pharmacology, Faculty of Medicine, Universitas Sumatera Utara, Medan, Indonesia

*Address all correspondence to: bas28al@gmail.com

References

[1] Whitfield KC, Bourassa MW, Adamolekun B, et al. Thiamine deficiency disorders: Diagnosis, prevalence, and a roadmap for global control programs. Ann N Y Acad Sci. 2018. DOI:10.1111/nyas.13919

[2] Fattal-Valevski A. Thiamine (vitamin B 1). Complement Health Pract Rev. 2011;16(1):12-20. DOI:10.1177/1533210110392941

[3] Frank RAW, Leeper FJ, Luisi BF. Structure, mechanism and catalytic duality of thiamine-dependent enzymes. Cell Mol life Sci. 2007;64(7):892-905.

[4] Donnino MW, Andersen LW, Chase M, et al. Randomised, double-blind, placebo-controlled trial of thiamine as a metabolic resuscitator in septic shock: A pilot study. Crit Care Med. 2016;44(2):360.

[5] Smithline HA, Donnino M, Greenblatt DJ. Pharmacokinetics of high-dose oral thiamine hydrochloride in healthy subjects. BMC Clin Pharmacol. 2012;12(1):4. DOI:10.1186/1472-6904-12-4

[6] Alemanno F. Thiamine (Vitamin B1). In: Biochemistry for Anesthesiologists and Intensivists. Cham: Springer International Publishing; 2019:139-159. DOI:10.1007/978-3-030-26721-6_12

[7] Williams RD, Mason HL, Smith BF, Wilder RM. Induced thiamine (vitamin B1) deficiency and the thiamine requirement of man: Further observations. Arch Intern Med. 1942;69(5):721-738. DOI:10.1001/archinte.1942.00200170003001

[8] Osiezagha K, Ali S, Freeman C, et al. Thiamine deficiency and delirium. Innov Clin Neurosci. 2013;10(4):26-32.

[9] Information N. Thiamin 1,2. 2017;(5):395-397. DOI:10.3945/an.116.013979.395

[10] Lonsdale D. A review of the biochemistry, metabolism and clinical benefits of thiamin(e) and its derivatives. Evidence-based Complement Altern Med. 2006;3(1): 49-59. DOI:10.1093/ecam/nek009

[11] Kalliyath A, Korula S, Mathew A, Abraham S, Isac M. Effect of preoperative education about spinal anesthesia on anxiety and postoperative pain in parturients undergoing elective cesarean section: A randomised controlled trial. J Obstet Anaesth Crit Care. 2019;9(1):14. DOI:10.4103/joacc.joacc_35_18

[12] Wrenn KD, Murphy F, Slovis CM. A toxicity study of parenteral thiamine hydrochloride. Ann Emerg Med. 1989;18(8):867-870. DOI:10.1016/S0196-0644(89)80215-X

[13] Wishart DS, Feunang YD, Guo AC, et al. DrugBank 5.0: A major update to the DrugBank database for 2018. Nucleic Acids Res. 2018;46(D1):D1074-D1082. DOI:10.1093/nar/gkx1037

[14] Galvin R, Bråthen G, Ivashynka A, Hillbom M, Tanasescu R, Leone MA. EFNS guidelines for diagnosis, therapy and prevention of Wernicke encephalopathy. Eur J Neurol. 2010;17(12):1408-1418. DOI:10.1111/j.1468-1331.2010.03153.x

[15] Rao J, Oz G, Seaquist ER. Regulation of cerebral glucose metabolism. Minerva Endocrinol. 2006;31(2): 149-158.

[16] Butterworth RF, Kril JJ, Harper CG. Thiamine-dependent enzyme changes in the brains of alcoholics: Relationship to the Wernicke-Korsakoff syndrome. Alcohol Clin Exp Res. 1993;17(5):1084-1088. DOI:10.1111/j.1530-0277.1993.tb05668.x

[17] Ishii K, Sarai K, Sanemori H, Kawasaki T. Concentrations of thiamine

and its phosphate esters in rat tissues determined by high-performance liquid chromatography. J Nutr Sci Vitaminol (Tokyo). 1979;25(6):517-523. DOI:10.3177/jnsv.25.517

[18] Haas RH. Thiamin and the brain. Annu Rev Nutr. 1988;8(1):483-515. DOI:10.1146/annurev.nu.08.070188.002411

[19] Bâ A. Metabolic and structural role of thiamine in nervous tissues. Cell Mol Neurobiol. 2008;28(7):923-931. DOI:10.1007/s10571-008-9297-7

[20] Itokawa Y, Schulz RA, Cooper JR. Thiamine in nerve membranes. BBA – Biomembr. 1972;266(1):293-299. DOI:10.1016/0005-2736(72)90144-7

[21] Matsuda T, Cooper JR. Thiamine as an integral component of brain synaptosomal membranes. Proc Natl Acad Sci U S A. 1981;78(9 II):5886-5889. DOI:10.1073/pnas.78.9.5886

[22] Moskowitz A, Andersen LW, Huang DT, et al. Ascorbic acid, corticosteroids, and thiamine in sepsis: A review of the biologic rationale and the present state of clinical evaluation. Crit Care. 2018;22(1):1-7. DOI:10.1186/s13054-018-2217-4

[23] Costa NA, Gut AL, de Souza Dorna M, et al. Erratum: Corrigendum to "serum thiamine concentration and oxidative stress as predictors of mortality in patients with septic shock" (journal of critical care (2014) 29(2) (249-252) (S0883944113004723) (10.1016/j.jcrc.2013.12.004)). J Crit Care. 2016;36:311. DOI:10.1016/j.jcrc.2016.07.001

[24] Corcoran TB, O'Neill MP, Webb SAR, Ho KM. Inflammation, vitamin deficiencies and organ failure in critically ill patients. Anaesth Intensive Care. 2009;37(5):740-747. DOI:10.1177/0310057x0903700510

[25] Attaluri P, Castillo A, Edriss H, Nugent K. Thiamine deficiency: An important consideration in critically ill patients. Am J Med Sci. 2018;356(4):382-390.

[26] Donnino MW, Carney E, Cocchi MN, et al. Thiamine deficiency in critically ill patients with sepsis. J Crit Care. 2010;25(4):576-581. DOI:10.1016/j.jcrc.2010.03.003

[27] Hwang SY, Park JE, Jo IJ, et al. Combination therapy of vitamin C and thiamine for septic shock in a multicentre, double-blind, randomised, controlled study (ATESS): Study protocol for a randomised controlled trial. Trials. 2019;20(1):1-8. DOI:10.1186/s13063-019-3542-x

Chapter 11

Role of Dietary Supplements in Prevention of Renal Stones: An Update

Akshata Sangolli, Shridhar C. Ghagane and Rajendra B. Nerli

Abstract

Kidney stone disease is an oldest known and widespread medical condition characterised by its high prevalence in all over the world. Literature suggests that around 9–12% of population in industrialised countries have kidney stone disease in their lives with the 30–50% of reoccurrence rate. Because of high prevalence, recurrent and unpredictable nature of stone formation and its predominance mainly in adults contributes to the substantial impact on society, individual and health care system. In light of these trends, it's imperative to use optimum preventive strategies to reduce the burden of kidney stone disease on individual and society. The aetiology of kidney stone disease is a multifactorial and it's related to diet, environmental factors, genetics, metabolic syndromes and various life style factors. Its noteworthy that dietary and life style modification are the major contributors in the prevention of kidney stone reoccurrence. Dietary interventions aim to reduce the urinary abnormalities known to promote lithogenesis. Therefore, modification in the dietary factors is appealing way to patients and physicians in the treatment and prevention of stone recurrence as it is relatively inexpensive and safe. So, the present chapter is focusing on the role of dietary supplements in prevention of renal stones.

Keywords: kidney stones, dietary factors, prevention, health

1. Introduction

Kidney stone disease or renal calculi is a serious medical condition though not life-threatening disorder. In medical terms it referred as urolithiasis or nephrolithiasis where "Lith" meaning stone [1]. Renal stone formation is an oldest and widespread disease in the world affecting human beings. Its prevalence in the Europe is around 7–9%, Asia 1–5% and in North America 6–12%. The lifetime prevalence of renal stones in India is 5–11% [2]. This prevalence represents threefold increment and 5–6% absolute increment in last 20–30 years. Increased in the number of cases is reported in all groups irrespective of gender, racial and ethnic variation [3]. An alteration in normal mineral content of urine is the main cause for lithiasis [4]. Urinary components play a vital role in stone formation as they will be in their metastable state with several pre-existing substance which can crystalize to form calculi. These substances if exists in super saturation level makes urine unstable and will lead to crystallisation of excess of solutes [5].

Kidney stone may be found with different shapes, sizes and colours depending on their composition. Smaller stones may pass in the urine without any symptoms, but often stones grow in their size and develop a level of discomfort while passing through urine. In some case it may cause a severe pain if the surface of the stone is rough or it may require medical intervention as bigger stone cannot pass through the urinary system [6]. Other complications such as urinary tract infections, sever pain, or decline in the renal function may be associated with urolithiasis [7]. If neglected it may lead to the substantial damage to the kidney [6]. Since significant number of patients may have to undergo surgical interventions for the treatment, the management of the kidney stone has become considerably expensive [8]. Because of high prevalence, recurrent and unpredictable nature of stone formation and its predominance mainly in adults contributes to the substantial impact on society, individual and health care system [9]. In light of these trends, it's imperative to use optimum preventive strategies to reduce the burden of kidney stone disease on individual and society. Thus, awareness regarding importance of preventive measures particularly on consumption of healthy diet certainly help to reduce the cost of hospitalisation and will increase the compliance in general.

According to literature survey various preventive measures are available to reduce the risk of kidney stone formation such as life style modification, high water intake, less consumption of salt and modification in dietary habits. As diet has shown its strong association with stone formation, changes in dietary habits may help to reduce the burden of stone formation. Less awareness on effect of food on stone formation is one of the main reasons for increase prevalence. So, more studies have to be focused effect of various diets at molecular level to understand the actual mechanism behind stone formation. Thus, present chapter is focusing on role of various diets in prevention of kidney stone disease.

2. Types of kidney stones

Chemical constitutes of urine is the main factor for variation in chemical composition of the stones. These variations may be associated with other risk factors such as environmental factors, diet, climate and life style habits. Based on this, urinary stones can be classified in to five major classes.

- **Calcium stones:** Calcium stone can exist as calcium oxalate or calcium phosphate or calcium carbonate. Among all the types these are the prominent ones constituting around 70–80% of the total stone forms. Calcium oxalate can exist as calcium oxalate monohydrate (COM) or dihydrate (COD). COM is the more stable form and common compared to the COD. Various factors may contribute to calcium stone such as hypercalciuria, hyperoxaluria, hyperuricosuria etc. [10].

- **Uric acid or urate stones:** Uric acid stone accounts approximately for 5–10% of all the types of stones. High purine content of diet such as animal protein is main risk factor for the formation of uric acid stone. Other risk factors are low urine volume, hyperuricosuria, and low urinary pH [11].

- **Magnesium ammonium phosphate stones or struvite stone:** Also called as infection stones or triple phosphate stones. These stone occur at the extent of 12–15%. The main cause for the struvite stone is urinary tract infections where

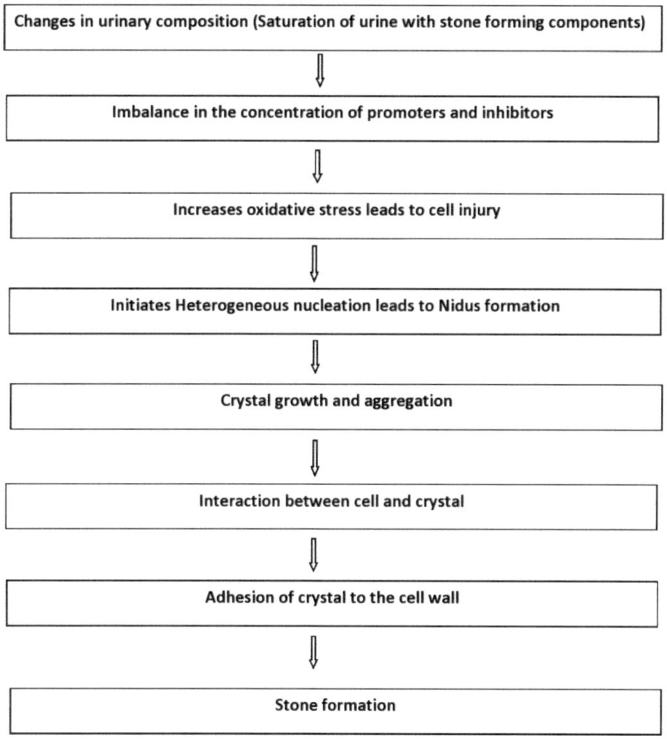

Figure 1.
Steps for mechanism of stone formation.

urease produced by microbes split urea into ammonia and carbon dioxide. This makes the urine alkaline and makes phosphate insoluble at its high pH. Thus, phosphate gets precipitated on the ammonia leading to the stone formation. Its more common in female compared to male [12].

- **Cystine stones:** The occurrence of cysteine stone is less than 3% among all the types of the stones. It occurs as a genetic disorder with defect in cystine amino acid transportation, which results in excess excretion of cystine in urine referred as cystinuria. It is an autosomal recessive disorder resulting in impaired renal tubular absorption of cystine and excretion of cystine in urine. As cystine is insoluble in urine results in the formation of stones in kidney [13].

- **Drug-induced stones:** These stone accounts for only 1% of total stones. Drugs such as triamterene, atazanavir, and sulfa drugs induce stone formation. Lithogenic drugs and their metabolites may get deposited to form nidus or on already existing stone. Some drugs may also interfere in purine and calcium oxalate metabolism and may lead to the stone formation (**Figure 1**) [14].

3. The aetiology of kidney stones

The aetiology of kidney stone formation is multifactorial. There are various risk factors associated with formation of kidney stones such as age, gender, ethic and family background, life style habits, environmental variations, occupation and dietary habits. Variations in these risk factors initiates super saturation of urine which may cause changes in the morphology of kidney, change in urine flow, urinary tract infections and metabolic abnormalities (**Figure 1**) [1].

- **Age:** The most vulnerable group for the urolithiasis is 20–60. The incidence of kidney stones is increased with age, thus middle-aged people are very prone to get kidney stone disease [15]. One of the most common reason which could be related to this age group is less fluid intake, dehydration, stress at the work place and unhealthy life styles. In some population the age distribution is different in males and females; male is affected after the age of 60 whereas females are affected at the age of 45–50 [16].

- **Gender:** According to most of the literature survey kidney stone disease is more common in males compared to female with the ratio ranging from 1.5 to 5. It may be associated with the changes in the dietary habits and testosterone promote the stone formation in males [17]. However, in recent decades this ratio is narrowed in many countries, which says that even females are more affected and prevalence is increase in both genders. The main reason for this could be standard living habits, high calorie food consumption and variations in occupations [18].

- **Climate:** Geographic and climatic changes are one of the major risk factors for urolithiasis, specifically temperature, seasonal variations, atmospheric pressure, humidity. High prevalence is seen in tropical and subtropical countries than in frigid zones. Considering seasons incidence will be high in summer compared to spring and winter. This may be associated with the concept of higher temperature leading to loss of water through the body fluids and dehydration which may lead to excretion of concentrated urine may become cause for formation of stone [19].

- **Dietary habits:** Diet also plays an important role in formation of stones in kidney. Diet rich with high amount of oxalate is the main precursor for stone formation. In addition, diet containing high amount sodium, protein, calcium also acts as risk factors for stone formation. This is the main reason for the raising trends of kidney stone disease in many of the Asian countries [20]. Oxalates present in the food gets metabolised in liver and Calcium in intestine combine with excess of oxalate and lead to formation of insoluble calcium oxalate stone [21]. Excessive consumption of animal meat will lead to increased uric acid concentration and results in hyperuricosuria, is the main risk factor for the uric acid stone formation. In addition, pH of the urine plays an important role in lithiasis as appropriate pH favours crystal precipitation, initial step of stone formation [22, 23]. Along with diet less fluid intake is also a main cause for stone formation. Water with high fluoride content may also become risk factor for stone formation. The fluoride in the intestine favours the absorption of oxalate and promotes excess of oxalate excretion and formation of calcium fluoride in urine [24]. Excretion of high concentration of magnesium in urine is also one of the causes for stone formation in kidney [25]. On other hand one of the studies conducted by Chandrajith et al., have not found any association between hardness of water and urolithiasis [26].

- **Occupation:** According to studies conducted on stone formation and occupation, sedentary life style is one of the risk factors for urolithiasis. Some studies reports that there is a positive association between more physical work and kidney stone disease. The risk of kidney stone formation is more in case of people expose to the sunlight or high temperature for longer period like farmers, miners' drivers etc. than people working at room temperature [27]. People working in these conditions may consume less fluid and they will be more prone to have dehydration, which may lead to excretion of concentrated urine and lead to urolithiasis. However, there are some studies which have shown negative relation between stone formation and occupation [28].

- **Genetics:** Genetic factor also contribute to the renal stone formation, especially cysteine stones are formed by mutation in the gene SLC3A1 and ALC7A9. In addition, even in case of uric acid stones some mutations seen in SLC2A9 and SLC22A12. In some cases, calcium oxalate stones are formed because of the deficiency of enzymes such as glyoxylate reductase/hydroxy pyruvate reductase (GRPHR), alanine glyoxylate aminotransferase. As a result, synthesis and excretion of oxalate is increased leading to calcium oxalate stone formation [29].

- **Racial distribution:** The association of kidney stone disease with different racial background is still controversial. In Asian population the association is been reported by certain studies. Whereas studies conducted in Iran have not shown a significant association between racial difference and prevalence of urolithiasis. In general terms, dietary habits, life style changes and gene of various races are the main key factors for the variations [30, 31].

4. Importance of diet

Intake of healthy and fresh diet plays an important role in maintaining the health status of human kind. Balanced diet has been immensely accepted all over the world owing to the increased awareness regarding the maintenance of health status among people. Balanced diet is comprised of all essential nutrients, which are required for good health of human beings [32]. Nutrients present in food provides energy to perform vital functions of life and also helps for growth and differentiation of cells. These nutrients can be classified into micro nutrients and macro nutrients based on the requirements. Micro nutrients are the one which are requited in a smaller quality which includes vitamins and minerals whereas macro nutrients are the one which are required in a larger quantity which includes carbohydrates, lipids and proteins [33]. All these nutrients are present in the food materials like cereals, pulses, vegetables, fruits, meat and dairy products. Among these products high amount of protein is found in meat. The concentration of protein may vary in different kind of animal meat like beef, mutton, chicken fish etc. whereas carbohydrate will be present in cereals, potatoes, milk and lipids will be rich in nuts, peanuts, ghee, oil, butter etc. [34].

5. Variations in the dietary habits

Variation in food habits arises from the people's origin and it is modified by resource of the respective place or origin. The major resources having impact on the food culture are climate, Land, soil, water, cultural and religion of the habitat [35]. The era of globalisation has changed the eating and life style habits which has shown a very strong impact on the health of human beings. Urban areas of most of the countries, have embraced more processed and packed food, which have led to increased obesity and body mass in people. Diversified food habits have been seen in various parts of the world and also within the countries itself with different geographical areas. These food variations are the main factor for variations in prevalence for urolithiasis at different geographical places [35].

6. Role of dietary habits in stone formation

Among various risk factors of urolithiasis, food is considered as one of the important modifiable risk factors in the kidney stone disease. According to the study conducted by Maalouf et al. [36] states that a load of protein diet in food will lead to increased calcium excretion. This may lead to increased risk for kidney stone formation. High amount of protein diet induces acid load in the body because of production of protons during the metabolism by sulphur containing amino acids and also there will be increased calcium excretion followed by high protein diet [36]. In addition to this high intake of animal protein leads to increased concentration of calcium, oxalate, uric acid and phosphorus in the urinary tract. There are various mechanisms involved which may lead to increased concentration of these substances in the urine and may lead to the formation calculi [37]. High intake of carbohydrates and lipids also have shown similar effects on the urinary composition. Furthermore, less intake of fruits and vegetables may also act as risk factor for urolithiasis even though some of them will be rich in oxalates [38]. High intake of sodium in the form of excess of common salt is noticed in case of many industrialised countries lead to more calcium deposition in kidney. The role of magnesium and vitamin C in the kidney stone formation is still not clear. Few studies have reported with no significant association with stone formation, whereas the effect of these nutrients on urinary composition shows its role in urolithiasis [39]. Thus, most of the studies support the fact that there is a relation between dietary habits and kidney stone formation, although contradictory results are also available.

Different food items	Content of food	Role in urolithiasis	Reference
Milk, cheese, dark green vegetables, yogurt, calcium fortified beverages etc.	Dietary calcium	Decreases the risk of calcium oxalate stone formation	[40]
White meat poultry, lean beef, eggs, beans, etc.	Protein	High load of acid in the kidney increases risk of kidney stone formation Increase urinary excretion of calcium	[41]
Canned food, corn meal, Black eyed beans, beets etc.	Sodium	High level of urinary calcium	[42]
Green leafy vegetables, beets, berries, chocolates, cranberries	Oxalate	Increased oxalate absorption from the intestine lead to high amount of excretion	[40]
Citrus fruits, peppers, strawberries, blackcurrants, broccoli etc.	Vitamin C	High oxalate excretion in urine	[43]
Energy drinks, soft drinks, carbonated drinks, coffee etc.	Carbonated beverages	High level of oxalate excretion in urine	[44]

Table 1.
Role of various food stuffs on kidney stone formation.

In the present chapter we are discussing in detail about the role of diet in KSD, so that we could summarise important preventive dietary habits for the urolithiasis (**Table 1**).

7. Impact of fluid on urolithiasis

Less fluid intake is one of the major risk factors in the stone formation, whereas adequate amount of urine excretion will eventually reduce the saturation of urine. To achieve 2 L/day of urine excretion fluid intake should be higher at the range of 2–3 L/day as water will be lost for extra renal functions like sweating, breathing and perspiration [45]. Along with volume of fluid intake, the quality and composition of fluid or water is also equally responsible for stone formation and it can be considered as a modifiable risk factor. Many studies have reported that high amount of fluid intake will eventually reduce the risk of urolithiasis [46].

Apart from water, other beverages like soda, tea, coffee and aerated or carbonated drinks consumption will also be having their impact on stone formation. High intake of sugar sweetened soda rich with fructose increases the risk of urolithiasis, as fructose promotes synthesis of uric acid and increases the excretion of uric acid, calcium and oxalate in urine [47]. Fructose rich food, makes the cells to utilise excess ATP for its uptake and such monosaccharides reduces phosphate concentrations within the cells and induces production of uric acid which leads to hyperuricosuria. So, in hyperuricosuria patients, it's advisable to reduce the intake of fruits, beverages and fruit juices rich with fructose content [48]. Another study conducted by Shuster et, al. reported that intake of carbonated/aerated beverages is also one of the risk factors for formation of calculi as these beverages contains high amount of phosphoric acid [49]. On the contrary some studies also evidenced that citric soda content of the aerated drinks has a capacity to reduce the risk of stone formation by increasing the excretion of citrate. However low energy aerated drinks have not shown significant association in large cohort studies, suggesting that more than carbonated content fructose content of the drinks is the main culprit for urolithiasis [50]. A study conducted by Ferraro et al. [51] reports that beverages such as tea, coffee reduces the risk of calculi formation as studies have noticed that caffeine intake is associated with increased urinary output. Still excess consumption of tea, coffee is not advisable as it may interfere with other metabolic reactions which influence changes in the blood pressure. Hot beverages such as beer, alcohol have shown controversial results for their association with stone formation. In the study conducted by Ferraro et al., these drinks have shown to reduce the risk of stone formation as they reduce the activity of antidiuretic hormone and helps to excrete excess amount of diluted urine [51]. In contrast to this, a study conducted by Borghi et al. reports that alcohol intake should be avoided in case of urolithiasis patients as it will be rich in purine and it may cause hyperuricosuria [46]. Another study conducted by Rodgers et al., found that magnesium and calcium content of mineral water acts as protective in case of calcium oxalate stones [52]. Studies conducted on effect of various fruits juices have shown its impact on stone formation. Fresh lemon juice involved in the excretion of citrate and reduces excretion of calcium in urine. In concern with non-citrus fruits the results of the studies are still controversial as some fruits have shown beneficiary effect whereas some have not shown any significant association with urolithiasis [53]. Thus, these studies suggests that all fluids are not having same effect on urolithiasis. So, it's advisable to reduce the intake of sweetened beverages and high citrate content drinks as they have shown unfavourable outcomes in KSD patients.

8. Effect of carbohydrate diet on urolithiasis

A study conducted by Nouvenne et al. reports that high carbohydrate food intake has shown increased excretion of calcium in urine compared to healthy individuals because carbohydrates decrease the calcium reabsorption in the renal tubules [54]. In contradictory to this some studies have reported that increased glucose concentration in diet has enhanced the calcium absorption in intestine. Various epidemiological studies have been conducted to find out the relation between insulin action and calculi formation specifically with uric acid stones. The insulin renal receptors show imbalance in acid handling which results in impaired excretion of ammonia in urine leading to excretion of acidic urine which favours precipitation of uric acid crystals leading to uric acid stones. This could one of the main reasons for high prevalence of urolithiasis in metabolic syndrome cases [55]. In addition to this high fructose intake has shown a strong association with formation of kidney stones as it enhances excretion of citrate and calcium in urine which favours the formation of stones in kidney. Increased fructose intake may cause insulin resistance and it may become trailing step for the formation of uric acid stone as it decreases the urinary pH and lead to uric acid stone formation [56]. Another study conducted by Curhan et al. reports that intake of sucrose has also shown its association with stone formation as high sucrose may increase the urinary excretion of calcium which is not dependent on the calcium intake (**Figure 2**) [57].

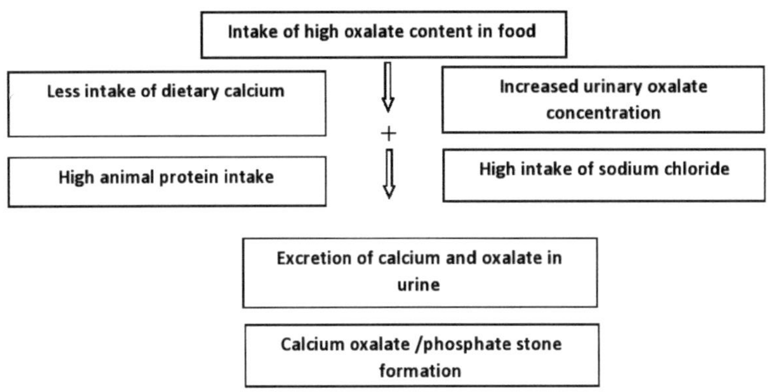

Figure 2.
Effect of various food items in calcium oxalate/phosphate stone formation.

9. Effect of protein diet on urolithiasis

Various studies have shown a strong association of protein intake specifically animal protein with kidney stone formation. Animal protein will be rich in purines and after its degradation it produces uric acid [58]. These contains amino acids such as tyrosine, tryptophan and glycine and degradation of these amino acids produces oxalate which is the main component of calcium stones [59]. Increased oxalate content causes calcium and citrate resorption, renal acid excretion and increased urinary excretion of calcium which ultimately cause kidney stones [60]. Animal protein such as meat, poultry, fish shows unfavourable effect as their intake leads to uricosuria, calcinuria and phosphaturia and also reduces urinary pH. All these conditions increase the risk of precipitation of substances like calcium and uric acid and lead to calculi. Formation [61]. According a study conducted by Kerstetter et al. reports that in a normal healthy individual for every 20–25 g of increase in dietary animal protein, will rise the urinary calcium by

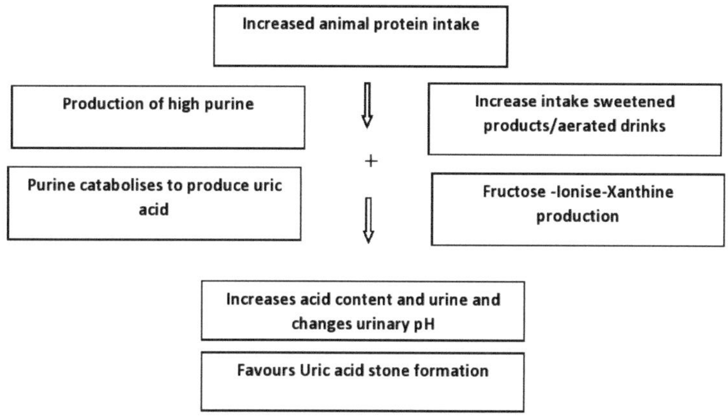

Figure 3.
Effect of various food items in uric acid stone formation.

30–35 mg/day [62]. The final outcome of high animal protein intake by keeping volume of the urine constant is super saturation of urine with calcium oxalate and uric acid which are the main risk factors for stone formation. The underlying mechanism for uricosuria and phosphaturia is related to high content of these substances in animal protein. In addition to this increased calcinuria and change in urinary pH is mainly attributed by sulphureted amino acids such as methionine and cysteine which produces hydrogen ions leading to subclinical acidosis [63]. The reason for oxaluria is still not clear yet according to studies conducted, presence of oxalate in animal proteins produced by amino acids such as tryptophan, tyrosine increases endogenous production of oxalate which may create favourable environment for formation of stones [64]. Its notable that effect of vegetable proteins on urinary composition is different from those of animal proteins. A study conducted by Breslau et al. [65] observed that intake of exclusively vegetarian diet with vegetable proteins leads to less excretion of calcium, phosphate and more oxalate, citrate in urine with less acids as vegetable proteins contains different quantity of sulphates, purines, oxalates and fibres. On a whole, this suggests that vegetable proteins are less harmful for urolithiasis specifically in context with uric acid stones (**Figure 3**) [65].

10. Effect of lipid diet on urolithiasis

According to some observational studies there is an association between lipid intake and stone formation in kidney. A study conducted by Khan et al. [66] observed significant changes in the concentration of lipids in urine among stone formers and healthy individuals. The altered lipid content in the membrane enhances nucleation and retention of calcium oxalate crystals which are the initial steps of calculi formation [66]. Another study conducted by Naya et al. [67] reported that there is a relation between urinary lipids and crystal formation as it correlates with urinary oxalate excretion. This association is more evident in case of arachidonic acid content of diet as it increases absorption of oxalate from intestine and increases its clearance from kidney [67]. Another study conducted by Baggio et al. also supports these results as they evidence a high concentration of arachidonic acid in red blood cell membranes and plasma of urolithiasis patients [68]. Still some contradictory studies are also available which says that there is no correlation between lipid diet and urolithiasis [69].

11. Effect of milk and milk products on urolithiasis

The requirement of calcium to body is satisfied by calcium rich food stuffs such as milk (100–120 mg/dl), cheese (approx. 500–600 mg/100 g) and yogurt (approx. 100 mg/100 g). The amount of calcium intake affects the level of calciuria in kidney stone patients and also in healthy individuals. Specifically, the absorption of calcium will be on higher side in kidney stone formers. This is suggesting that increasing in dietary calcium intake than its normal range have its impact on calcium stone formation. This was considered to be a risk factor for many ears and physicians used to suggest to avoid calcium rich food in urolithiasis patients [70]. But according to most of the recent research work, the dietary calcium is no longer involved in formation stone as reduction in the dietary calcium did not show its impact on reduction of calciuria [71] and avoiding dietary calcium may lead imbalance in calcium concentrations and cause certain complications like osteoporosis or osteopenia over a long period. In addition to this when there is a restriction on dairy products, patients may compensate its protein by consuming high quantity of animal protein which may show its own complications in long terms [72]. According to the results of most of pathophysiological studies reported so far, the risk of stone formation is less in subjects who consume high quantity of calcium than in those people who consume less amount of calcium irrespective of gender. A randomised study carried out over 5-year period reports that intake of less calcium diet by reducing milk related products is less significant in preventing calculi reoccurrence than a normal calcium intake with low animal protein diet. Hence, in view of these results it's necessary to understand that no idiopathic stone formers should be advised for less intake of milk and milk product which may lead to complications because of hypocalcaemia [73].

12. Effect of sodium chloride and potassium on urolithiasis

A significant relation between calcium stone formation and salt intake was first showed in a cohort study conducted by Curhan et al. [74]. Some studies conducted in further years did not succeed in confirming these results. A study conducted by Sabto et al. reports that daily intake of around 20–25 mmol of sodium will increase the calcium excretion in urine by 0.5–0.7 mmol/day, thus suggesting greatest impact of sodium on urinary calcium [75]. Salt present in the food stuff inhibits the tubular reabsorption of calcium thus leading to increased excretion of calcium in urine. In addition to this sodium chloride also inhibit the excretion of citric acid in urine, which is one of the significant risk factors for urolithiasis. The mechanism by which sodium decreases citric acid levels in urine is still unclear [76, 77]. Although there are no studies available to prove the fact that less sodium chloride intake will decrease the risk of calculi formation, but there are some studies which reports that beneficiary effect of low animal protein can be enhance by taking less sodium chloride. A well-balanced diet is with only required amount of sodium chloride will eventually help in preventing renal calculi [73]. Along with sodium even potassium is also involved in regulation of urinary calcium in human body. According to a study conducted by Muldowney et al., potassium deprivation was associated with increased calcium excretion in case of healthy individuals with normal diet with normal content of sodium chloride [78]. Another study conducted by Knight et al. also noticed in their study that sodium and potassium are involved in increasing urinary pH and its volume which are initial stages of cysteine stone formation [79].

13. Effect of fruits and vegetable on urolithiasis

The role of fruits and vegetables in kidney stone formation is always been a controversial as these have shown both beneficial effects as well as harmful effects in case of Kidney stone disease. As fruits and vegetables are one of the important dietary sources of oxalate and this absorbed oxalate will be excreted in urine. If urine gets saturated with oxalate content it may become a risk factor for urolithiasis but however this oxalate is not withstanding [80]. Whereas there are some studies which have shown beneficiary effects of fruits and vegetables on urolithiasis as they contain high amount of magnesium and potassium and less amount of animal protein and sodium chloride. In addition, some fruits and vegetables also give alkaline therapy to the urine composition with their high content of bicarbonate and citric acid [81]. So it's important to note that not all the vegetables and fruits are harmful for urolithiasis as very few will be rich in oxalate such as spinach, beets, nuts wheat bran etc. which significantly results in oxaluria [82]. Along with effect of dietary oxalate content, the absorption rate of oxalate may vary person to person, as a study conducted by 80 showed that around 9–12% of idiopathic urolithiasis patients have shown increased oxaluria because of their increased intestinal absorption rate by 15–30% [83]. Another study conducted by Lemann et al. [84] reported that intake of fruits and vegetables can enhance magnesium excretion which is one of the important inhibitors for calcium crystallisation. And also favours the dissolution of uric acid by changing pH of urine [84]. By considering all above-mentioned factors, we suggest physicians to recommend intake of fruits and vegetable in their day today life to all the type of stone formers with a note of restricting foods having increase oxaluric activity to avoid calcium oxalate stone formation. Elimination of fruit and vegetables from diet of normal subjects causes unfavourable changes in urinary composition and may become risk factor for stone formation as their deficiency may significantly increase in super saturation of urine for calcium oxalate and calcium phosphate [85]. Considering above mentioned facts, we encourage physicians to advise their patients to consume fruits and vegetables regularly with restriction of vegetables showing hyperoxaluric effect to avoid increment in urinary oxalate content.

14. Effect of vitamins on urolithiasis

The role of vitamins in the formation of stone is still uncertain. But according to literature review, vitamins with higher risk of causing urolithiasis are ascorbic acid (vitamin C), pyridoxine (vitamin B6) and calcitriol (vitamin D). A study conducted by Broadus et al., reported that the subjects with increased calcium excretion had high levels of vitamin D in their blood sample which lead to increased absorption of calcium in the intestine. Excluding some special cases, it's not suggestible to give supplementations of vitamin D particularly with combination of calcium to kidney stone patients [86]. High intake of vitamin C (ascorbic acid) has become a widespread practice all over the world as ascorbic acid helps in wound healing and preventing degenerative diseases. Vitamin C is a precursor of oxalate and it may increase excretion of oxalate in urine which a risk factor for calculi formation [87]. Intake of vitamin C around 1300–1500 mg/day is acceptable, if intake increases more than 1500 mg/day it will lead to initiation of crystal formation in urine [88]. Among vitamin B complex Vitamin B6 paly a vital role in reduction of risk of stone formation. Vitamin B6 (Pyridoxine) involves in the metabolism of oxalate, so deficiency of pyridoxine may lead to increased production of endogenous oxalic acid. Intake of Vitamin B6 around 40–50 mg/day in diet will eventually help to reduce excretion of oxalic acid in urine and reduce the risk of urolithiasis [89, 90]. In a summary, with regards to vitamins adequate amount of all the vitamins should be consumed through the diet as they a play important role in metabolism of vital biomolecules and also helps to maintain good health status of an individual. Kidney stone patients should avoid excess in take of ascorbic acid, and vitamin D supplementations as these have been reported as risk factors for urolithiasis. The patients can be advised to take good amount of pyridoxine as it is considered to reduce the risk of stone formation.

15. Conclusion

The prevalence of kidney stone disease has increased in recent years as a result of modification in eating and life style habits. Changes in urinary composition and urinary saturation is the initial step for stone formation. So, focusing on reducing urinary saturation may help to reduce the initiation of urolithiasis. Various preventive measures are available which could reduce the burden of stone disease. Among all, dietary interventions show promising results in reducing the risk of stone formation as diet shows its direct impact on urinary composition. Among various types of stones, the most prominent stones such as calcium oxalate/phosphate and uric acid stones shows direct association with diet. According to our review of literature diet containing animal protein will increase urinary uric acid concentration which favours the uric acid stone formation. Diet with high oxalate content will increases urinary oxalate and combine with calcium to form calcium oxalate stones. Less fluid intake is one of the major risk factors for urolithiasis as fluid will help to dilute urine and reduce the saturation of urine. Considering these facts avoiding the foods with increased risk of stone formation and consuming balanced diet in kidney stone formers will help to reduce the reoccurrence and eventually help to reduce the prevalence of disease.

16. Summary

Urolithiasis is a highly prevalent disease with its increased rate in recent years across the world. A change in food habits and intake of high calorie food is one of the main reasons for increased prevalence of kidney stone disease. The main aim of focusing on dietary interventions is to reduce urinary lithogenic risk factors such as increased calciuria, uricosuria, phosphaturia and low urinary pH. According to literature survey it's advisable to cut down high intake of animal protein and excess salt intake as animal protein increases uric acid concentration in urine and salt will increase mineral content of urine. Intake of high calorie food should also be reduced as it's involved in increasing saturation of urine and it may also lead to other health complications such as metabolic syndromes. High intake of aerated or carbonated drinks should be avoided in kidney stone formers as they contain high amount of sugar (Fructose). Along with reduction in consumption of above-mentioned food items it's equally important to consume food items which helps to alkaline the urine and reduce the risk of stone formation. According the literature survey it's advisable to consume good amount of green leafy vegetables with less oxalate content. Adequate amount of vitamins have to be taken in diet as their absence may lead to some deficiency manifestations in an individual. Excess consumption of vitamins such as vitamin D and C should be avoided as they may increase the risk of stone formation. In addition to diet, intake of high quantity of water will help to dilute the urine sample and reduce urinary saturation. As some food items still shows contradictory results on stone formation, so more studies have to be conducted in this regard considering higher population in order to establish the relation of these food items in stone formation which will eventually help to reduce the burden of stone formation.

Funding

None.

Conflict of interest

The authors declare conflict of interest as none.

Author details

Akshata Sangolli[1], Shridhar C. Ghagane[2,3,4]* and Rajendra B. Nerli[3]

1 Department of Biochemistry, JN Medical College, KLE Academy of Higher Education and Research (Deemed to be University), Belagavi, Karnataka, India

2 KAHER's Dr. Prabhakar Kore Basic and Applied Science Research Center [BSARC], V.K. Institute of Dental Sciences Campus, India

3 Department of Urology, JN Medical College, KLE Academy of Higher Education and Research (Deemed-to-be-University), Belagavi, Karnataka, India

4 Division of Urologic-Oncology, Department of Urology, KLES Dr. Prabhakar Kore Hospital and Medical Research Centre, Belagavi, Karnataka, India

*Address all correspondence to: shridhar.kleskf@gmail.com

References

[1] Sohgaura A, Bigoniya P. A review on epidemiology and etiology of renal stone. American Journal of Drug Discovery and Development. 2017;**7**(2): 54-62

[2] Faridi MS, Singh KS. Preliminary study of prevalence of urolithiasis in North-Eastern city of India. Journal of Family Medicine and Primary Care. 2020;**9**(12):5939

[3] Thongprayoon C, Krambeck AE, Rule AD. Determining the true burden of kidney stone disease. Nature Reviews Nephrology. 2020;**16**(12):736-746

[4] Corbetta S, Baccarelli A, Aroldi A, Vicentini L, Fogazzi GB, Eller-Vainicher C, et al. Risk factors associated to kidney stones in primary hyperparathyroidism. Journal of Endocrinological Investigation. 2005; **28**:122-128

[5] Grases F, Costa-Bauza A, Prieto RM. Renal lithiasis and nutrition. Nutrition Journal. 2006;**5**(1):1-7

[6] Moe OW. Kidney stones: Pathophysiology and medical management. Lancet. 2006;**367**:333-344

[7] ElSheemy MS, Shouman AM, Shoukry AI, El Shenoufy A, Aboulela W, Daw K, et al. Ureteric stents vs percutaneous nephrostomy for initial urinary drainage in children with obstructive anuria and acute renal failure due to ureteric calculi: A prospective, randomised study. BJU International. 2015;**115**(3):473-479

[8] Chung DY, Cho KS, Lee DH, Han JH, Kang DH, Jung HD, et al. Impact of colic pain as a significant factor for predicting the stone free rate of one-session shock wave lithotripsy for treating ureter stones: A Bayesian logistic regression model analysis. PLoS One. 2015;**10**(4): e0123800

[9] Saigal CS, Joyce G, Timilsina AR, Urologic Diseases in America Project. Direct and indirect costs of nephrolithiasis in an employed population: Opportunity for disease management? Kidney International. 2005;**68**(4):1808-1814

[10] Alelign T, Petros B. Kidney stone disease: An update on current concepts. Advances in Urology. 2018;**2018**

[11] Saljoughian M. The management of urolithiasis. US Pharm. 2020;**45**(9): 34-36

[12] Griffith DP. Struvite stones. Kidney International. 1978;**13**(5):372-382

[13] Ahmed K, Dasgupta P, Khan MS. Cystine calculi: Challenging group of stones. Postgraduate Medical Journal. 2006;**82**(974):799-801

[14] Dursun M, Otunctemur A, Ozbek E. Kidney stones and ceftriaxone. European Medical Journal of Urology. 2015;**3**(1):68-74

[15] Kale SS, Ghole VS, Pawar NJ, Jagtap DV. Inter-annual variability of urolithiasis epidemic from semi-arid part of Deccan Volcanic Province, India: Climatic and hydrogeochemical perspectives. International Journal of Environmental Health Research. 2014; **24**(3):278-289

[16] Zeng Q, He Y. Age-specific prevalence of kidney stones in Chinese urban inhabitants. Urolithiasis. 2013; **41**(1):91-93

[17] Huang WY, Chen YF, Carter S, Chang HC, Lan CF, Huang KH. Epidemiology of upper urinary tract stone disease in a Taiwanese population: A nationwide, population-based study. The Journal of Urology. 2013;**189**(6): 2158-2163

[18] Silva GR, Maciel LC. Epidemiology of urolithiasis consultations in the Paraíba Valley. Revista do Colégio Brasileiro de Cirurgiões. 2016; **43**:410-415

[19] Basiri A, Shakhssalim N, Khoshdel AR, Ghahestani SM, Basiri H. The demographic profile of urolithiasis in Iran: A nationwide epidemiologic study. International Urology and Nephrology. 2010;**42**(1):119-126

[20] Ganesamoni R, Singh SK. Epidemiology of stone disease in Northern India. In: Urolithiasis. London: Springer; 2012. pp. 39-46

[21] Vinarova L, Vinarov Z, Tcholakova S, Denkov ND, Stoyanov S, Lips A. The mechanism of lowering cholesterol absorption by calcium studied by using an in vitro digestion model. Food & Function. 2016;**7**(1): 151-163

[22] Guasch-Ferré M, Bulló M, Babio N, Martínez-González MA, Estruch R, Covas MI, et al. Mediterranean diet and risk of hyperuricemia in elderly participants at high cardiovascular risk. Journals of Gerontology Series A: Biomedical Sciences and Medical Sciences. 2013;**68**(10):1263-1270

[23] Alatab S, Pourmand G, Howairis E, El Fatih M, Buchholz N, Najafi I, et al. National profiles of urinary calculi. Iranian Journal of Kidney Diseases. 2016;**10**(2)

[24] Singh PP, Dhing S, Bhatnagar R, Kothari S, Dhar V. Evidence suggesting that high intake of fluoride provokes nephrolithiasis in tribal populations. Urological Research. 2001;**29**(4): 238-244

[25] Afaj AH, Sultan MA. Mineralogical composition of the urinary stones from different provinces in Iraq. The Scientific World Journal. 2005;**5**:24-38

[26] Chandrajith R, Wijewardana G, Dissanayake CB, Abeygunasekara A. Biomineralogy of human urinary calculi (kidney stones) from some geographic regions of Sri Lanka. Environmental Geochemistry and Health. 2006; **28**(4):393-399

[27] Hussein NS, Sadiq SM, Kamaliah MD, Norakmal AW, Gohar MN. Twenty-four-hour urine constituents in stone formers: A study from the northeast part of Peninsular Malaysia. Saudi Journal of Kidney Diseases and Transplantation. 2013; **24**(3):630

[28] Lee YH, Huang WC, Tsai JY, Lu CM, Chen WC, Lee MH, et al. Epidemiological studies on the prevalence of upper urinary calculi in Taiwan. Urologia Internationalis. 2002;**68**(3):172-177

[29] Vasudevan V, Samson P, Smith AD, Okeke Z. The genetic framework for development of nephrolithiasis. Asian Journal of Urology. 2017;**4**(1):18-26

[30] Portis AJ, Hermans K, Culhane-Pera KA, Curhan GC. Rapid communication: Stone disease in the Hmong of Minnesota: Initial description of a high-risk population. Journal of Endourology. 2004;**18**(9):853-857

[31] Muslumanoglu AY, Binbay M, Yuruk E, Akman T, Tepeler A, Esen T, et al. Updated epidemiologic study of urolithiasis in Turkey. I: Changing characteristics of urolithiasis. Urological Research. 2011;**39**(4):309-314

[32] Eze NM, Maduabum FO, Onyeke NG, Anyaegunam NJ, Ayogu CA, Ezeanwu BA, et al. Awareness of food nutritive value and eating practices among Nigerian bank workers: Implications for nutritional counseling and education. Medicine. 2017;**96**(10)

[33] Kihara J, Sileshi GW, Nziguheba G, Kinyua M, Zingore S, Sommer R.

Application of secondary nutrients and micronutrients increases crop yields in sub-Saharan Africa. Agronomy for Sustainable Development. 2017;**37**(4): 1-4

[34] De Smet S, Vossen E. Meat: The balance between nutrition and health. A review. Meat Science. 2016;**120**:145-156

[35] Wahlqvist ML. Regional food culture and development. Asia Pacific Journal of Clinical Nutrition. 2007;**16**:2

[36] Maalouf NM, Moe OW, Adams-Huet B, Sakhaee K. Hypercalciuria associated with high dietary protein intake is not due to acid load. The Journal of Clinical Endocrinology & Metabolism. 2011;**96**(12):3733-3740

[37] Breslau NA, Brinkley L, Hill KD, PAK CY. Relationship of animal protein-rich diet to kidney stone formation and calcium metabolism. The Journal of Clinical Endocrinology & Metabolism. 1988;**66**(1):140-146

[38] Meschi T, Nouvenne A, Ticinesi A, Prati B, Guerra A, Allegri F, et al. Dietary habits in women with recurrent idiopathic calcium nephrolithiasis. Journal of Translational Medicine. 2012;**10**(1):1-8

[39] Taylor EN, Stampfer MJ, Curhan GC. Dietary factors and the risk of incident kidney stones in men: New insights after 14 years of follow-up. Journal of the American Society of Nephrology. 2004;**15**(12):3225-3232

[40] Mitchell T, Kumar P, Reddy T, Wood KD, Knight J, Assimos DG, et al. Dietary oxalate and kidney stone formation. American Journal of Physiology-Renal Physiology. 2019; **316**(3):F409-F413

[41] Tracy CR, Best S, Bagrodia A, Poindexter JR, Adams-Huet B, Sakhaee K, et al. Animal protein and the risk of kidney stones: A comparative metabolic study of animal protein sources. The Journal of Urology. 2014;**192**(1):137-141

[42] Park SM, Jee J, Joung JY, Cho YY, Sohn SY, Jin SM, et al. High dietary sodium intake assessed by 24-hour urine specimen increase urinary calcium excretion and bone resorption marker. Journal of Bone Metabolism. 2014; **21**(3):189-194

[43] Ferraro PM, Curhan GC, Gambaro G, Taylor EN. Total, dietary, and supplemental vitamin C intake and risk of incident kidney stones. American Journal of Kidney Diseases. 2016;**67**(3): 400-407

[44] Saldana TM, Basso O, Darden R, Sandler DP. Carbonated beverages and chronic kidney disease. Epidemiology (Cambridge, Mass.). 2007;**18**(4):501

[45] D'Alessandro C, Ferraro PM, Cianchi C, Barsotti M, Gambaro G, Cupisti A. Which diet for calcium stone patients: A real-world approach to preventive care. Nutrients. 2019; **11**(5):1182

[46] Borghi L, Meschi T, Amato F, Briganti A, Novarini A, Giannini A. Urinary volume, water and recurrences in idiopathic calcium nephrolithiasis: A 5-year randomized prospective study. The Journal of Urology. 1996;**155**(3): 839-843

[47] Fox IH, Kelley WN. Studies on the mechanism of fructose-induced hyperuricemia in man. Metabolism. 1972;**21**(8):713-721

[48] Taylor EN, Curhan GC. Fructose consumption and the risk of kidney stones. Kidney International. 2008; **73**(2):207-212

[49] Shuster J, Jenkins A, Logan C, Barnett T, Riehle R, Zackson D, et al. Soft drink consumption and urinary

stone recurrence: A randomized prevention trial. Journal of Clinical Epidemiology. 1992;**45**(8):911-916

[50] Eisner BH, Asplin JR, Goldfarb DS, Ahmad A, Stoller ML. Citrate, malate and alkali content in commonly consumed diet sodas: Implications for nephrolithiasis treatment. The Journal of Urology. 2010;**183**(6):2419-2423

[51] Ferraro PM, Taylor EN, Gambaro G, Curhan GC. Soda and other beverages and the risk of kidney stones. Clinical Journal of the American Society of Nephrology. 2013;**8**:1389-1395

[52] Rodgers AL. Effect of mineral water containing calcium and magnesium on calcium oxalate urolithiasis risk factors. Urologia Internationalis. 1997;**58**(2): 93-99

[53] Gettman MT, Ogan K, Brinkley LJ, Adams-Huet B, Pak CY, Pearle MS. Effect of cranberry juice consumption on urinary stone risk factors. The Journal of Urology. 2005;**174**(2):590-594

[54] Nouvenne A, Meschi T, Guerra A, Allegri F, Prati B, Borghi L. Dietary treatment of nephrolithiasis. Clinical Cases in Mineral and Bone Metabolism. 2008;**5**(2):135

[55] Li H, Klett DE, Littleton R, Elder JS, Sammon JD. Role of insulin resistance in uric acid nephrolithiasis. World Journal of Nephrology. 2014;**3**(4):237

[56] Cox CL, Stanhope KL, Schwarz JM, Graham JL, Hatcher B, Griffen SC, et al. Consumption of fructose-but not glucose-sweetened beverages for 10 weeks increase circulating concentrations of uric acid, retinol binding protein-4, and gamma-glutamyl transferase activity in overweight/obese humans. Nutrition & Metabolism. 2012;**9**(1):1-0

[57] Curhan GC, Willett WC, Knight EL, Stampfer MJ. Dietary factors and the risk of incident kidney stones in younger women: Nurses' Health Study II. Archives of Internal Medicine. 2004; **164**(8):885-891

[58] Giannini S, Nobile M, Sartori L, Carbonare LD, Ciuffreda M, Corrò P, et al. Acute effects of moderate dietary protein restriction in patients with idiopathic hypercalciuria and calcium nephrolithiasis. The American Journal of Clinical Nutrition. 1999;**69**(2): 267-271

[59] He Y, Chen X, Yu Z. The change of human Na+/dicarboxylate co-transporter 1 expression in the kidney and its relationship with pathogenesis of nephrolithiasis. Zhonghua Yi Xue Za Zhi. 2001; **81**(17):1066-1069

[60] Reddy ST, Wang CY, Sakhaee K, Brinkley L, Pak CY. Effect of low-carbohydrate high-protein diets on acid-base balance, stone-forming propensity, and calcium metabolism. American Journal of Kidney Diseases. 2002;**40**(2):265-274

[61] Goldfarb S. Dietary factors in the pathogenesis and prophylaxis of calcium nephrolithiasis. Kidney International. 1988;**34**(4):544-555

[62] Kerstetter JE, O'Brien KO, Insogna KL. Low protein intake: The impact on calcium and bone homeostasis in humans. The Journal of Nutrition. 2003;**133**(3):855S-861S

[63] Alpern RJ, Sakhaee K. The clinical spectrum of chronic metabolic acidosis: Homeostatic mechanisms produce significant morbidity. American Journal of Kidney Diseases. 1997;**29**(2):291-302

[64] Jaeger P, Robertson WG. Role of dietary intake and intestinal absorption of oxalate in calcium stone formation. Nephron Physiology. 2004;**98**(2):64-71

[65] Breslau NA, Brinkley L, Hill KD, Pak CYC. Relationship of animal

protein-rich diet to kidney stone formation and calcium metabolism. Journal of Clinical Endocrinology and Metabolism. 1988;**66**:140-146

[66] Khan SR, Glenton PA, Backov R, Talham DR. Presence of lipids in urine, crystals and stones: Implications for the formation of kidney stones. Kidney International. 2002;**62**(6):2062-2072

[67] Naya Y, Ito H, Masai M, Yamaguchi K. Association of dietary fatty acids with urinary oxalate excretion in calcium oxalate stone-formers in their fourth decade. BJU International. 2002;**89**(9):842-846

[68] Baggio B, Budakovic A, Nassuato MA, Vezzoli G, Manzato E, Luisetto G, et al. Plasma phospholipid arachidonic acid content and calcium metabolism in idiopathic calcium nephrolithiasis. Kidney International. 2000;**58**(3):1278-1284

[69] Taylor EN, Stampfer MJ, Curhan GC. Fatty acid intake and incident nephrolithiasis. American Journal of Kidney Diseases. 2005; **45**(2):267-274

[70] Lemann J Jr, Adams ND, Gray RW. Urinary calcium excretion in human beings. New England Journal of Medicine. 1979;**301**(10):535-541

[71] Coe FL, Favus MJ, Crockett T, Strauss AL, Parks JH, Porat A, et al. Effects of low-calcium diet on urine calcium excretion, parathyroid function and serum 1, 25 (OH) 2D3 levels in patients with idiopathic hypercalciuria and in normal subjects. The American Journal of Medicine. 1982;**72**(1):25-32

[72] Hess B. Low calcium diet in hypercalciuric calcium nephrolithiasis: First do no harm. Scanning Microscopy. 1996;**10**(2):21

[73] Borghi L, Schianchi T, Meschi T, Guerra A, Allegri F, Maggiore U, et al. Comparison of two diets for the prevention of recurrent stones in idiopathic hypercalciuria. New England Journal of Medicine. 2002;**346**(2):77-84

[74] Curhan GC, Willett WC, Speizer FE, Spiegelman D, Stampfer MJ. Comparison of dietary calcium with supplemental calcium and other nutrients as factors affecting the risk for kidney stones in women. Annals of Internal Medicine. 1997; **126**(7):497-504

[75] Sabto J, Powell MJ, Breidahl MJ, Gurr FW. Influence of urinary sodium on calcium excretion in normal individuals: A redefinition of hypercalciuria. Medical Journal of Australia. 1984;**140**(6):354-356

[76] Wills MR. The interrelationships of calcium and sodium excretions. Clinical Science. 1969;**37**:621-630

[77] Sakhaee K, Harvey JA, Padalino PK, Whitson P, Pak CYC. The potential role of salt abuse on the risk for kidney stone formation. The Journal of Urology. 1993;**160**:310-312

[78] Muldowney FP, Freaney R, Moloney MF. Importance of dietary sodium in the hypercalciuria syndrome. Kidney International. 1982;**22**:292-296

[79] Knight J, Jiang J, Wood KD, Holmes RP, Assimos DG. Oxalate and sucralose absorption in idiopathic calcium oxalate stone formers. Urology. 2011;**78**(2):475-4e9

[80] Marangella M, Bianco O, Martini C, Petrarulo M, Vitale C, Linari F. Effect of animal and vegetable protein intake on oxalate excretion in idiopathic calcium stone disease. British Journal of Urology. 1989;**63**(4):348-351

[81] Curhan GC, Willett WC, Rimm EB, Stampfer MJ. A prospective study of dietary calcium and other nutrients and the risk of symptomatic kidney stones.

New England Journal of Medicine. 1993;**328**(12):833-838

[82] Massey LK, Roman-Smith H, Sutton RA. Effect of dietary oxalate and calcium on urinary oxalate and risk of formation of calcium oxalate kidney stones. Journal of the American Dietetic Association. 1993;**93**(8):901-906

[83] Hesse A, Schneeberger W, Engfeld S, Von Unruh GE, Sauerbruch T. Intestinal hyperabsorption of oxalate in calcium oxalate stone formers: Application of a new test with [13C2] oxalate. Journal of the American Society of Nephrology: JASN. 1999;**10**:S329-S333

[84] Lemann J Jr, Bushinsky DA, Hamm LL. Bone buffering of acid and base in humans. American Journal of Physiology-Renal Physiology. 2003;**285**(5):F811-F832

[85] Meschi T, Maggiore U, Fiaccadori E, Schianchi T, Bosi S, Adorni G, et al. The effect of fruits and vegetables on urinary stone risk factors. Kidney International. 2004;**66**(6):2402-2410

[86] Broadus AE, Insogna KL, Lang R, Ellison AF, Dreyer BE. Evidence for disordered control of 1, 25-dihydroxyvitamin D production in absorptive hypercalciuria. New England Journal of Medicine. 1984;**311**(2):73-80

[87] Tiselius HG, Almgård LE. The diurnal urinary excretion of oxalate and the effect of pyridoxine and ascorbate on oxalate excretion. European Urology. 1977;**3**:41-46

[88] Chai W, Liebman M, Kynast-Gales S, Massey L. Oxalate absorption and endogenous oxalate synthesis from ascorbate in calcium oxalate stone formers and non-stone formers. American Journal of Kidney Diseases. 2004;**44**(6):1060-1069

[89] Curhan GC, Willett WC, Speizer FE, Stampfer MJ. Intake of vitamins B6 and C and the risk of kidney stones in women. Journal of the American Society of Nephrology. 1999; **10**(4):840-845

[90] Curhan GC, Willett WC, Rimm EB, Stampfer MJ. A prospective study of the intake of vitamins C and B6, and the risk of kidney stones in men. The Journal of Urology. 1996;**155**(6): 1847-1851

Chapter 12

Novel Treatment Strategy against COVID-19 through Anti-Inflammatory, Antioxidant and Immunostimulatory Properties of the B Vitamin Complex

Quratulain Babar, Anwar Ali, Ayesha Saeed and Muhammad Farrukh Tahir

Abstract

The immune system is influenced by vitamin B complex: B1, B2, B3, B5, B6, B7, B9 and B12. The B complex insufficiency can cause significant impairment in cellular and immune function and trigger an inflammatory response. There is currently no certified SARS-CoV-2 treatment or a credible vaccine, but strengthening the immune system with B vitamins could go a long way in preventing and treating COVID-19 patients. Thus, a complete and nutritious diet must be followed before approved drugs and potential good vaccine research results are available to boost the normal functioning of the immune system. In order to activate adaptive and inborn immune responses, reduce cytokine levels such as proinflammatory cytokines, decrease oxidative stress, preserve endothelial homogeneity, improving pulmonary function, prevents hypercoagulable conditions and shortening the length of hospital stay; B-Complex vitamins play a significant role. Thus, the role of B complex in patients with COVID-19 needs to be evaluated and additional non-drug B vitamins can be used in existing treatments.

Keywords: vitamins, immunity, COVID-19, inflammation, cytokines, anti-oxidant

1. Introduction

The severe acute respiratory syndrome coronavirus 2 SARS-CoV-2 virus is the cause of COVID-19. The World Health Organisation reported that COVID19 became an international global health crisis in January 2020 and that COVID19 has become a global epidemic, causing more than 20 million infections and 7 million deaths by March 2020. COVID-19 clinical signs include arthromyalgia, cough, diarrhoea, fevers, headache, lethargy, multiorgan disorder, severe interstitial pneumonia and septic coagulomas [1]. The severe development of COVID-19 also leads to a cytokine storm and excessively pro-inflammatory cytokine secretion [2]. In 2002–2004 and 2012–2014, epidemics of similar β-caronaviral virus (SARS) and Middle East Respiratory Syndrome (MERS), have previously taken place [3, 4].

IntechOpen

2. Food supplements to improve immune quality, antioxidants and anti-inflammation

Today, most states all around globe are working to develop corona vaccines, with a number entering human studies while most of them are being studied and developed in various stages. Furthermore, there are no specialised COVID-19 drugs, nor are there any meaningful statistics on the impacts of nutritional additives on COVID-19 risk or seriousness at both national and international levels. Developing new antiviral medications for COVID-19 is a major challenge that requires significant time and resources in the development and evaluation of this product. Several scores of proof signify, in particular in people with insufficient food resources as well as through their free radical scavenging, anti-inflammatory or viricidal capabilities, that many supplements from various vegetables, fruits, spices, herbs and root sources can decrease the hazard or intensity of a diverse variety of viral infections. These nutrients may be recycled to reduce the disease effects of SARS-CoV-2 infection. Thus the utilisation of natural compounds, together with the treatment for COVID-19, can propose new preventative and treatment support. The positive effects of certain nutrients are discussed in the following section [5].

3. Vitamin B and its biological roles

Vitamin B complex are the main vitamin of the brain function, eyes, gastrointestinal tract, liver, hair, muscular tone, nervous system, skin, and are critical to the health of the nervous system. These vitamins next to each other help to detoxify the organ, promote good metabolism, release enzymes from the food, stabilise the functions of your nervous system, provide cells with plenty of oxygen, maintain healthy skin and hair, protect faulty vision and have also been utilised in weaknesses [6]. **Table 1** provides an overview of the vitamin B complex with its cellular roles, scientific name, recommended male and female dose.

4. Vitamin B and pandemic

Due to anti-inflammatory and immunomodulatory properties of vitamin C and vitamin D in this era of pandemic they are getting much attention. Low vitamin D and C levels lead to coagulopathy and suppression of the immune system causing lymphocytopenia. The data showed that in corona virus patients with low levels of vitamin D have high mortality rate. In corona virus patients, the consumption of vitamin C increases the oxygenation index [9]. Accordingly, a lack of vitamin B can seriously alter the function of a cell and immune system that leads to hyperhomocysteinemia inflammation. Vitamin B must be stressed because it plays a key role in proper immune function, energy metabolism and cell function [10]. Vitamin B helps to reduce inflammation, strengthens respiratory functioning, preserves endothelial homogenity, inhibits hypercoagulation, activate innate and adaptive immune responses properly, and can decrease hospitalisation for long periods of time [11]. Thus, the role of B complex in patients with COVID-19 needs to be evaluated and additional non-drug B vitamins can be used in existing treatments.

5. Food supplements to counteract COVID-19

Phase 1 is critical from the point of view of prevention because individuals are carriers and can unintentionally propagate the infection. The organisation and

DOI: http://dx.doi.org/10.5772/intechopen.100251

Vitamin	Scientific name	Coenzymes	Groups transferred	For women daily recommended dose	For men daily recommended dose	Physiological/cellular role
B1	Thiamine	Thiamin pyrophosphate (TPF)	Aldehydes	1.1 mg	1.2 mg	Energy releasing, tissue growth
B2	Riboflavin	Flavin Mononucleotide (FMN)	H Atoms	1.1 mg	1.3 mg	Energy releasing, nucleotide synthesis
B3	Niacin	Nicotinamide Adenine (NAD+), Nicotinamide adenine dinucleotide phosphate (NADP+)	H atoms	14 mg	16 mg	Energy releasing, lipid breakdown and synthesis
B5	Pantothenic acid	Coenzyme A (CoA)	Acyl groups	5 mg	5 mg	Energy production from food stuff Synthesis of fatty acids
B6	Pyridoxine	Pyridoxal-5-phosphate (PLP), pyridoxine-5-phosphate (PNP), pyridoxamine-5′-phosphate (PMP)	Amino groups	1.3 mg	1.3 mg	Amino acid and glycogen breakdown
B7	Biotin	Biotin	CO_2	30 μg	30 μg	Fatty acid, glucose and leucine synthesis
B9	Folic acid	Tetrahydrofolate (THF)	One carbon group except CO_2	400 μg	400 μg	Amino acid and nucleotide synthesis
B12	Cobalomine/ Cyanocobalomin	methylcobalamine	Methyl groups, hydrogen atoms	2.4 μg	2.4 μg	Amino acid metabolism Nucleotide synthesis Breakdown of fatty acids Folic acid regeneration

Table 1.
Summary of Vitamin B complex [7, 8].

mounting of specific adaptive immune responses in persons and the utilisation of antivirals in phase 1 is vital for preventing entry of virus, disease progression and replication to phase 2. Global strategies can thus include administering external antivirals and food supplements that increase immune levels. In addition to retaining the overall status of the patient, phase 2 of the infection can be adapted for treatment to protect damage and malfunction to tissue in the course of the treatment by using nutritional supplements that can repress continuing oxidative stress, acute inflammation and cytokine storms. In short, in order to improve the immune response in phase one and eliminating it in the second phase, approaches to counter SARS-CoV-2 are effective, in addition to symptomatic treatment [5, 12].

6. Pathogenesis of COVID-19

The information provided shows that infection pathogenesis can be divided into two components. Phase 1: an asymptomatic phase of detectable viruses or not. Phase 2: High viral load symptomatic phase [13]. After binding the S protein into the ACE2 receptors and then initiating cellular transmembrane protease, serine 2 (TMPRSS2) and then the virus enters the airway epithelium. After the virus enters the host, innate interferon (IFN) immune response will be inhibited or delayed [14]. Ubiquitation and breakdown of RNA sensor molecules interrupts with downstream signalling [5, 15]. After an impairment of the IFN viral replication system, activation is generated of the granulocytes macrophages and monocytes that lead to the description of the "cytokine storm." The activation of proinflammatory cytokines, which involves interleukin IL-12, IL-8, IL-6 and IL-1 are described, involves massive secretions of the tumour necrosis factor (TNF-α). Tissue fibrosis, pneumonia and hyperinflammation of tissues is associated with this [16]. Research has suggested that oxidative stress is involved in COVID-19 pathogenesis. The evidence seems to indicate that SARS-CoV-2 actually causes oxidative stress by improving reactive oxygen species (ROS) production and indirectly suppressing host defence [17]. Moreover, granulocytosis also contributes to superoxide ions, a kind of of ROS

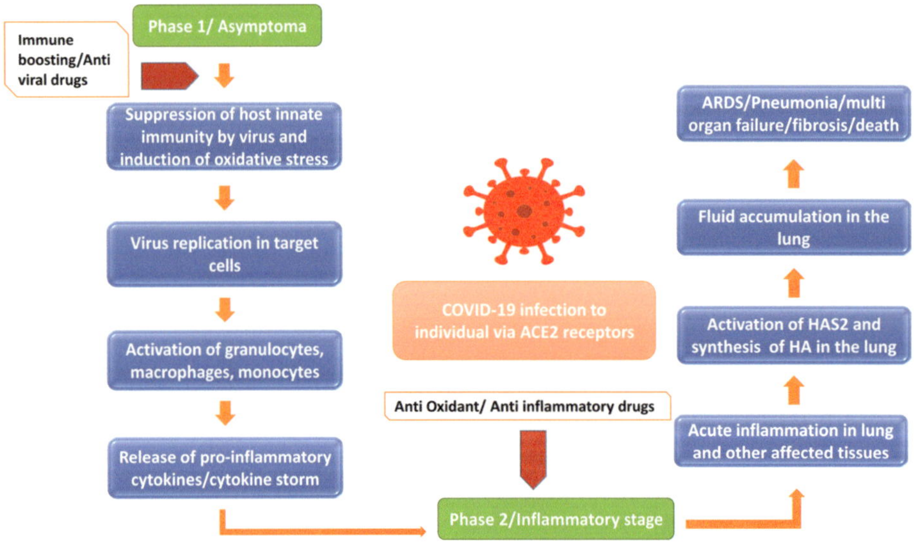

Figure 1.
Schematic representation of pathogenesis of COVID-19. Phase 1 which is asymptomatic includes dysregulation of host innate immune system, elevation of oxidative stress and phase 2 is acute inflammatory harmful phase [12, 24].

and an extra development of proinflammatory cytokines in response to SARSCoV-2 infections [16, 18, 19]. High levels of cytokines also cause HA-synthase-2 (HAS2) endothelial inductions in alveolar and fibroblast epithelial cell (Type 2) [20–23]. Lungs of patients with corona virus have increased cytokine (IL-1, TNF-α) inflammatory levels. This is related to higher HAS2 activity and successive corona virus lung pathophysiology. The clinical and scientific study results above therefore recommend that corona virus pathogenesis consists of two stages: Stage 1, dysregulation of host innate immune system, elevation of oxidative stress and phase 2 is acute inflammatory harmful phase (**Figure 1**).

7. Cross-discussion of immunity, inflammation and oxidative stress with vitamin B complex

In 1936 R.R. Williams and his coworkers defined their chemical structure and were able to synthesise the Vitamin B1 (thiamin) as the oldest vitamin. Vitamin B1 has an impact on anti-inflammatory characteristics, cytochrome C release, mitochondrial membranes, oxidative stress-induced, NF-kappaβ and protein kinases, P38-MAPK. Over expression of proinflammation cytokines like TNF, IL-1, IL-6, and arachidonic acid products, nervous system malfunction, T-cycle infiltration, neuroinflammation, expression CD40 by the microglia and CD40L, causing the loss of astrocytes, beriberi, CL2 chemokine over expression all are the outcomes of deficient vitamin B1 [25].

Therapies with vitamin B complex reduced proinflammatory expression and enhanced anti - proliferative cytokine activity, thereby making a contribution to neuroinflammatory resolution. Macrophages are usually grouped into two major subtypes: (i) macrophages (M1) that are involved as principal phagocytic cells in the inflammatory sites; and (ii) macrophages (M2) which carry out the process of tissue reshaping following inflammatory cellular activity. At the same time, B vitamins reduced the macrophage count for M1 and improved the macrophage count of M2. Thus, B vitamins have the potency for neuroinflammatory and neuroregenerational treatments and could be an excellent remedy for human peripheral nerve injury (PNI) [26].

Vitamin B-6 played an important role during the last decades in the mechanism for inflammatory and antioxidant activities [27]. PLP may interact with peroxy radicals and sequester free radicals and, through its group of hydroxyls, prevent lipid peroxidation on the pyridine ring [28]. PLP plays the role as a coenzyme in the manufacturing, throughout inflammation, of cytokines as well as other multipeptide intermediaries [29]. Therefore insufficient vitamin B-6 may diminish its antioxidant potential directly or interfere with inflammatory reactions [30].

8. COVID-19 patients: thiamine in hypoxia

Thiamine inadequacy impairs inflammatory profile through neuroinflammation by affecting cardiovascular system [31]. As the corona virus needs antibodies, mainly T cells, thiamine deficiencies can result in insufficient antibody responses and consequently in more serious symptoms. Thus, the correct immune responses to corona virus infection are effective to assist with sufficient thiamine levels. Moreover, COVID-19 symptoms seem to be much related to the disease of altitude and pulmonary edema of high altitude. By inhibiting carbonic anhydrase enzymes and consequently increasing oxygen level, for prevention of high altitude sickness and pulmonary edema acetazolamide is commonly prescribed. Thiamine also works as an inhibitor of the carbon anhydrase enzyme, hence the potential of hypoxia limitation and lowered hospitalisation at high concentrations of thiamine administered in early COVID-19 people. Research is still needed on the possibility of helping to heal COVID-19 patients by administering of increased thiamine doses [11].

9. Riboflavin-UV for inhibition of COVID-19 replication

Riboflavin with UV light is responsible for irreparable destruction to nucleic acids such as RNA and DNA, which makes it impossible to replicate microbial pathogens. The effectiveness of Riboflavin and UV light against MERS-CoV was demonstrated and it may also be of assistance with SARS-CoV-2 [32]. In fact, riboflavin-UV reduced the SARS-CoV-2 infective titre below the human blood and plasma and platelet identification limit [33]. It may ameliorate the risk of COVID-19 transfusion and reduce other pathogenic organisms in blood products in patients critically ill with COVID-19.

10. Vitamin B3 (Nicotinamide, Niacin) for blocking cytokine storm in inflammation

Niacin is a component of NAD and NADP, which are both essential during chronic systemic inflammatory responses [34]. NAD^+ works as a coenzyme in a broad variety of metabolic pathologies, and its elevated concentrations are crucial for dealing with a range of scenarios. NAD+ has immunomodulatory effects that are proven to affect proinflammatory cytokines throughout initial inflammatory periods [35, 36]. Current proof shows that IL-6 can regulate inflammatory storm in COVID-19 patients [37]. Niacin also decreases neutrophil infiltration and has an anti-inflammatory impact in ventilator-induced patients. Nicotinamide and niacin in hamsters minimise injury to the lung tissue [38]. Nicotinamide also minimises replication of virus and reinforces body mechanisms of protection. It can be used as an additional treatment for COVID-19 patients with a view to the lung protective and immune bolstering role of niacin [39].

11. Vitamin B5 (pantothenic acid) as anti-inflammatory agent

Vitamin B5 has a variety of functions, such as lipids and triacylglycerols reduction, improve wounds recovery, reduces inflammation, and enhances mental health [10]. Although there is scarce research which shows vitamin B5's effects on the immune system, it is a feasible vitamin for scientific investigations.

12. PLP supplementation as immunodulatory and anti-inflammatory agent

Pyridoxal 5′ phosphate (PLP) is an active component of pyridoxine and is a cofactor vital for several inflammatory disorder processes that contribute to immune disorder. In chronic inflammatory diseases, PLP has a reverse relation with plasma TNF-α and IL-6. Throughout inflammation, PLP use enhances its depletion, which means that COVID-19 can be deficient in patients with high inflammatory response. In patients with type-2 diabetes and cardiovascular disorder, groups at greater risk for relatively poor COVID-19 outcomes were reported with low levels of PLP [40, 41]. Elevated incidence of coagulopathy between many COVID-19 patients was also observed as a result of immune disregulation. In a recent manuscript, PLP intake was associated to ameliorate COVID-19 symptoms by regulate immune functions, lower pro-inflammatory cytokines, preserve endothelial stability and avoid hypercoagulability [42]. Indeed, it was found 30 years ago that platelet accumulation and blood coagulation defects were reduced by PLP amounts [43]. Vitamin B6 (along with B2 and B9) prediculated IL 10, a potent anti-inflammatory and immune suppressant cytokine which can disable monocytes and prevent T cells as well as cells present with the antigen [44]. Patients with COVID-19 often frequently react to the virus by implementing a pro-inflammatory T cell response and secretion. PLP may help dampen the cytokine storm and inflammatory processes suffered by a few patients with COVID-19.

13. Vitamin B9 (folic acid, folate) as furin inhibitor

For DNA, synthesis of protein and adaptive immune reaction, folate seems to be essential vitamin. Furin is a bacterial and viral infections-related enzyme and a good potential goal for infection treatment. Folic acid has recently been noted for being capable of inhibiting furin by avoiding binding of the spike protein SARS-CoV-2 and trying to prevent cell input and virus retention. Follic acid was thus recommended to be helpful for COVID-19-associated early phase respiratory disease management [44]. A latest publication shows the strength and stability of folic acid and its derivatives tetrahydrofolic acid and 5 methyl tetrahydrofolic acid, via structure-based molecular docking, in connection with SARS-CoV-2. Follic acid can thus be used as a treatment strategy for COVID-19 management [45].

14. Vitamin B12 (cobalmin) as antioxidant and gut modulatory agent

For the synthesis of erythrocytes, safety of the nervous system, myelin production, angiogenesis and rapid production of DNA, vitamin B12 is critical. Adenosyl-, h ydroxo- and methyl cobalamin are active forms of vitamin B12. As modulator for intestinal flora and low B12 concentrations, vitamin B12 increases homocysteine

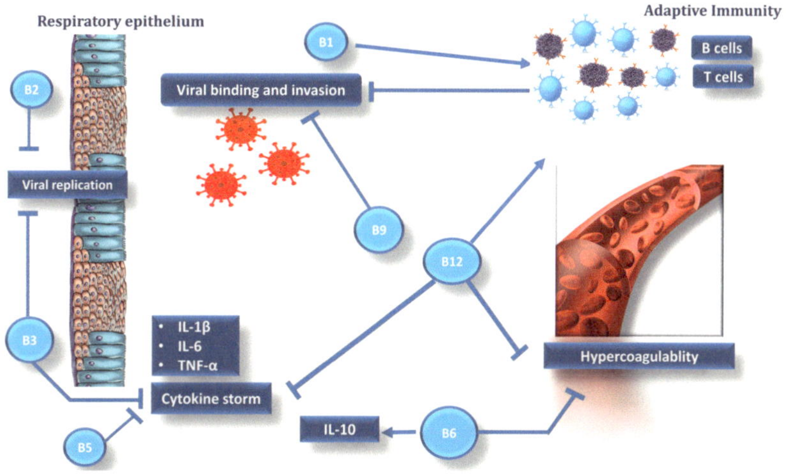

Figure 2.
Summary of Vitamin B complex as anti-COVID-19 agent for inhibition of viral replication, viral binding and invasion, cytokine storm and hypercoagulabity [9].

and methyl malonic acid, leading to increasing inflammatory process, ROS and oxidative stress [35]. Reduced immune response, endothelial dysfunction, myelin sheathing integrity interruption, megaloblastic anaemia, platelets and coagulation activation and are caused by hyperhomocysteinemia [46–48]. SARS-CoV-2 may perhaps interact with the metabolic activities of vitamin B12 which may affect microbiological bowel propagation. Provided that symptoms such as vasoconstriction, increased oxidative stress, cascade-activation of clotting, lactate dehydrogenase, renal and pulmonary vascular disorder and hyperhomocysteinemia are feasible [47, 49]. Furthermore, B12 insufficiency can lead to CNS, gastrointestinal and respiratory and abnormalities [48]. Remarkably, a new study shows that additional methylcobalamine may minimise damage to the organs and symptoms associated with COVID 19 [50]. A Singapore diagnostic research demonstrates that the intensity of COVID-19 in patients receiving magnesium, vitamin D (1000 IU) and vitamin B12 supplements (500 μg), reduced considerably the need for COVID-19 symptoms [51]. **Figure 2** shows a summary of the anti-viral vitamin B complex.

15. Conclusion and future perspective

Besides building and maintaining a stronger immune system, vitamin B can actually prevent or alleviate the symptoms of COVID-19 and / or treat infectious diseases with SARSCoV2. Inadequate nutritional conditions are much more susceptible to infections. Therefore, a balanced diet is needed to enhance immunity. Complementary or safe and cost-effective treatments are needed to reverse abnormal activation of the immune system that can lead to cytokine storms and as an antiplatelet agent. An adequate intake of vitamins is essential for the normal functioning of the body and the boosting of the immune system. B vitamins help control the immune response by decreasing pro-inflammatory cytokines and inflammatory conditions, minimising respiratory and gastrointestinal disease, preventing hypercoagulability, and possibly improving outcomes and shortening the hospital stay for COVID-19 patients.

Acknowledgements

Authors acknowledge the role of Food and Nutrition Society Gilgit Baltistan, Pakistan for providing access to journals and data bases.

Conflict of interest

Authors declare no conflict of interest.

Notes/thanks/other declarations

Authors pay special thanks to everyone for contribution.

Abbreviations

COVID-19	Coronavirus disease of 2019
SARS-COV-2	Severe acute respiratory syndrome coronavirus 2

ACE2	Angiotensin-converting enzyme 2
CCL2	C-C Motif Chemokine Ligand 2
CD40	Cluster of differentiation 40
HAS2	Hyaluronan Synthase 2
IFN	Interferon
IL	Interleukin
MAPK	Mitogen-activated protein kinase
MCG	Microgram
MERS	Middle East respiratory syndrome
NAD	Nicotinamide adenine dinucleotide
NADP	Nicotinamide adenine dinucleotide phosphate
NFKB	Nuclear Factor kappa-light-chain-enhancer of activated B cells
P38	Type of mitogen-activated protein kinases
PLP	Pyridoxal-5′-phosphate
TMPRSS2	Transmembrane Serine Protease 2

Author details

Quratulain Babar[1], Anwar Ali[2]*, Ayesha Saeed[1] and Muhammad Farrukh Tahir[1]

1 Department of Biochemistry, Government College University, Faisalabad, Pakistan

2 Department of Epidemiology and Health Statistics, Xingya School of Public Health, Central South University, Changsha, China

*Address all correspondence to: 206908003@csu.edu.cn

References

[1] Zu, Z.Y., et al., *Coronavirus disease 2019 (COVID-19): a perspective from China*. 2020. **296**(2): p. E15-E25.

[2] Chi, Z., et al., *The Cytokine Release Syndrome (CRS) of Severe COVID-19 and Interleukin-6 Receptor (IL-6R) Antagonist Tocilizumab man be the Key to Reduce the Mortality*. 2020.

[3] Zumla, A., D., and S.J.T.L. Perlman, *Middle East respiratory syndrome*. 2015. **386**(9997): p. 995-1007.

[4] Schoeman, D. and B.C.J.V.j. Fielding, *Coronavirus envelope protein: current knowledge*. 2019. **16**(1): p. 1-22.

[5] Mrityunjaya, M., et al., *Immune-boosting, antioxidant and anti-inflammatory food supplements targeting pathogenesis of COVID-19*. 2020. **11**.

[6] Zhuang, H. and M.M. Earth, *The physiological roles of vitamins in vegetables*, in *Postharvest physiology and pathology of vegetables*. 2002, CRC Press. p. 398-418.

[7] Robinson, F.A., *The vitamin B complex*. 1951: Рипол Классик.

[8] Schellack, G., P. Harirari, and N.J.S.P.J. Schellack, *Vitamin B-complex deficiency, supplementation and management*. 2019. **86**(3): p. 23-29.

[9] Shakoor, H., et al., *Be well: A potential role for vitamin B in COVID-19*. 2021. **144**: p. 108-111.

[10] Mikkelsen, K. and V. Apostolopoulos, *Vitamin B1, B2, B3, B5, and B6 and the immune system*, in *Nutrition and immunity*. 2019, Springer. p. 115-125.

[11] Zhang, L. and Y.J.J.o.m.v. Liu, *Potential interventions for novel coronavirus in China: A systematic review*. 2020. **92**(5): p. 479-490.

[12] Jin, Y., et al., *Virology, epidemiology, pathogenesis, and control of COVID-19*. 2020. **12**(4): p. 372.

[13] Shi, Y., et al., *COVID-19 infection: the perspectives on immune responses*. 2020, Nature Publishing Group.

[14] Channappanavar, R. and S. Perlman. *Pathogenic human coronavirus infections: causes and consequences of cytokine storm and immunopathology*. in *Seminars in immunopathology*. 2017. Springer.

[15] Li, X., et al., *Molecular immune pathogenesis and diagnosis of COVID-19*. 2020. **10**(2): p. 102-108.

[16] Chew, B.P. and J.S.J.T.J.o.n. Park, *Carotenoid action on the immune response*. 2004. **134**(1): p. 257S–261S.

[17] Spears, J.W. and W.P.J.T.V.J. Weiss, *Role of antioxidants and trace elements in health and immunity of transition dairy cows*. 2008. **176**(1): p. 70-76.

[18] Ertesvag, A., et al., *Retinoic acid stimulates the cell cycle machinery in normal T cells: involvement of retinoic acid receptor-mediated IL-2 secretion*. 2002. **169**(10): p. 5555-5563.

[19] Bono, M.R., et al., *Retinoic acid as a modulator of T cell immunity*. 2016. **8**(6): p. 349.

[20] Bell, T.J., et al., *Defective lung function following influenza virus is due to prolonged, reversible hyaluronan synthesis*. 2019. **80**: p. 14-28.

[21] Wang, D., et al., *Clinical characteristics of 138 hospitalised patients with 2019 novel coronavirus–infected pneumonia in Wuhan, China*. 2020. **323**(11): p. 1061-1069.

[22] Xu, Z., et al., *Pathological findings of COVID-19 associated with acute*

respiratory distress syndrome. 2020. **8**(4): p. 420-422.

[23] Hallgren, R., et al., *Accumulation of hyaluronan (hyaluronic acid) in the lung in adult respiratory distress syndrome*. 1989. **139**(3): p. 682-7.

[24] Mason, R.J., *Pathogenesis of COVID-19 from a cell biology perspective*. 2020, Eur Respiratory Soc.

[25] Spinas, E., et al., *Crosstalk between vitamin B and immunity*. 2015. **29**(2): p. 283-288.

[26] Ehmedah, A., et al., *Vitamin B complex treatment attenuates local inflammation after peripheral nerve injury*. 2019. **24**(24): p. 4615.

[27] Bilski, P., et al., *Vitamin B6 (pyridoxine) and its derivatives are efficient singlet oxygen quenchers and potential fungal antioxidants*. 2000. **71**(2): p. 129-134.

[28] Ohta, B.K. and C.S.J.J.o.t.A.C.S. Foote, *Characterisation of endoperoxide and hydroperoxide intermediates in the reaction of pyridoxine with singlet oxygen*. 2002. **124**(41): p. 12064-12065.

[29] Friso, S., et al., *Low circulating vitamin B6 is associated with elevation of the inflammation marker C-reactive protein independently of plasma homocysteine levels*. 2001. **103**(23): p. 2788-2791.

[30] Taysi, S.J.C.n., *Oxidant/antioxidant status in liver tissue of vitamin B6 deficient rats*. 2005. **24**(3): p. 385-389.

[31] Mikkelsen, K., A.J.N. Vitamin, and Immunity, *B1, B2, B3, B5, and B6 and the Immune System*. 2019: p. 115-125.

[32] Ragan, I., et al., *Pathogen reduction of SARS-CoV-2 virus in plasma and whole blood using riboflavin and UV light*. 2020. **15**(5): p. e0233947.

[33] Keil, S.D., et al., *Inactivation of severe acute respiratory syndrome coronavirus 2 in plasma and platelet products using a riboflavin and ultraviolet light-based photochemical treatment*. 2020. **115**(6): p. 495-501.

[34] Boergeling, Y. and S.J.T.F.j. Ludwig, *Targeting a metabolic pathway to fight the flu*. 2017. **284**(2): p. 218-221.

[35] Mikkelsen, K., et al., *The effects of vitamin B on the immune/cytokine network and their involvement in depression*. 2017. **96**: p. 58-71.

[36] Mikkelsen, K., L. Stojanovska, and V.J.C.m.c. Apostolopoulos, *The effects of vitamin B in depression*. 2016. **23**(38): p. 4317-4337.

[37] Liu, B., et al., *Can we use interleukin-6 (IL-6) blockade for coronavirus disease 2019 (COVID-19)-induced cytokine release syndrome (CRS)?* 2020. **111**: p. 102452.

[38] Nagai, A., et al., *Effects of nicotinamide and niacin on bleomycin-induced acute injury and subsequent fibrosis in hamster lungs*. 1994. **20**(4): p. 263-281.

[39] Mehmel, M., N. Jovanović, and U.J.N. Spitz, *Nicotinamide riboside—the current State of research and therapeutic uses*. 2020. **12**(6): p. 1616.

[40] Merigliano, C., et al., *The relationship between vitamin B6, diabetes and cancer*. 2018. **9**: p. 388.

[41] Lengyel, C.O., et al., *Nutrient inadequacies among elderly residents of long-term care facilities*. 2008. **69**(2): p. 82-88.

[42] Desbarats, J., *Pyridoxal 5′-phosphate to mitigate immune dysregulation and coagulopathy in COVID-19*. 2020.

[43] Van Wyk, V., H. Luus, and A.d.P.J.T.r. Heyns, *The in vivo effect in*

humans of pyridoxal-5′-phosphate on platelet function and blood coagulation. 1992. **66**(6): p. 657-668.

[44] Mikkelsen, K., et al., *Anti-tumour effects of vitamin B2, B6 and B9 in promonocytic lymphoma cells.* 2019. **20**(15): p. 3763.

[45] Kumar, V., S. Kancharla, and M.K.J.V. Jena, *In silico virtual screening-based study of nutraceuticals predicts the therapeutic potentials of folic acid and its derivatives against COVID-19.* 2021. **32**(1): p. 29-37.

[46] Nemazannikova, N., et al., *Is there a link between vitamin B and multiple sclerosis?* 2018. **14**(2): p. 170-180.

[47] Sabry, W., et al., *Vitamin B12 deficiency and metabolism-mediated thrombotic microangiopathy (MM-TMA).* 2020. **59**(1): p. 102717.

[48] Wolffenbuttel, B.H., et al., *The many faces of cobalamin (vitamin B12) deficiency.* 2019. **3**(2): p. 200-214.

[49] Grangé, S., et al., *Adult-onset renal thrombotic microangiopathy and pulmonary arterial hypertension in cobalamin C deficiency.* 2015. **386**(9997): p. 1011-1012.

[50] dos Santos, L.M.J.J.G.B. and P. Sciences, *Can vitamin B12 be an adjuvant to COVID-19 treatment?* 2020. **11**(3): p. 001-005.

[51] Tan, C.W., et al., *A cohort study to evaluate the effect of combination Vitamin D, Magnesium and Vitamin B12 (DMB) on progression to severe outcome in older COVID-19 patients.* 2020.